Synthesis of Biocomposite Materials

Chemical and Biological Modifications of Natural Polymers

Edited By

Yukio Imanishi, D.Eng.
Department of Polymer Chemistry
Kyoto University
Kyoto, Japan

CRC Press
Boca Raton   Ann Arbor   London   Tokyo

Library of Congress Cataloging-in-Publication Data

Synthesis of biocomposite materials : chemical and biological
 modifications of natural polymers / editor, Yukio Imanishi.
    p.   cm.
 Includes bibliographical references and index.
 ISBN 0-8493-6771-9
 1. Polymeric composites.   I. Imanishi, Y. (Yukio), 1934-
TA418.9.C6S95   1992
620.1'92—dc20   91-33746
                                                                                    CIP

© 1992 by CRC Press, Inc.

International Standard Book Number 0-8493-6771-9

Library of Congress Card Number 91-33746

Printed in the United States of America 1 2 3 4 5 6 7 8 9 10

Printed on acid-free paper

PREFACE

Biomaterials have been used for artificial-organ materials and bioreactor materials, and have gained an important position for enhancement of human welfare. Natural polymers such as plasma proteins and polysaccharides are usually biocompatible, and they have been used to improve biocompatibility of synthetic polymers by hybridization. On the other hand, natural polymers such as enzyme proteins are biofunctional but not very stable for industrial use. Chemical modification or mutation of enzyme proteins has been employed to create bioreactor catalysts and biofunctional materials. Remarkable progress in chemical modification of natural polymers for synthesis of biomaterials has made it an exciting time to organize this volume.

This volume summarizes the research works that have been devoted to create useful biofunctional materials by chemical modification of natural polymers in the near past and to forecast development in the near future.

This volume begins with the introduction of research results on the hybridization of natural polymers with synthetic polymers to make synthetic polymers biocompatible (nonthrombogenic) or to make natural polymers (proteins) free of antigenicity. These hybridized materials are useful in clinical test and therapy.

The second chapter of this volume describes examples of the use of biologically active polymers for construction of time-specific and site-specific drug delivery systems by chemical modification. The synthesis of superhormones on the basis of the membrane-compartment concept is described. Also, the use of enzymes as a sensor of physiological conditions (glucose level) in combination with an insulin-release system is described.

The third chapter of this volume describes current and future development of protein modification and mutation to create biofunctional materials. Emphasis is placed on enzyme solubilization into organic solvent by grafting of vinyl polymers and on chemical or genetic mutation of proteins with photofunctional nonnatural α-amino acid. The former will lead to creation of new bioreactor catalysts, and the latter to photo-energy-transducing protein devices.

The final chapter of this volume describes the synthesis of bioactive materials by immobilization of cell-adhesion or cell-growth proteins on synthetic polymer membrane. These hybridized materials are useful for cell culture under serum-free conditions and for creation of truly biocompatible materials.

In summary, this volume offers timely and useful suggestions for the design and synthesis of biofunctional materials.

Yukio Imanishi
Kyoto, Japan
September, 1991

THE EDITOR

Yukio Imanishi, D.Eng., is Professor of Radiation Polymer Chemistry in the Department of Polymer Chemistry, Kyoto University, Kyoto, Japan, working in the field of polymer synthesis and biofunctional materials since 1961. His research interests lie in biomedical materials, membrane-active polypeptides, signal-responsive membrane systems, and, recently, protein engineering.

Dr. Imanishi obtained his D.Eng. at the Department of Polymer Chemistry, Faculty of Engineering, Kyoto University, Kyoto, Japan (1965). He has published more than 200 research papers, review papers, and books on polymer synthesis and biofunctional materials.

CONTRIBUTORS

Yukio Imanishi
Department of Polymer Chemistry
Kyoto University
Kyoto, Japan

Yji Inada
Department of Materials Science
  and Technology
Toin University of Yokohama
Yokohama, Japan

Yoshihiro Ito
Department of Polymer Chemistry
Kyoto University
Kyoto, Japan

Shunsaka Kimura
Department of Polymer Chemistry
Kyoto University
Kyoto, Japan

Hideo Kise
Institute of Materials Science
University of Tsukuba
Tsukuba, Ibaraki, Japan

Y. Kodera
Department of Materials Science
  and Technology
Toin University of Yokohama
Yokohama, Japan

A. Matsushima
Department of Materials Science
  and Technology
Toin University of Yokohama
Yokohama, Japan

H. Nishimura
Department of Materials Science
  and Technology
Toin University of Yokohama
Yokohama, Japan

TABLE OF CONTENTS

Introduction

I. INTRODUCTION

Y. Imanishi

TABLE OF CONTENTS

1. INTRODUCTION

I. SCOPE OF BIOCOMPOSITE MATERIALS

Biocomposite materials are defined as materials synthesized by conjugation of biologically active substances with synthetic or nonbioactive polymers. They comprise an important part of biofunctional materials.

Life engineering as a field of life sciences consists of medical and biological engineering. Medical engineering enhances human welfare by developing innovative therapy methods such as artificial organ transplantation and signal-responsive drug delivery. Biological engineering also enhances human welfare by developing innovative production methods such as biological production and bioreactor systems. The evolution of both engineering fields becomes possible through the development of key materials in each field. Biospecific and biosimulation materials are vital to the support of medical and biological engineering, respectively. These relationships are depicted in Figure 1.

Biospecific materials function at the materials/biological system interface by controlling the specific recognition function of the biological system including proteins, cells, and biological tissues. Typical examples of biospecific materials are the biocompatible materials used for artificial organs.

Biosimulation materials possess exquisite and efficient functionalities similar to the biological system. Artificial enzymes, artificial cells, and artificial muscles are typical examples of biosimulation materials.

Biospecific and biosimulation materials cover the whole field of biofunctional materials. Biofunctional materials are synthesized from the view point of functionality design. The functionality design is based on determination of the polymer structure that realizes the desired functionality and property of materials, and on exploration of the appropriate method of polymer synthesis, polymer reaction, and polymer modification that yields the designed polymer structure. In these processes, chemical modification of natural polymers to give biospecific or biosimulating functionalities is very important.

II. FUNCTIONALITY DESIGN OF BIOSPECIFIC MATERIALS

In this section, the functionality design of biocompatible materials will be described. The biocompatibility required for artificial organ materials includes blood compatibility and tissue compatibility. At an early stage of blood/material contact, the material is required to possess nonthrombogenicity in a narrow sense of the blood compatibility. In the case of a long-term implantation of material in a living body, the material/growth cell interaction should be taken into account. In the following sections, the devices for acquiring a short-term nonthrombogenicity and a long-term biocompatibility will be described.

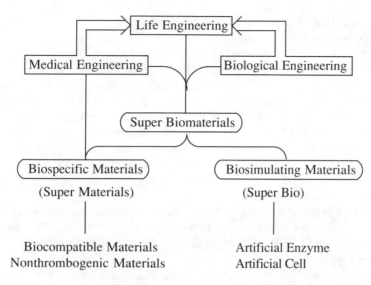

FIGURE 1. Super biomaterials supporting life engineering and human welfare.

A. MATERIALS POSSESSING HETEROGENEOUS SURFACE MORPHOLOGY

Among synthetic polymers in clinical use, polyetherurethaneurea (PEUU) is relatively nonthrombogenic. It has been considered that the relatively high nonthrombogenicity of PEUU comes from the heterogeneous morphology appearing on the membrane surface. On the surface of the PEUU membrane, a domain structure develops by microphase separation of nonpolar and flexible regions (consisting of polyether chains) from polar and rigid regions (consisting of urethane and urea linkages). The domain structure suppresses adsorption and denaturation of plasma proteins and subsequent adhesion and activation of platelets, leading to suppression of blood coagulation.[1]

This explanation is incomplete but has been used as a guideline for the design of nonthrombogenic materials, for example, graft polymerization of appropriate monomers on the surface of matrix membrane that is activated by chemical treatment or irradiation in advance.[2] It has been reported that the nonthrombogenicity of a PEUU membrane can be remarkably improved by the surface-grafting of polyoxyethylene chains.[3]

B. CONJUGATION OF BIOLOGICALLY ACTIVE SUBSTANCES

Despite continuing efforts to synthesize nonthrombogenic materials, it has generally been accepted that the acquisition of a perfect nonthrombogenicity of synthetic materials is, in practice, impossible. Then, hybridization of synthetic materials with a biological component has been considered useful in synthesizing highly nonthrombogenic materials. The biological component includes physiologically active substances of biological origin and living cells from various organs.[4,5]

1. Conjugation of Coagulation-Inhibiting Factors

Investigations on the design and synthesis of nonthrombogenic materials along these lines will be thoroughly reviewed in Chapter 2.1.

Heparin has been most frequently used as an anticoagulant in conjugation with synthetic polymers.[6,7] The method of heparin conjugation can be summarized in three ways: first, solubilization of heparin in organic solvent by chemical treatment and blending with or coating on a polymer membrane; second, electrostatic binding of heparin with a cationic matrix membrane; third, immobilization of heparin to a chemically functional matrix membrane. In all of these methods, heparin inactivates thrombin through the activation of antithrombin III (AT III) either by release into the blood or by remaining in an immobilized state.

A vinyl polymer carrying sulfonic acid groups which activated AT III, poly(vinyl sulfonic acid), was graft-polymerized to a PEUU membrane which was glow-discharged in advance, and the graft membrane showed a high nonthrombogenicity.[8]

A tripeptide, Val-Pro-Arg, mimics the reaction site of fibrinogen by thrombin and is named synthetic thrombin substrate. The synthetic thrombin substrate was immobilized on a PEUU membrane, and the composite membrane was found to prolong the thrombin time in blood coagulation.[9]

2. Conjugation of Reagent-Suppressing Platelet Activation

Adhesion and activation of platelets on the material play an important role in thrombus formation. For reagents suppressing platelet activation, adrenalin-shielding reagent drugs participating in the prostaglandin metabolic system, those affecting the cyclic nucleotide production, and those participating in the Ca^{2+}-regulating mechanism are considered. These platelet tranquilizers were immobilized on various polymer membranes to yield highly nonthrombogenic materials. An excellent inhibition effect on platelet adhesion and activation of PGI_2-immobilized polystyrene beads has been reported by Ebert et al.[10]

It has been pointed out that adhesion and activation of platelets are suppressed on heparin-immobilized materials, too.[6,7]

3. Conjugation of Fibrinolytic Enzymes

Urokinase, streptokinase, fibrinolysine, and brynolase have been used as fibrinolytic enzymes for immobilization onto polymer membranes. Some of the enzyme-immobilized membranes have been used clinically. In the clinical use of immobilized enzymes, the stability of enzymes, in particular, the inactivation of enzymes by AT III or α_2-macroglobulin, is a significant problem. Encapsulation of urokinase in a polymeric network is a successful solution to this problem.[11] In this case, the enzyme maintains the intrinsic activity, because it is physically bound in the network and not chemically damaged. The enzyme is also protected by the network from invasion of macromolecular destroyers.

Binary immobilization of antithrombogenic reagents also has been conducted. Examples are prostaglandin/heparin complex immobilization[12] and urokinase/heparin coimmobilization.[13] With reference to a polytetrafluoroethylene membrane coated with a heparin/collagen complex, an early thrombus formation is inhibited by the action of heparin, and a long-term endothelialization is accelerated by collagen.[14]

4. Acceleration of Endothelialization

For long-term implantation of artificial organ materials, an appropriate regeneration of vascular wall tissue is desirable. Herring et al.[15] succeeded in synthesizing a cell hybridization-type artificial vascular tissue. They seeded autologal vascular endothelial cells to the fibrin layer formed on a polymer membrane by preclotting (seeded graft). However, in this method, the survival of collected cells is very difficult.

Next, an idea of "cultured graft" was proposed, in which endothelial cells are adhered and grown in advance on the surface of a porous membrane to cover it completely with a layer of endothelium.[16,17] In this method, the adhesion and growth of cells are enough to exert cell functions. However, a steady supply of living cells and an immune response are still serious problems.

In order to obtain cell hybridization materials without encountering these problems, the design and synthesis of polymer membranes, on which living cells easily adhere and proliferate upon implanting in a living body, are strongly desired. The present author and co-workers are very interested in the synthesis of novel composite materials that enhance adhesion and growth of cells. One of the cell-adhesion-accelerating materials is synthesized by the immobilization of the core peptide of cell-adhesion proteins, that is, the Arg-Gly-Asp sequence.[18] The background and usefulness of the biocomposite materials will be described in details in Chapter 5.1.

The other approach synthesizes the cell-growth-accelerating materials by immobilizing cell-growth proteins such as insulin, transferrin, and epidermal growth factor (EGF). Successful results of growth factor immobilization prompted the author's team to investigate the effect of coimmobilization of the cell-growth factor and the cell-adhesion factor. All of these experimental results will be thoroughly reviewed in Chapter 5.2.

III. FUNCTIONALITY DESIGN OF BIOSIMULATION MATERIALS

In this section, functional design in the synthesis of materials mimicking specific functionalities of proteins, such as enzymes, will be described. In order to use proteins for materials, modification of proteins is necessary to be resistant against changes in temperature, pH, and ionic strength of the environment. When proteins are used as molecular materials constituting a molecular system, they should possess functionalities of acceptance, amplification, and transmission of external signals.

A. SYNTHETIC APPROACH

It does not appear that the synthetic approach toward biosimulation of materials has progressed to a sufficiently sophisticated level to produce perfection of functionality. Artificial enzymes have been designed toward the construction of specific and efficient sites for substrate binding. Utilization of cyclic host compounds has been an important approach among many attempts along this line of investigation. Cyclodextrins, cyclic peptides, cyclophanes, and crown ethers have been used as cyclic host compounds.

In particular, it has been reported that β-cyclodextrin carrying a pyridoxalamine substituent is an excellent model for transaminase.[19] A remarkably high stereospecificity was observed in the conversion of keto acid to amino acid that was catalyzed by β-cyclodextrin carrying pyridoxamine and another amino group.[20]

B. INNOVATION OF PROTEIN FUNCTIONALITY BY CHEMICAL MODIFICATION

An artificial enzyme, flavopapain, has been synthesized in which a flavin coenzyme was connected to a thiol group involved in the catalytic site of papain.[21] The flavopapain possesses the substrate-binding site of papain, in which *N*-alkyl dihydronicotinamides are specifically bound. The oxidation reactions of *N*-alkyl dihydronicotinamide catalyzed by flavopapain are faster by some hundred times than nonenzymatic reactions. Another example of an artificial enzyme based on glyceroaldehyde 3-phosphate dehydrogenase has been reported by the same author.[22] This sort of chemical modification of enzyme functionality is named "chemical mutation". Synthesis of a membrane-active hydrolytic enzyme by the modification of gramicidin S is an example of new enzyme production by a method analogous to chemical mutation.[23] Chemical mutation of enzymes for the creation of novel catalysts, along with genetic mutation, will be comprehensively reviewed in Chapter 4.2.

The product of chemical modification of proteins with synthetic polymers has been named "hybrid protein".[24] Hybridization of enzymatic proteins with lipophilic synthetic polymers solubilizes the proteins into an organic solvent, enables chemical reactions that do not occur in an aqueous solution to proceed, and thus extends the utilization of enzymes to bioreactor catalysts. This aspect of hybrid proteins will be extensively reviewed in Chapter 4.1.

The enzyme hybridization with synthetic polymers also is effective in preventing antigenicity of the enzyme in medical application. This aspect of hybrid proteins will be thoroughly reviewed in Chapter 2.2.

Substances in a living body are metabolized by specific catalysis of enzymes. In other words, enzymes are sensors for specific substrates. If the substrate is related to a disease and the enzyme is coupled with a drug or a hormone by a linkage sensitive to the enzyme reaction, the hybrid enzyme is a molecular device usable for a drug delivery system that is an innovative therapeutic method. This aspect of hybrid proteins will be thoroughly reviewed in Chapter 3.2.

There are many kinds of polypeptide hormones produced in a living body. They act as messengers of biological signals. They are accepted by specific receptor proteins present in the cell membrane and transmit the signal to the nucleus of the cell. Polypeptide hormones consist of a message segment and an address segment. The former carries biological signals and interacts with receptor proteins. The latter has different degrees of membrane affinities and determines selectivity of the receptor or receptor subtype. We can replace the address segment with synthetic polymers, resulting in intensification or toleration of the hormone effect and varied selectivities of the receptor. This aspect of hybrid proteins will be comprehensively reviewed in Chapter 3.1.

C. PROTEIN ENGINEERING

A new research field, protein engineering, has been created by the progress of genetic engineering as the realization of a dream that novel proteins are produced at will by modification of a specific site in an amino acid sequence. Many new functionality proteins have been synthesized by using protein engineering technology. Typical examples are genetic modifications of antibodies. As described previously, construction of a specific and efficient substrate-binding site is essential for the synthesis of artificial enzymes, and cyclic compounds are useful for the substrate-binding site. Another useful candidate for the substrate-binding site is the antibody. The specific antigen recognition by an antibody is a very attractive property to be used in the design of an artificial enzyme. Mutant proteins in which an antibody is used have been synthesized by the recombinant DNA method. Typical examples are the synthesis of an antibody with nuclease activity by connecting immunoglobulin Fab to nuclease and the synthesis of a chimera protein in which immunoglobulin Fab is connected to the terminal part of a protein responding to oncogene c-myc.[25] A fibrinolytic fused protein specific to fibrin has also been synthesized by gene technology.[26] Investigations on humanized monoclonal antibodies have progressed in the field of protein engineering involving antibodies.[27,28,29] The recent trend in the creation of novel proteins by protein engineering will be comprehensively reviewed in Chapter 4.2, along with chemical mutation of enzymes.

The utilization of antibodies in the creation of new functional proteins leads to the opening of a new scientific field, antibody engineering. The new scientific field will be a topic in a forthcoming book of this series.

REFERENCES

1. **Okano, T. and Kataoka, K.**, Molecular design of antithrombogenic polymers, in *Biomaterial Science, Part 2*, Tsuruta, T. and Sakurai, Y., Eds., Nankodo, Tokyo, 1982, 57.
2. **Fischer, J. P., Becker, U., von Halasz, S.-P., Muck, K.-F. Puschner, Rosinger, S., Schmidt, A., and Suhr, H. H.**, The preparation and characterization of surface-grafted plastic materials designed for the evaluation of their tissue and blood compatibility, *J. Polym. Sci. Polym. Symp.*, 66, 443, 1979.
3. **Liu, S.-Q., Ito, Y., and Imanishi, Y.**, Synthesis and non-thrombogenicity of polyurethanes with poly(oxyethylene) side chains in soft segment regions, *J. Biomater. Sci. Polym. Ed.*, 1, 111, 1989.
4. **Wilson, J. E.**, Hemocompatible polymers: preparation and properties, *Polym. Plast. Technol. Eng.*, 25, 233, 1986.
5. **Ito, Y. and Imanishi, Y.**, Blood compatibility of polyurethanes, *CRC Crit. Rev. Biocompatibility*, 5, 45, 1989.
6. **Wilson, J. E.**, Heparinized polymers as thromboresistant biomaterials, *Polym. Plast. Technol. Eng.*, 16, 119, 1981.
7. **Ito, Y.**, Antithrombogenic heparin-bound polyurethanes, *J. Biomater. Appl.*, 2, 235, 1987.
8. **Ito, Y., Iguchi, Y., Kashiwagi, T., and Imanishi, Y.**, Synthesis and nonthrombogenicity of polyetherurethaneurea film grafted with poly(sodium vinyl sulfonate), *J. Biomed. Mater. Res.*, in press.
9. **Ito, Y., Liu, L.-S., and Imanishi, Y.**, *In vitro* non-thrombogenicity of a thrombin-substrate-immobilized polymer surface by the inhibition of thrombin activity, *J. Biomater. Sci. Polym. Ed.*, 5, 123, 1991.
10. **Ebert, C. D., Lee, E. S., and Kim, S. W.**, The antiplatelet activity of immobilized prostacyclin, *J. Biomed. Mater. Res.*, 16, 629, 1982.
11. **Liu, L. S., Ito, Y., and Imanishi, Y.**, Biological activity of urokinase immobilized to cross-linked poly(2-hydroxyethyl methacrylate), *Biomaterials*, in press.
12. **Jacobs, H. A., Okano, T., and Kim, S. W.**, Antithrombogenic surfaces: characterization and bioactivity of surface immobilized PGE$_1$-heparin conjugate, *J. Biomed. Mater. Res.*, 23, 611, 1989.
13. **Kusserow, B. K., Larrson, R., and Nichols, J.**, The urokinase-heparin bonded synthetic surface. An approach to the creation of a prosthetic surface possessing composite antithrombogenic and thrombolytic properties, *Trans. Am. Soc. Artif. Intern. Organs*, 17, 1, 1971.
14. **Kodama, M.**, The artificial blood tube with a trilayer structure (in Japanese), *Kagaku to Kogyo (Chem. Chem. Ind.)*, 40, 480, 1987.
15. **Herring, M. B., Gardner, A., and Glover, J. L.**, A single-staged technique for seeding vascular grafts with autogenous endothelium, *Surgery*, 84, 498, 1978.
16. **Takagi, A.**, Delayed seeding of cultured endothelial cells to synthetic vascular prosthesis, *Jpn. J. Artif. Organs*, 15, 23, 1986.
17. **Moroboshi, Y., Suga, M., Maeyama, T., Ookuma, T., Sasaki, H., Ichiki, M., Sato, S., Kinebuchi, C., Kaneda, H., Saito, K., Oouchi, H., and Mori, S.**, Development of prosthetic vascular graft covered with autogenous canine endothelial cells — basic research, *Jpn. J. Artif. Organs*, 17, 1609, 1988.
18. **Ito, Y., Kajihara, M., and Imanishi, Y.**, Materials for enhancing cell adhesion by immobilization of cell-adhesive peptide, *J. Biomed. Mater. Res.*, in press.
19. **Breslow, R., Czarnik, A. W., Lauer, M., Leppkes, R., Winkler, J., and Zimmerman, S.**, Mimics of transaminase enzymes, *J. Am. Chem. Soc.*, 108, 1969, 1986.
20. **Tabushi, I., Kuroda, Y., Yamada, M., and Higashimura, H.**, A-(modified B$_6$)-B-[ω-amino(ethylamino)]-β-cyclodextrin as an artificial B$_6$ enzyme for chiral amino transfer reaction, *J. Am. Chem. Soc.*, 107, 5545, 1985.

21. **Slama, J. T., Oruganti, S. R., and Kaiser, E. T.,** Semisynthetic enzymes: synthesis of a new flavopapain with high catalytic efficiency, *J. Am. Chem. Soc.,* 103, 6211, 1981.
22. **Hilvert, D., Hatanaka, Y., and Kaiser, E. T.,** A highly active thermophilic semisynthetic flavoenzyme, *J. Am. Chem. Soc.,* 110, 682, 1988.
23. **Yagi, Y., Kimura, S., and Imanishi, Y.,** Interaction of gramicidin S analogs with lipid bilayer membrane, *Int. J. Peptide Protein Res.,* 36, 18, 1990.
24. **Inada, Y., Ed.,** *Protein Hybrid,* Kyoritsu Publ. Co., Tokyo, 1987.
25. **Neuberger, M. S., Williams, G. T., and Fox, R. O.,** Recombinant antibodies possessing novel effector functions, *Nature (London),* 312, 604, 1984.
26. **Pennica, D., Halmes, W. E., Kohr, W. J., Harkins, R. N., Vehar, G. A., Ward, C. A., Bennett, W. F., Yelverton, E., Seeburg, P. H., Heyneker, H. L., and Goeddel, D. V.,** Cloning and expression of human tissue-type plasminogen activator cDNA in *E. coli, Nature (London),* 301, 214, 1983.
27. **Jones, P. T., Dear, P. H., Foote, J., Neuberger, M. S., and Winter, G.,** Replacing complimentarity-determining regions in a human antibody with those from a mouse, *Nature (London),* 321, 522, 1986.
28. **Verhoeyen, M., Milstein, C., and Winter, G.,** Reshaping human antibodies: grafting an antilysozyme activity, *Science,* 239, 1534, 1988.
29. **Riechmann, L., Clark, M., Waldmann, H., and Winter, G.,** Reshaping human antibodies for therapy, *Nature (London),* 332, 323, 1988.

Modification of Biologically Active
Substances for Biocompatible Materials

2.1 ANTICOAGULANT HYBRIDIZATION WITH SYNTHETIC POLYMERS

Yoshihiro Ito

TABLE OF CONTENTS

I. INTRODUCTION

Biomaterials, especially those for artificial organs, require medicofunctionality and biocompatibility as shown in Figure 1. Medicofunctionality is the functional characteristic of individual organs, such as selective permeability, adsorption-separation ability, carrier transportation ability, optical sensitivity, and mechanical properties. Biocompatibility is specific to biomedical materials and is required for all kinds of biomedical materials that contact living components. Generally, biocompatibility is divided into blood com-

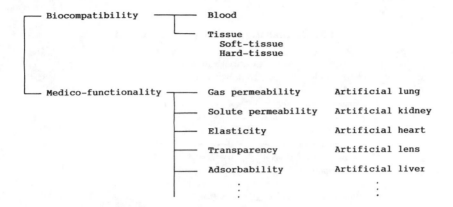

FIGURE 1.   Properties required for biomaterials.

patibility and tissue compatibility, of which, the former poses serious problems for the materials contacting blood. This is because thrombus formation on the surface not only affects functionality of the artificial organs, but also affects the living body by the flown thrombus.

When foreign material is brought into contact with the blood, plasma proteins are first adsorbed onto the material, followed by activation of clotting factors, adhesion and activation of platelets, and finally formation of the fibrin network, leading to thrombus formation. There are also biological systems such as fibrolysis which dissolve thrombi and the complement system which activates leukocytes. These systems are interrelated with elementary reactions of the thrombogenic system.

Some books concerned with blood compatibility have already been published.[1-3] This chapter will describe the design of synthetic anticoagulants, hybridization of synthetic reagents, bioactive reagents, or cells with synthetic polymers, after a short explanation of blood components and blood-material interaction.

A. BLOOD COMPONENTS

Blood is a complex fluid composed of formed elements (red blood cells, white blood cells, and platelets) suspended in a liquid portion (plasma) as shown in Table 1. Whole blood consists of about 45% formed elements and about 55% plasma. The plasma itself consists of a complex solution of proteins, inorganic molecules, and electrolytes.

1. Formed Elements

Red blood cells — The red blood cell (erythrocyte) is a biconcave disk, with a diameter *in vivo* of about 8 μm. Red cells are formed in the bone marrow and have a mean survival time of about 3 months in humans. In human males the average normal red cell count is about 5.4 million cells per cubic millimeter. Mechanical trauma or changes in osmotic pressure can cause rupture of the cells with liberation of hemoglobin (hemolysis).

TABLE 1
The Composition of Circulating Blood

Cell	Concentration (number/mm³)	Vol% in blood
Erythrocytes	$4—6 \times 10^6$	45
Leukocytes		
Neutrophils	$1.5—7.5 \times 10^3$	
Eosinophils	$0—4 \times 10^2$	
Basophils	$0—2 \times 10^2$	
Lymphocytes	$1—4.5 \times 10^3$	1
Monocytes	$0—8 \times 10^2$	
Platelets	$250—500 \times 10^3$	

Protein	Concentration (mg/100 ml serum)	
Albumin	3500—5500	
IgG	800—1800	
Fibrinogen	200—450	
α_1-Lipoprotein	290—770	
α_1-Antitrypsin	200—400	
β-Lipoprotein	200—650	
Transferrin	200—400	55
IgA	90—450	
IgM	60—250	
IgE	~300	
α_2-Macroglobulin	160—370	
Haptoglobin		
1-1 Type	100—220	
2-1 Type	160—300	
2-2 Type	120—260	
At least 100 known minor plasma proteins	<100	

White blood cells — White blood cells or leukocytes can be divided into two main groups: granulocytes (or polymorphonuclear leukocytes) which make up 60 to 70% of the white cells, and nongranular leukocytes, which make up 20 to 30% of the total number of white cells. Neutrophilic leukocytes compose from 50 to 65% of the granulocytes and are about 8 μm in diameter. These granulocytes exhibit the property of phagocytosis (ingestion of materials). Basophilic and acidophilic leukocytes make up the remaining granulocytes. The nongranular leukocytes include lymphocytes (6 to 8 μm in diameter) and monocytes (9 to 12 μm in diameter). Monocytes also exhibit phagocytosis.

Platelets — Platelets or thrombocytes are not cells but fragments of giant bone marrow cells (megakaryocytes). They are disk-shaped under normal physiological conditions, are found at a concentration of about 300,000/µl in human blood, with a diameter of 2 to 4 µm. Platelets contain protein-rich granules (dense and α-granules). When activated by physiological or foreign stimuli, platelets can undergo spatial transformation and degranulation leading to the release of platelet aggregating agents such as ADP or serotonin.

2. Proteins

Many blood proteins have been implicated in hemostasis and thrombosis, as shown in Table 1. The plasma proteins are divided into the following groups: proteins which provide nutrients to cells (albumin, lipoproteins, etc.); those involved in the transport of the other chemicals including hormones, minerals, and intermediates (transferrin, ceruloplasmin, etc.); and those concerned with defense against disease (the proteins of the clotting sequence and fibrinolysis, immune and complement systems).

B. BLOOD-BIOMATERIAL INTERACTIONS (THROMBUS FORMATION)

The first event that occurs after blood contacts a polymer surface, following instantaneous redistribution of interfacially bound water and ions, is the absorption of plasma proteins and the formation of a protein layer at the blood-polymer interface. It is now generally recognized that thrombogenesis (a result of blood-material interaction and one of the major obstacles in the use of artificial organs and prosthetic devices contacting the blood) is mediated in most, if not in all, instances, by a layer of plasma protein adsorbed to the artificial surface. The mechanisms of thrombogenesis are considered to involve interactions of multiple components at multiple levels as illustrated in Figure 2.

1. Protein Adsorption

All the effects indicated in Figure 2 stem from the adsorbed protein layer. Therefore, in the cause of research into hemocompatible materials, attempts have been made mostly to establish the relationship between the physicochemical properties and the adsorptional and hemocompatible characteristics of the surface.[4,5] Andrade and Hlady[6] surveyed plasma protein components greater than 1 mg/ml in concentration, which they called "the big twelve". They considered their size, concentration, diffusion coefficient, structure and function, and methods of estimating their "surface denaturability" by using bulk solution measures of denaturation and conformational change. They suggested that the role of carbohydrate moieties in plasma proteins had some bearing on their adsorption properties, and that lipoproteins, because of their lipid-phase transition and conformational lability at body temperature, tended to dominate the adsorption process, particularly on mobile elastomeric polymer surfaces. The adsorption of lipoprotein was investigated by Jayalumari et al.[7] The importance of enzymes released by cell damage was also referred to by Mulzer and Brash.[8]

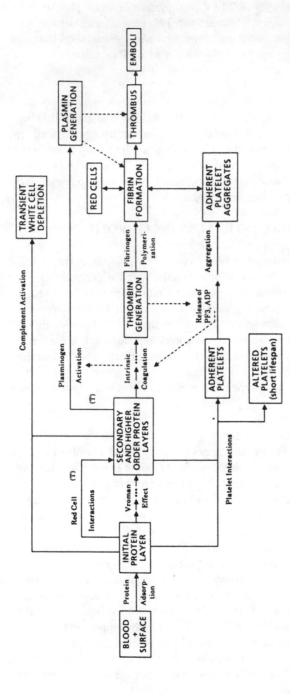

FIGURE 2.   Sequence of events during blood-material interactions. The notation T indicates tentative. (From Brash, J. L., *Ann. N.Y. Acad. Sci.*, 516, 206, 1987. With permission.)

The conformational analysis of adsorbed proteins has been performed by various spectroscopic techniques.[9-12] FT-IR and CD techniques revealed a loss of the α-helix and the gain of a β-structure in adsorbed proteins. Sanada et al.[10] showed that hydrogen-bonding formation of the material surface reduced conformational change of adsorbed proteins.

A computer simulation to determine the interaction potential energy between the protein and the polymer surface was conducted by Lu and Park.[13] The adsorption of four proteins — lysozyme, trypsin, immunoglobulin F_{ab}, and hemoglobin — on five polymer surfaces — polystyrene, polyethylene, polypropylene, poly(hydroxyethyl methacrylate), and poly(vinyl alcohol) — was studied. Lu and Park found that the interaction energy was dependent on the orientation of the protein on the polymer surfaces. The energy varied from -850 to $+600$ kJ/mol with an average of about -155 kJ/mol. The average interaction energies of the four proteins with poly(vinyl alcohol) were always lower than those with the other polymers.

Among the plasma proteins, fibrinogen is considered to be an important component of the adsorbed protein film and to be an essential player in the early events of platelet activation,[14] which will be described in the next section. An early feature of activation of platelets by soluble agonists such as ADP or thrombin is the exposure of specific membrane glycoproteins (GPIIb-IIIa), to which fibrinogen molecules bind with high affinity. According to current dogma, the divalent fibrinogen molecule is necessary for the formation of connecting bridges between contiguous platelets, perhaps in conjunction with other "adhesive proteins", such as the von Willebrand factor, thrombospondin, or fibronectin. Salzman and Lindon's group[15] showed that the interaction of platelets with fibrinogen molecules adsorbed on surfaces in their "native" conformation might itself be a stimulus to platelet activation, as shown in Figure 3.

Composition of the adsorbed protein layer varies from surface to surface, and it seemed likely that variations in thrombogenicity among surfaces stem from protein layer differences. An aspect of the layer that has recently been recognized is its evolutionary character. It has been hypothesized that proteins may replace one another at the surface in an ordered sequence, and considerable evidence is now available to support at least a partial adoption of this view. These transient phenomena have been designated the "Vroman effect" in recognition of Vroman's seminal contributions in this regard. The effect has been investigated by many researchers.[16-19]

The effect of shear rate is an important problem in protein adsorption. Pitt and Cooper[20] reported the effect of wall shear rate upon albumin adsorption onto a polyurethaneurea, Biomer, using FTIR-ATR spectroscopy. They showed that increasing wall shear rate during adsorption did not significantly affect the adsorption kinetics, but decreased the rate of protein adsorption. These results suggested that albumin at higher shear rates was more tightly bound to the surface. Comparison of the data with simple mathematical models suggested that more than one layer of protein is adsorbed in 2 h.

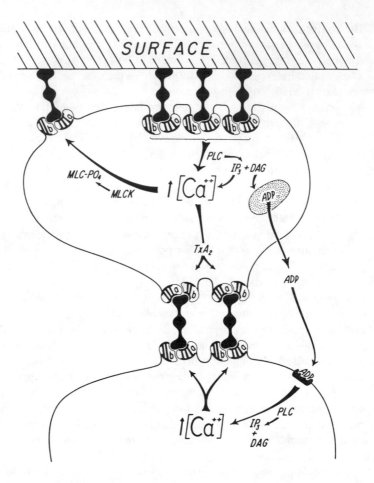

FIGURE 3.   Hypothetical scheme of platelet activation by adhesion to fibrinogen-coated surfaces. Adsorbed fibrinogen, conformationally relatively intact, reacts with glycoprotein IIb-IIa platelet receptors, masked and of low affinity, which by activation of phospholipase C and resulting generation of inositol triphosphate and diacylglycerol leads to elevation of the ionized cytoplasmic calcium concentration. This, in turn, activates myosin light-chain kinase leading to exposure of glycoprotein receptors of higher affinity, resulting in further fibrinogen binding and to adherence of additional platelets and formation of a platelet aggregate. The response may be enhanced by ADP secreted from the adherent platelets and by activation of calcium-sensitive phospholipase A_2 with resultant generation of thromboxane A_2. (From Salzman, E. W., Lindon, J., McManama, G., and Ware, J. A., *Ann. N.Y. Acad. Sci.,* 516, 184, 1987. With permission.)

The effects of adsorbed protein into cellular events were observed by many researchers. It was observed, for example, that fibronectin, which promotes platelet release, was bound more tightly to hydrophobic than to hydrophilic areas. This, however, was accompanied by a decrease in platelet releasing activity.[21] Ito et al.[22] found a relationship between conformational change of adsorbed proteins and platelet adhesion. The increase in the β-structure of the protein led an increase of platelet adhesion.

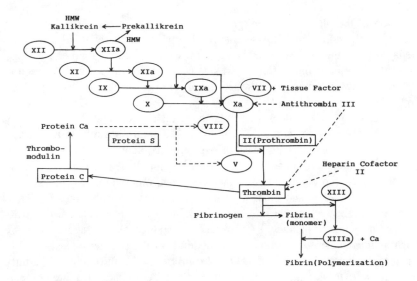

FIGURE 4. The coagulation system and protective systems; →, suppression.

2. Coagulation System Activation

Blood coagulation can be initiated either by activation of the intrinsic pathway or activation of the extrinsic pathway.[23] The intrinsic pathway is initiated by the contact system. On the other hand, the extrinsic pathway is initiated by activation of factor X. These are shown in Figure 4.

The contact system comprises four proteins that exist in plasma in the zymogen or procofactor form, namely, factor XII (XII, Hageman factor), prekallikrein (PK, Fletcher factor), factor XI (XI, PTA), and high molecular weight kininogen (HMWK). Each of these proteins is required to initiate, amplify, and propagate surface-mediated defense reactions. Each of these proteins is converted to the corresponding active enzyme or cofactor during contract activation. The zymogens, XII, PK, and XI, are converted. The activation of the contact system of plasma proteolysis is probably initiated by the binding of plasma XII to a negatively charged surface where autoactivation occurs; i.e., the inert zymogen XII is converted to an active serine protease, XIIa, presumably by a trace amount of XIIa. In contrast, the substrates of XIIa and the zymogens PK and XI cannot be activated on a surface in the absence of HMWK. PK and XI exist in bimolecular complexes with the contact system procofactor, HMWK, which serves to place them on the surface for cleavage by factor XIIa. Thus, of the four proteins involved in the initiation of blood coagulation at artificial surfaces, only two interact directly with the surface. There are presently many publications concerning the adsorption of these proteins.[19,24]

3. Platelet Adhesion

Platelet deposition on artificial surfaces is a characteristic feature of platelet function *in vivo* and *in vitro*, and a vital component of the physiological function of the platelet as a building block in the formation of a hemostatic plug. Platelet adhesion follows protein adsorption on artificial surfaces. The behavior of platelets at surfaces is a complex phenomenon in which rheology controls the kinetic interactions, and several biochemical and cellular factors determine the outcome of the interactions. These factors have been categorized by Coller as shown in Table 2.[25]

After the platelets adhere, granular contents are released. With relatively weak stimuli, only the contents of the dense granules are released, whereas stronger stimuli also cause release of α-granule contents. The dense granules contain serotonin and adenosine diphosphate while the α-granules contain the platelet-specific proteins (platelet factor 4, β-thromboglobulin); platelet-derived growth factor; and other proteins including fibrinogen, factor V, and von Willebrand factor. These release-inducing agents also activate the prostaglandin synthetic pathway of platelets and induce platelet aggregation by various mechanisms, some of which depend on the platelet-release mechanism, and others which do not. ADP, thromboxane A_2, thrombin, and a fatty acid are known as platelet-activating factors (PAF). Platelet aggregation is preceded by a change in platelet shape from discoid to spiny spheres. This shape change is mediated by the contractile microtubular system. Platelet aggregation requires a platelet membrane glycoprotein receptor, GPIIb-IIIa, which is thought to be the receptor for fibrinogen. Platelet-platelet bridging may also involve the platelet α-granule proteins fibronectin and thrombospondin. The platelet activation process is shown in Figure 3.

Platelet activation is accelerated by thrombin generation. Stimulated platelets have an enhanced capacity to facilitate the interactions between coagulation factors. This platelet coagulation capacity is probably a manifestation of a rearrangement of the platelet surface lipoprotein which accompanies the change in platelet shape.

Thus far, several investigations have been concerned with the number and deformation degree of platelets adhering to various artificial surfaces.[26-28] Recently, some studies have dealt with platelet interactions in terms of various methods. The shape change of platelets adhered on polymers has been discussed from the standpoint of cytoskeleton reorganization by some researchers.[29,30] For example, platelets were treated with cytoskeleton breakers; then adhesion and serotonic release were investigated. The results are shown in Figure 5. Platelet release reaction caused by contacting materials has also been investigated.[31] Kang et al.[31] reported that serotonin release was induced to a similar extent by stimulation upon adhesion, while β-thromboglobulin only sometimes was released (Figure 6). Variation in the extent of stimulation by material surfaces was argued.

4. Complement System Activation

In the interaction of blood with vascular grafts, most attention has generally been directed to the coagulation system and platelets, as thrombosis is a major cause of vascular graft failure. However, complement activation by

TABLE 2
Major Determinants of Platelet Behavior at Surfaces

I. Flow
    A. Rate
    B. Geometry
    C. Shear
    D. Turbulence
    E. Effect of other cellular elements

II. The reactive surface
    A. Natural
        1. Blood vessel location
        2. Nature of injury
        3. Exposed mediator(s)
            a. Collagen
            b. Immobilized von Willebrand factor
            c. Noncollagenous microfibrils
            d. Fibronectin
            e. Thrombospondin
    B. Artificial
        1. Fibrinogen deposition
        2. Fibrinogen transformation

III. The platelets
    A. Receptors
        1. For fluid-phase agonists
        2. For inhibitors
        3. For glycoproteins (GPIa, GPIb, GPIIb-IIIa, GPIV)
    B. Metabolic pathways
        1. Arachidonic acid (cyclooxygenase, thromboxane synthetase, lipoxygenase)
        2. Phospholipase C/diglyceride lipase: diacyglycerol/inositol trisphosphate/protein kinase C
        3. Cyclic AMP generation
        4. Calcium metabolism
    C. Effector pathways
        1. Contractile mechanism
        2. Release reaction
        3. Procoagulant activity
    D. Content of adhesion- and aggregation-augmenting agents
        1. Dense bodies (ADP, epinephrine, serotonin)
        2. α-Granules (fibrinogen, fibronectin, thrombospondin, von Willebrand factor)

IV. Soluble protein cofactors
    A. Fibrinogen
    B. von Willebrand factor

V. Activators
    A. Fluid phase
        1. Singly
            a. ADP
            b. Thrombin
            c. Epinephrine
            d. Serotonin
            e. TXA_2 (PGG_2, PGH_2)
            f. Platelet-activating factor

TABLE 2 (continued)
Major Determinants of Platelet Behavior at Surfaces

            2.   In combination
    B.   Solid phase
            1.   Collagen
            2.   Immobilized von Willebrand factor
            3.   Immobilized fibrinogen
            4.   Immobilized fibronectin
            5.   Immobilized thrombospondin
VI.   Inhibitors
        A. PGI_2
        B. PGE_1
        C.   Adenosine

vascular graft materials may have systemic effects such as pulmonary leukostasis, pulmonary inflammation, and anaphylaxis. Additionally, adherent leukocytes may contribute to vascular graft thrombosis by augmenting coagulation and stimulating platelet aggregation by liberation of the platelet activating factor, as well as by direct platelet/neutrophil interaction.[32-36] Of further interest is the possible role that complement activation and leukocytes may play in the organization and healing of vascular grafts as well as in the species-dependent phenomenon of vascular graft endothelialization.[34,35]

The complement system consists of approximately 18 proteins which together comprise roughly 2 to 4% of the plasma proteins. These proteins of the complement system function in one of two ways, via the classical or the alternative pathways. It is generally accepted that the complement activation observed during hemodialysis proceeds by the alternative pathway.[37] The first phase, initiation, begins in C3. The active site of this reactive species consists of an activated thiolester that can undergo nucleophilic attack. Consequently, this reactive form of C3 can be deposited upon and become covalently attached to biomaterials.

The role of complement activation in thrombogenesis on the surface of hydrophilic monomer-graft copolymerized polyethylene (PE) tubes has been investigated.[38,39] *N*-Vinylpyrrolidone (NVP)-grafted tubes were activated in an *in vitro* complement system of canine serum; but no activation occurred in 2-hydroxyethyl methacrylate (HEMA)-grafted tubes. The relative patent time for NVP-grafted tubes implanted in canine peripheral veins was shorter than that for HEMA-grafted tubes, and adhesion of numerous leukocytes was observed on the luminal surfaces of the NVP-grafted tubes. Decomplementation by prior administration of the cobra venom factor elongated the relative patent time for only NVP-grafted tubes and also inhibited the adhesion of leukocytes onto them. These results suggest that complement activation participates in thrombus formation on the polymer surfaces in canine veins. Similar results were obtained by Payne and Horbett.[40] The mechanism of complement-induced thrombus formation is proposed in Figure 8.

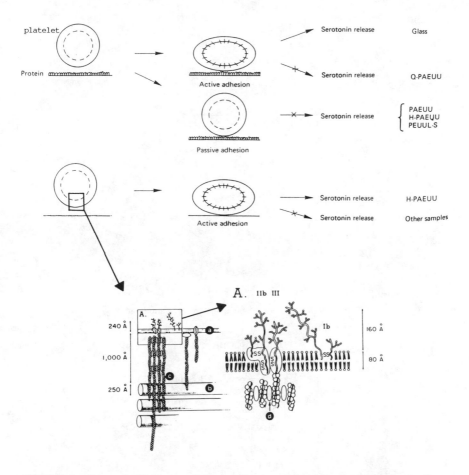

FIGURE 5.　Schematic illustration of the participation of cytoskeleton proteins in platelet-material interaction.

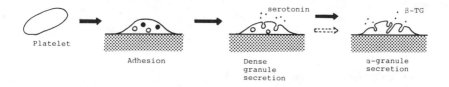

FIGURE 6.　A model for the activation mechanism of platelets upon adhesion to materials; →, stimulus transmitted and ⇢, stimulus not transmitted.

5. Protective System Activation

Thrombogenesis is normally modulated by a number of efficient protective mechanisms to maintain the endothelial hemostatic balance (protective systems are shown in Figure 5).[41] The protective mechanisms include the normal intact endothelium, the inhibitors of activated coagulation factors, the hepatic

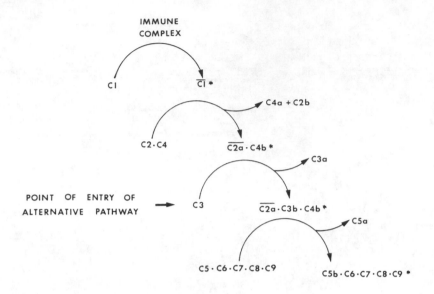

FIGURE 7.   The complement system activation pathway.

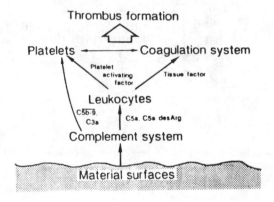

FIGURE 8.   Schematic diagram of the process of complement-induced thrombus formation on material surfaces. (From Hayashi, K., Fukumura, H., and Yamamoto, N., *J. Biomed. Mater Res.*, 24, 1385, 1990. With permission.)

clearance of activated coagulation factors, and the fibrinolytic system.

Normal endothelial cells have nonthrombogenic properties to maintain blood flow *in vivo*. Endothelial cells synthesize prostaglandin I_2 or prostacyclin (a powerful inhibitor of platelet aggregation); synthesize a lipoxygenase product, 13-HODE, which inhibits platelet adhesion; and synthesize glycosaminoglycans, which inhibit blood coagulation by enhancing the activities of antithrombin III and heparin cofactor II.

There are circulating inhibitors in blood. Antithrombin III inhibits thrombin and activated factors X, IX, XI, and XII. The rates of antithrombin III with these enzymes are markedly increased by heparin. Protein C is a vitamin K-dependent serine protease zymogen and is activated by thrombomodulin in the endothelial cell surface. The activated protein C is a potent anticoagulant; its anticoagulant properties are exerted by proteolytically inactivating activated factors V and VIII. Protein S is a vitamin K-dependent plasma protein that acts as a cofactor for activated protein C and is required for the maximal expression of the anticoagulant effect of protein C. Heparin cofactor II is a plasma protein that inactivates thrombin by forming a covalent complex with the enzyme. The reaction between thrombin and heparin cofactor II is an order of magnitude slower than that between thrombin and antithrombin III. Activated coagulation factors are inactivated by the liver in the human body by removal from the blood.

The fibrinolytic system consists of a plasma and a cellular component. The plasma fibrinolytic system is the conversion of a β-globulin, plasminogen, to the active proteolytic enzyme plasmin. There are two distinct classes of activators (tissue-type and urokinase-like) as well as a fast-acting plasminogen activator inhibitor. Once plasminogen is activated to plasmin it can hydrolyze fibrin and various plasma coagulation proteins, including fibrinogen. There is considerable evidence that some fibrinolytic activity is derived from proteolytic enzymes released from neutrophils.

C. OVERVIEW OF CURRENT BLOOD-COMPATIBLE MATERIALS

1. Polyurethane

Polyurethane is known as a blood-compatible elastomer. As described by Lelah and Cooper,[2] and Ito and Imanishi,[42] the polyurethanes are widely used in medicine, for example as aortic balloon pumps, catheters, and housings for artificial hearts. Thus, the hemocompatibility of polyurethane has been investigated by many researchers.[43,44]

Pelzer and Heimburger[27] compared blood compatibility of some materials by assaying platelet factor 4 release, platelet adsorption, and fibrinogen adsorption. Polyurethane induced slight *in vitro* platelet factor 4 release and adsorbed significantly lower concentrations of fibrinogen as well as platelets from human blood.

Brothers et al.[45] compared the blood compatibility of polyurethane and polytetrafluoroethylene. Qualitative examination of representative sections of polyurethane conduits demonstrated thick inner capsules with numerous small islands of graft material surrounded by macrophages and bands of mature fibrous tissue, in contrast to the thinner neointima and limited anastomotic pannus ingrowth observed in expanded polytetrafluoroethylene grafts.

Bakker et al.[46] described that the biocompatibility of the polyurethanes and polyether polyester copolymer was better than that of both silicone rubbers and polypropylene oxide.

Hydrophilic and hydrophobic microporous polyurethane (Mitrathane R) was synthesized, and the hemocompatibility was investigated.[47-49] The hydrophilic type has a microporous surface, and displays good mechanical characteristics, and the behavior *in vitro* is excellent. The *in vivo* results were, however, disappointing. A hyperplastic reaction appeared to have been initiated on the polyurethane, for which no rationale was given. On the other hand, the hydrophobic grafts were patent for 6 months.

The microporous, hydrophilic polyetherurethaneurea (PEUU) vascular graft was compared with expanded polytetrafluoroethylene (PTFE) in the canine carotid artery by Martz et al.[50] They reported that at implantation times ranging from 4 h to 6 months, all the PEUU grafts were occluded while of the PTFE grafts, only those implanted for 1 week and 6 months were blocked. Histopathologic analysis of the expanded grafts and their capsules revealed an ongoing inflammatory reaction at the anastomotic sites of the PEUU grafts.

The driving force behind the introduction of polyetherurethanes into biomedical devices is their desirable bulk properties. These include elasticity, strength, durability, ease of fabrication, and variety of bulk compositions which can be synthesized. However, the surface structure and biocompatibility of segmented polyurethane are still topics of much discussion.[51-53]

Castner et al.[54] used X-ray photoelectron spectroscopy and contact angle methods to examine the surfaces of a homologous series of polyetherurethane samples before and after cleaning. The samples have similar surface compositions, although they differ in bulk composition. Critical surface tension values are all similar and high (45 to 46 dyn/cm). The similarity in the surfaces of the polyurethanes, despite the differences in their mechanical properties, demonstrates that surface properties do not necessarily reflect bulk properties. Soap washing and methanol-acetone extraction of the polyurethane films resulted in surfaces more representative of the bulk compositions of the polyurethanes. Analysis of intravenous catheters confirmed that they are lubricated with polydimethylsiloxane, a common practice in the medical device industry. The significant extractable compositions of polyetherurethaneurea were observed by other researchers.[52,53] The role of cleanliness, contamination, and preparation procedures of inorganic biomaterials has been reviewed by Kasemo and Lausmaa.[55]

Block or segmented copolymers have a surface activity such that the air-facing surface is enriched with the lower surface free-energy component in order to minimize the interfacial free energy between the solid and the air as shown in Figure 9. Several researchers demonstrated the surface differences between hydrated and dry states by means of X-ray photoelectron spectroscopy and dynamic contact angle measurements.[56,57]

2. Silicone

Silicone also has a long history as a biomedical polymer and has preferentially been used for special catheters and shunts which require good compatibility. This is because the polymer exhibits an extremely high chemical stability and hydrophobicity. Poor mechanical properties can be improved by mixing with fillers, which, however, decrease the blood compatibility.

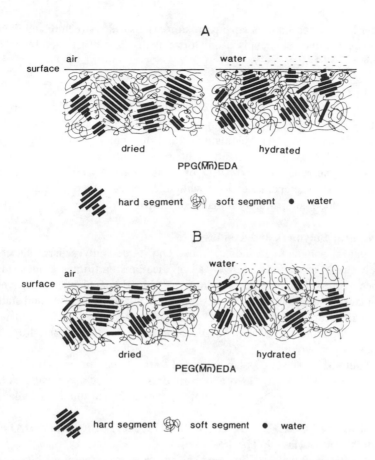

FIGURE 9.   Schematic representation of the surface structure of a (A) hydrophobic and (B) hydrophilic soft segment containing segmented polyetherurethane in the hydrated and dehydrated states. (From Takahara, A., Jo, N. J., and Kajiyama, T., *J. Biomater. Sci., Polym. Ed.,* 1, 17, 1989. With permission.)

3. Polytetrafluoroethylene

Polytetrafluoroethylene (PTFE) is chemically more inert and displays a higher hydrophobicity than silicone. Expanded PTFE with its microporous structure is often used for vascular grafts and patches. Normally, vascular grafts made from PTFE become blood-compatible as a result of neointima ingrowth, but are also reported to exhibit relatively good antithrombogenicity if the micropores are filled with plenty of water.

Kottke-Marchant et al.[58] tested expanded polytetrafluoroethylene (ePTFE), crimped Dacron Bionit (DB), and preclotted Dacron Bionit (DB/PC). The observed platelet activation was striking for Dacron and especially preclotted Dacron, with ePTFE showing low levels of platelet activation. Platelet activation by Dacron was initially rapid and then leveled off, whereas the platelet activation with preclotted Dacron began more slowly, but reached much greater levels after 3 h of *in vitro* perfusion. The surface structure and *in vitro* cell response to commercially used Gortex is dramatically altered by solvent washing.[59]

4. LTI Carbon

A pyrolytic, low-temperature isotropic (LTI) carbon is also used by depositing it on substrates such as metals. This carbon has been described by Haubold et al.[60] as a good biomaterial, and this author claims a failure rate of less than 1 in 10,000 for prosthetic heart valves during 16 years of clinical use, since these devices are partly or totally coated with isotopic carbon deposited at low temperature along with silicone. However, why this carbon has good blood compatibility remains unknown.

5. Natural Polymers and Tissues

Natural polymers, such as cellulose and its derivatives, have also been widely used in various biomedical applications: artificial kidney membranes,[61,62] coating materials for drugs,[63,64] blood coagulants,[65] additives to pharmaceutical products,[65] supports for immobilized enzymes,[65] and stationary phases for optical resolution[66] and recently as blood anticoagulants,[67] blood-compatible materials,[68-71] tissue-compatible materials,[72] and drugs having antitumor activity.[73]

Natural tissues such as bovine carotid artery and human umbilical cord vein have been clinically used as thromboresistant vascular prostheses after treatment with aldehydes such as dialdehyde starch and glutaraldehyde.

D. DESIGN CONCEPT OF BLOOD-COMPATIBLE MATERIALS
1. Bulk and Surface Properties

It is believed that both bulk and surface properties determine blood compatibility. When material is fabricated as a medical device, the structural design is also an important rheological factor determining the blood compatibility. The importance of bulk properties, that is, mechanical properties, has been recognized. A mismatch caused by the variation in mechanical properties between vascular grafts and host arteries yielded local stress concentrations and flow disturbances which might result in suture failure, tearing of the host artery, local thrombus formation, and other complications.[74-82] Clinical experiences and chronic animal experiments on arterial grafts have indicated a fairly good correlation between compliance (elasticity) and patency rate.[83,84] A higher patency rate has been obtained for materials having a matching compliance to the host artery.[85] These results emphasize that the mechanical compatibility of materials has an important role in graft performance, and materials used for vascular grafts should be elastic and flexible enough to retain normal hemodynamics and strong and durable enough to resist overextension and fatigue failure.

Pordeyhimi and Wagner[86] undertook a statistical investigation of reported vascular graft failures over a period of 30 years. They emphasized the importance of mechanical behavior of artery/implant systems.[86]

2. Surface Design

Generally, there are three approaches in the design of antithrombogenic materials: (1) minimization of interactions of the material with blood components; (2) immobilization of biologically active substances such as heparin, urokinase, prostacyclin etc., which inhibit or suppress thrombus formation on the material; and (3) endothelialization of the material utilizing the biological function of living tissues. This chapter will discuss the progress and problems of each approach.

Approach (1) can be further divided: (a) minimization of adsorption of plasma proteins onto the material and (b) formation of layered protein that does not strongly interact with platelets.

Method (a) is related to several theories such as the minimum interfacial free-energy hypothesis proposed by Andrade,[87] the presence of optimum values of critical surface tension, proposed by Baier,[88] the presence of optimum negative charge density on the surface proposed by Costello et al.,[89] and a diffusion layer theory proposed by Ikada et al.[90] These theories aim at simplified treatment of interactions between a material and components of living tissue.

Baier's hypothesis[91] is as follows. For the desired production of material surfaces associated with the minimum of blood reactivity, minimum thrombosis, and biological adhesion, the cumulative results of this surface characterization approach strongly indicate that the most favorable surfaces are those exhibiting critical surface tensions in the range between 20 and 30 dyn/cm, independently of the intrinsic degree of hydrophilicity or hydrophobicity of those materials. Conversely, for those many situations in which strong biological anchorage, reaction, adhesion, or induction of adjacent cellular activity is desirable, the recommendation would be to develop materials having a critical surface tension in the range of 35 to 40 dyn/cm, with a high proportion of polar interactive sites present at their interfaces.

Ruckenstein and Gourisankar[92,93] proposed a surface energetic criterion of biocompatibility of foreign surfaces. This criterion, which was identified upon analysis of the surface interactions between a typical biological fluid (i.e., blood) and synthetic surfaces, was founded on the premise that a sufficiently low (but not very low) solid-biological fluid interfacial free energy of the order of 1 to 3 dyn/cm, was necessary to fulfill the dual requirements of maintaining a low thermodynamic driving force for the adsorption of fluid components on the solid surface as well as a mechanically stable solid-fluid interface. They presented the results shown in Figure 10 and indicated that UV irradiation enhanced the blood compatibility of poly([diphenyl] siloxane).

For method (b), the optimum hydrophobic-hydrophilic balance hypothesis proposed by Lyman et al.[94] is representative. This hypothesis was based on the fact that segmented polyetherurethanes which have been used as artificial

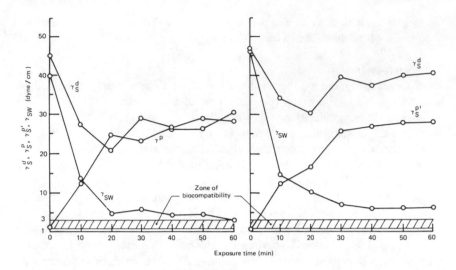

FIGURE 10.   Effect of UV irradiation (for various exposure times) on the surface free energy components of poly(diphenyl siloxane). The solid-water interfacial free energy is also plotted for different exposure times. The left side shows the values of γ_s^p and γ_s^d (evaluated by Kaelble's method), while the right side shows the values of γ_s^p and γ_s^d (evaluated by the Fowlkes-Hamilton method). (From Ruckenstein, E. and Gourisankar, S. V., *Biomaterials*, 7, 403, 1986. With permission.)

hearts are multiphase with a microphase-separated structure of polar and nonpolar segments, and that the excellent antithrombogenicity of segmented polyetherurethanes originates from the selective adsorption of albumin that is inactive to platelets. Some researchers stressed the dispersive component of surface-free over polar energy.[95,96] The hypothesis of optimal balance can also be cited between these components, as can the theory of the minimum sum of differences between dispersive and polar components of surface energies of polymer materials and blood for hemocompatible surfaces.[97,98]

The other hypothesis (b) explaining blood compatibility based on protein adsorption was provided by Okano et al.[99] This group observed that HEMA-St, HEMA-DMA, HEMA-EO, and HEMA-PO block copolymers showed promising antithrombogenic activities by suppressing activation and aggregation of platelets.[100,101] They considered that the high hemocompatibility of microphase-separated surfaces was caused by the high affinity of hydrophilic areas to serum albumin and of hydrophobic areas to fibrinogen, which resulted in surface energetic homogenicity. Accordingly, the regularity of the protein adsorptional layer provided for a lower probability of redistribution of membrane proteins of adherent platelets and decreased their morphological alteration.

As a possible mechanism involved in the suppressive effect of microdomain-structured surfaces on cellular contact-induced activation, Kataoka[102] proposed a mechanism of "capping control" as shown in Figure 11. Kataoka's

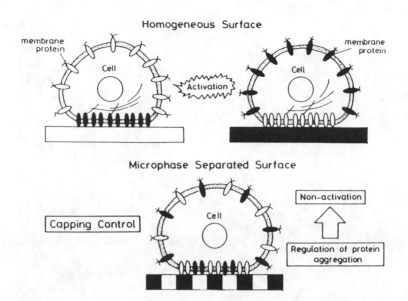

FIGURE 11. Schematic illustration of "capping control" mechanism. (From Kataoka, K., Sakurai, Y., and Tsuruta, T., *Macromol. Chem.* (Suppl. 9), 53, 1985. With permission.)

group[103,104] developed new adsorbents for adsorption chromatography of lymphocyte subpopulations using this mechanism.

Nojiri et al.[105] investigated the difference between these block copolymers and medical polyurethane derivatives. Polyurethane derivatives accumulated thick multilayers of adsorbed proteins (1000 to 2000 Å) after implantation. In contrast, block copolymers showed only a monolayer protein thickness (<200 Å). In addition, the pattern of protein distribution revealed high concentrations of fibrinogen and IgG, and less albumin adsorbed onto polyurethanes. In contrast, block copolymers showed a patchy protein distribution pattern with high concentrations of albumin and IgG, and relatively less fibrinogen. The grafts made of the polyurethane derivatives occluded within 1 month, while block copolymer grafts were patent for over 3 months. It was concluded that adsorbed monolayer patterns were more compatible than multilayered proteins. It was considered that the thin and stable adsorbed protein layer on the block copolymer was associated with the microdomain structures of the surface, and played an important role in long-term *in vivo* blood compatibility.

Recently, however, the microphase-separated structure at the surface has become suspect by some researchers.[106-108] They pointed out that the thermodynamically stable domain was concentrated on the surface. Figure 12 shows an electron microscope photograph of a polystyrene-polyisobutylene block copolymer. Only the polyisobutylene domain is seen at the surface. Okano's group revealed that the air-facing surface of PST-PHEMA block copolymer is covered by only polystyrene by using ESCA analysis (Figure 13).[109]

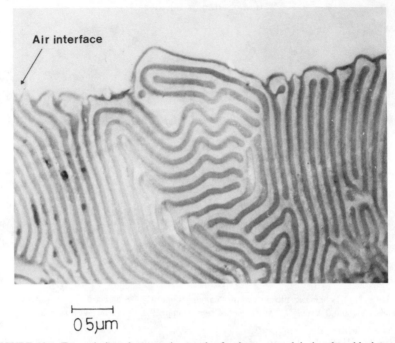

FIGURE 12. Transmission electron micrograph of polystyrene-polyisobutylene block copolymer. The polyisobutylene phase is stained. The air-facing surface is covered with the stained phase. (Photo courtesy of Dr. H. Hasegawa, Department of Polymer Chemistry, Kyoto University.)

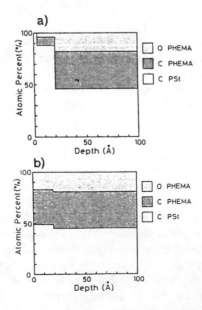

FIGURE 13. Depth profile obtained for poly(hydroxyethyl methacrylate)-polystyrene (a) block copolymer and (b) random copolymer, analyzed by X-ray photoelectron spectroscopy.

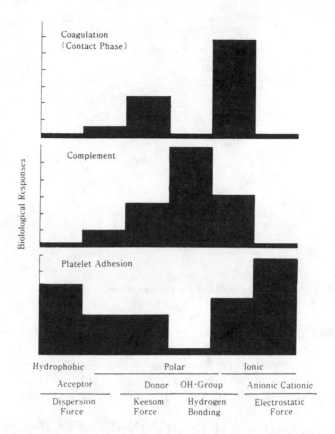

FIGURE 14. General biological responses on various material surfaces. (From Matsuda, T., *Polymers and Medicine,* Takemoto, K., Sunamoto, J., and Akashi, M., Eds., Mita Publ., Tokyo, 1989. With permission.)

Therefore, Cho et al.[110] investigated platelet adhesion onto cast film and a Langmuir-Blodgett (LB) film of block copolymers. On the surface of the cast film, adhering platelets extended a few pseudopods and maintained discoid integrity. By contrast, platelets adhering to the LB film were observed to extend many pseudopods. This result demonstrated that the highly ordered structure produced by the LB technique significantly enhanced platelet activity.

On the other hand, Yui et al.[111,112] emphasized the importance of controlling the crystalline-amorphous microstructure for the antithrombogenicity of polymer surfaces with a semicrystalline microstructure, using poly(propyleneoxide) segmented nylon 6,10. The domain size of the polymer was similar to that of polyurethane.

In approach (1), thermodynamic approach (a) and heterogenicity approach (b) were considered. Merrill[114] proposed some working hypotheses by categorizing the conventional biomaterials. Matsuda[113] summarized the biological response of biomaterials, as shown in Figure 14. Apparently, since there are

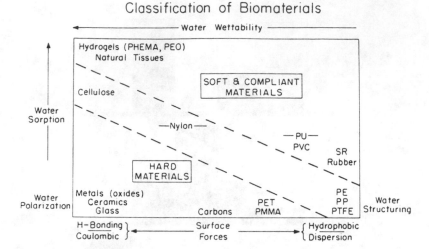

FIGURE 15. Classification of biomaterials. (From Hoffman, A. S., *J. Biomed. Mater. Res.*, 20, ix, 1986. With permission.)

many biological interactions between biocomponents and artificial materials, it would appear impossible to design blood-compatible material according to one simple concept. First, physicochemical approaches (1) will be described. Second, utilization of biological molecules (2) will be discussed. Finally, materials hybridized with cells (3) will be reviewed.

II. UTILIZATION OF NONBIOSPECIFIC ANTICOAGULANTS

A. HYDROPHILIZATION
1. Surface Treatments

Though Hoffman[115] pointed out that the terms "hydrophilic" and "hydrophobic" are too simple to describe the character of a biomaterial which is interacting with the biological environment (Figure 15), hydrophilization of surfaces is an important technique to improve biocompatibility of a material. Figure 16 shows that hydrophilization enhanced blood compatibility effectively. There are some published reports of plasma deposition methods for providing biomaterial surfaces with hydrophilic properties. Gombotz and Hoffman[116] described the plasma polymerization of alkyl alcohol and alkyl amine on polymer supports.

Marchant et al.[117] reported on the preparation of a hydrophilic polymer film by plasma polymerization of *N*-vinyl-2-pyrrolidone. Preliminary *in vitro* and *in vivo* biocompatibility studies indicated that the hydrophilic film was noncytotoxic and did not increase the inflammatory response when compared with negative controls. This result demonstrated that hydrophilization enhances not only blood compatibility but also tissue compatibility. Paul et al.[118]

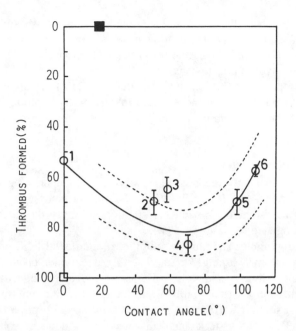

FIGURE 16. Relationship between the water-contact angle and thrombus formation on the polyurethane derivatives. (■), Heparinized polyurethane, (○), nonheparinized polyurethane, and (□), glass. 1, Polyoxyethylene grafted; 2, vinyl sulfonate-grafted; 3, acrylate grafted; 4, conventional; 5, polydimethylsiloxane grafted; and 6, fluoroalkyl-containing polyurethanes. (From Ito, Y. and Imanishi, Y., *CRC Crit. Rev. Biocompatibility*, 5, 45, 1989. With permission.)

have described the grafting of *N*-vinyl-*N*-methylacetamide hydrogels onto PVC for improving blood compatibility.

Since successful clinical application in 1960 by Wichterle,[119] hydrogels have been widely used as an important biomedical material. However, the main disadvantage of hydrogels is that they have poor mechanical properties after swelling. In order to eliminate this disadvantage, a hydrogel was grafted onto some base polymers with good mechanical properties using covalent bonding. Styrene-butadiene-styrene triblock copolymer (SBS), a microphase separation polymer with excellent mechanical properties, was used to make solvent-case films.[120]

As a kind of hydrophilization, Nagaoka and Akashi[121,122] synthesized a low-friction surface for medical devices. A low-frictional property of the surface (so-called slipperiness), is required for medical devices in contact with catheters or guide wires that are inserted into blood vessels, the urethra, or other parts of the body which have mucous membranes. If these devices do not have low friction, their introduction into the body is not only accompanied by pain, but there is also damage to the mucous membranes or the intima of the blood vessels, which may lead to infectious diseases or mural thrombus formation. One practical method involves coating with a hydrophilic polymer. Nagaoka and Akashi[121,122] developed epoxy groups containing hydrophilic copolymers that form stable, covalent bonds with amino groups coated on the substrate. On the coating surface, there were no lesions of the intima of the blood vessels and no thrombus formation.

However, hydrophilization has one downside. Ratner et al.[123] reported that not only platelet adhesion, but also the number of circulating platelets in the blood decreased with the increase in surface hydrophilicity. This indicates microemboli generation in circulating blood without thrombi formation on the surface.

2. Polyoxyethylene (POE) Graft

POE is increasingly regarded as a polymer with interesting blood-contacting properties. The low affinity of POE for proteins and other blood components has stimulated many investigators to study the interactions of blood and biomaterials based on POE.

The hydrophilicity and unique solubility properties of POE produce surfaces which are in a liquid-like state, with the polymer chains exhibiting considerable flexibility or mobility. The rapid movement of hydrated POE chains attached on a surface probably influences the microthermodynamics at the protein solution/surface interface and prevents adsorption or adhesion of proteins. POE has a steric stabilization effect in aqueous solution. It is thought that a repulsive force develops, due to a loss of configurational entropy of the surface-bound POE, when a protein or other particle approaches the POE surface. It appears that a POE surface in water has rapid motions and a large excluded volume compared to the less water soluble polyethers, thereby actively minimizing the adsorption of proteins. The POE-water interface also has a very low interfacial energy interface and thus a low driving force for protein adsorption. Proteins at or near a low interfacial energy interface will not be affected by any greater effects from the surface than they are from the bulk solution.

There has been much effort invested in preparing POE surfaces by a number of methods. Block copolymerization,[124] cross-linking to produce a POE network,[125,126] and surface treatments such as direct adsorption of high molecular weight POE[127,128] have been performed. Recent developments in the use of POE will be described.

Covalent grafting of POE has also been performed by Nagaoka and Nakao.[129] They called the POE graft "molecular cilia" as shown in Figure 17. Kinetically, the process by which plasma protein interacts with and is irreversibly adsorbed onto the surface of a material to form a foothold for platelet adhesion is largely influenced by the length of time the protein remains on the surface. On the surface of a POE-coated material, microflows of water are induced by the movements of hydrated POE chains; then because plasma protein is prevented from staying on the surface, absorption is inhibited. This is considered to be a molecular version of the way in which cilia function on some mucosal epithelia.

Takahara et al.[56] and Okkema et al.[130] investigated polyetherurethane

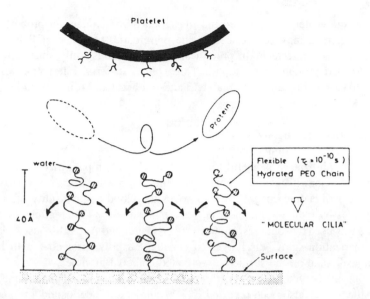

FIGURE 17. The interaction between blood components and hydrated POE chains on the surface. The rapid movement of the long POE chains influences the microhemodynamics at the blood-material interface more effectively than do the short POE chains. (From Nagaoka, S. and Nakao, A., *Biomaterials*, 11, 119, 1990.)

containing polytetramethylene glycol, polyoxyethylene, and a mixture of them in the main chain. On the other hand, Liu et al.[135] synthesized a new polyurethane with POE side chains in the soft segment. First, epoxized polybutadiene was synthesized and grafted with POEs. The grafting of POE to polyurethane suppressed adsorption and denaturation of plasma proteins and platelet adhesion. This investigation showed that protein conformation was not affected by the presence of POE chains.

van der Does and colleagues attempted to graft POE onto Pellethane® 2363 80A (Pell 80A).[132] They utilized two techniques. One was the use of grafting high molecular weight POE with dicumyl peroxide (DCP). Pell 80A substrates were dipped in a solution of POE and DCP. After drying, the POE/DCP-coated substrates were UV or heat treated in order to form a POE and Pell 80A network. The other technique was the use of controlled oxidation for grafting methoxy poly(ethylene glycol) 400 methacrylate (MPEGMA-400). The graft polymerization method was the same as that of 2-hydroxyethyl methacrylate onto polyurethanes as described by Feng et al.[132] Blood compatibility of polyurethane was improved by the grafted POE.[133,134]

POE-containing nonionic polymeric surfactants (block copolymers) were studied as a possible means of producing POE-rich surfaces by a simple coating treatment of a common hydrophobic medical material.[135-139]

Bots et al.[140,141] reported that UV-initiated cross-linked blends of poly(oxypropylene) and POE showed good blood compatibility. Surface analysis of these blends suggested that the good blood-contacting properties of these materials may be ascribed to the preferential presence of POE at the polymer-water interface. In protein adsorption studies, only small amounts of adsorbed proteins were found. Three blood material interaction tests *in vitro* have good results: low platelet adhesion and kallikrein generation, and high APTT value.

B. HYDROPHOBILIZATION
1. Surface Treatments
Hydrophilization is useful for obtaining blood compatibility as shown in Figure 16. Chawla[142] showed that plasma polymerized cyclic organosiloxanes provide a filler-free coating that improves the blood compatibility of silastic substrates by reducing platelet and leukocyte adhesion. Thereafter, the technique was used by some researchers.[142-144] Hasirci and Akovali[143] used plasma polymerization (glow discharge) to construct a polymeric coat with hexa-methylsiloxane on the surface of activated charcoal granules.

Plasma polymerization methods for modifying nonorganic biomaterial substrates have also been reported.[146-148] Hahn et al.[146] deposited a protective hydrocarbon coating for polar graphic electrodes used in the measurement of tissue oxygen concentration. Plasma-assisted techniques such as sputtering or spraying were investigated for modifying metal components of prostheses, particularly for hard-tissue devices.

2. Fluorination
Garfinkle and co-workers[149,150] modified small-diameter Dacron grafts with a plasma-deposited fluoropolymer, which markedly improved graft patency in an *ex vivo* femoral shunt model in the baboon. To understand the basis for this enhanced thromboresistance, studies of the adsorption of proteins from plasma were performed using radio frequency glow discharge (RFGD) treatment of poly(ethylene terephthalate) (PET). The adsorption of fibrinogen and albumin to the fluorocarbon-coated surfaces was comparable to the adsorption of the proteins to polytetrafluoroethylene (PTFE) and PET. However, the elutability of fibrinogen and albumin from the RFGD fluorocarbon surfaces with sodium dodecyl sulfate was much lower than that from PTFE or PET. Other RFGD treatments of PET, such as ethylene deposition or argon etching, did not reduce the extent of albumin elutability as dramatically as did the RFGD fluorocarbon treatments. Garfinkle et al.[149] considered that the strong albumin binding to RFGD fluorocarbon surfaces was exploited clinically to enhance the retention of albumin preadsorbed to blood-contacting surfaces to render them thromboresistant.

Yeh et al.[151] had plasma modified with silicone rubber tubing and expanded polytetrafluoroethylene graft materials with fluorocarbon. These surface modifications were found to impart improved blood compatibility as determined by reduced platelet consumption and acute thrombus formation

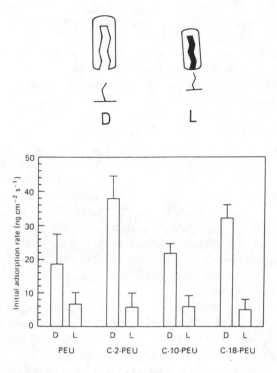

FIGURE 18. Initial adsorption rates of lipidized albumin (L) and delipidized albumin (D) onto alkyl grafted polyurethane surfaces. (From Pitt, W. G., Grasel, T. G., and Cooper, S. L., *Biomaterials*, 9, 40, 1988. With permission.)

in an arteriovenous shunt system in the baboon. They proposed a unique explanation for this high compatibility.

Fluoroalkyl groups have been introduced into polyurethane by several groups as reviewed by Ito and Imanishi.[42]

3. Alkyl Chain

The alkylation of blood-contacting polymers with C-18 linear alkyl chains reduces thrombus deposition at the blood/polymer interface. Munro et al.[152] proposed that the reduction of thrombus formation was a result of albumin adsorbing specifically and strongly to the alkyl chains, since albumin frees fatty acids[153,154] and binds the alkyl chain attached to a solid polymer substrate.[155] This specific adsorption was postulated to increase the amount of albumin on the surface and decrease the number of platelets on the surface available to aggregate into a thrombus. Figure 18 shows the initial adsorption of albumin derivatives. The delipidized albumin is referred to as "D-HSA", while the lipidized albumin is referred to as "L-HSA". The adsorption of L-HSA on the alkylated polyurethane surfaces was reduced remarkably.[156-160] The enhancement of albumin adsorption influenced other blood-material interactions. For example, fibrinogen adsorption-inhibiting and complement system passivation effects were observed.[161]

However, Salzman et al.[15] reported that platelet reactivity of poly(alkyl methacrylate)s increased with the length of the alkyl side chains. They considered that the increment of platelet activation was caused by fibrinogen adsorption without denaturation.

C. NEGATIVELY CHARGED SURFACES

It is known that polymers with hydrophilic and negatively charged surfaces have low adsorption of serum proteins and do not have strong interactions with blood cells.[89] Graft copolymerization is an effective method for improving the biocompatibility of polymers by introducing hydrophilicity and negative charges on the surface. Yamauchi[162] has grafted poly(vinyl chloride) (PVC) tubing with 2-hydroxyethyl-methacrylate for blood circulation applications. Singh et al.[163] grafted methacrylic acid onto PVC by a radiation-induced method. They reported that the weight of thrombus formed on the grafted PVC was less than that of PVC and glass and decreased with the increase in the graft level.

Nigretto et al.[164] investigated the effect of negative charge on the surface using dextran derivatives. The derivatives studied were dextran sulfonate, dextran methylcarboxylate, and derivatives thereof carrying methylbenzylamide and methylbenzylamide sulfonate groups in addition to the remaining methylcarboxylate groups. It was found that the contact activation capacities of the materials paralleled the course of their anticoagulant activities.

When human plasma is brought into contact with a negatively charged surface, proteolytic reactions which activate the intrinsic blood coagulation cascade occur. These reactions are connected to so-called contact activation as mentioned before. Four interacting proteins are involved in contact activation: factor XII, factor XI, prekallikrein (the zymogen of kallikrein), and the nonenzymatic cofactor high molecular weight kininogen (HMWK). However, the question as to whether this superficial process is specific (i.e., related to some biological recognition of certain functional groups) or not (i.e., related to any negatively charged surface) is still controversial.[165]

Miyamoto et al.[166] reported that though blood coagulation by positively charged particles was attributed to platelet activation, an enhancement of blood coagulation was also observed in the presence of erythrocytes, leukocytes, their cell membranes or negatively charged phospholipids, and phosphatidylserine instead of platelets. They concluded that hydrophilic and low-charged surfaces suppressed blood coagulation.

D. SYNTHETIC POLYMER-POLYMER HYBRIDIZATION FOR MULTIPHASE POLYMERS

Multiphase polymers in biomaterial applications are important. The representative example is polyurethane. Therefore, there are many ongoing efforts to synthesize new hybridized polymeric materials.

1. Polyurethane-Polydimethylsiloxane

Cooper's group[167] introduced polydimethylsiloxane into polyetherure-thane as the soft segment. They reported that polydimethylsiloxane-based materials showed higher levels of thrombus deposition than the polyethylene oxide-based polymers. Moreoever, they[167] described that despite the variation in the surface properties, the blood compatibility of these polymers was not significantly affected by the addition of the PDMS-containing polyols.

Other kinds of PDMS-containing polyurethane have been synthesized as reviewed by Ito and Imanishi.[42]

2. Polyurethane-Polyamide

Poly(amine-urethanes) with long repeating units have been prepared from 2-aminoethanol, α,ω-alkylene dicarboxylic acid chlorides, and 1,6-diiso-cyanatohexane.[169] These partially crystalline polymers undergo slow hydrolysis when suspended in buffered water, but were found to be more readily degraded by the enzyme subtilisin.

3. Polyurethane-Polyester

Polyurethane-Dacron was constructed by Pizzoferrato et al.[170] They found that polyethylene terephthalate (Dacron®) coated with urethane L 325 (Du Pont's Adriprene®) gave evidence of enhanced thromboresistance. It is considered that among substances released from platelets, β-thromboglobulin, platelet factor 4, and thromboxane B_2 are at very low levels in human plasma under physiological conditions and therefore can be used as markers of platelet release.[171,172] β-Thromboglobulin is a platelet-specific protein that has weak heparin neutralizing activity and inhibits PGI_2. Platelet factor 4 is a protein capable of neutralizing heparin. Thromboxane A_2 is predominantly produced from arachidonic acid in the platelets and is formed from PGH_2 in a reaction catalyzed by thromboxane synthetase. It undergoes nonenzymatic hydrolysis to a relatively inactive end product, thromboxane B_2. The assay of these three platelet-released substances showed that Dacron-urethane composites were better thromboresistors than any other materials conventionally used in vascular surgery.

Leake et al.[173] found that polyethylene terephthalate (Dacron®) coated with polyurethane L235 (Du Pont's Adriprene®) enhanced thromboresistance, by means of assays of three platelet-released substances, i.e., β-thrombo-globulin, platelet factor 4, and thromboxane B_2, as well as the quantification of platelet retention.

Nakagawa et al.[174,175] modified a polyetherurethane tube to render enough mechanical strength by incorporation of knitted polyester fibers for its skeleton and implanted the graft in dogs.

4. Polyurethane-Polylactide

Polyurethane-poly(L-lactide) mixtures were prepared by Gogolewski and Pennings.[176] Their current concept of the synthetic skin substitute consists of a microporous vapor-permeable polyetherurethane top layer and a separate bottom layer, composed of a biodegradable polyesterurethane elastomer network. This network is designed to degrade rapidly to nontoxic products, not having disadvantages such as low rate of degradation and release of the toxic, carcinogenic, aromatic diamine 4,4′-methylenedianiline upon degradation of the segmented polyurethane, associated with the polyurethane-poly(L-lactide) mixture.[177]

5. Polyurethane-Polypeptide

Kashiwagi et al.[178] synthesized polyetherurethaneurea-poly(γ-benzyl-L-glutamate) block copolymer. After the end groups of polyurethane were terminated with amino groups, the amino groups were used as initiators for polymerization of *N*-carboxyanhydride. Platelet adhesion was suppressed on the block copolymer, compared with each homopolymer. Polyurethanes containing peptide components have also been synthesized by Imanishi's group.[179]

6. Polypeptide-Synthetic Polymers

Since α-amino acid is a biocomponent and the block copolymer was believed to have high blood compatibility, various kinds of block polypeptide copolymers, such as vinylpolymer-polypeptide,[180,181] polydimethylsiloxane-polypeptide,[182,183] and polyether-polypeptide[184,185] and other polypeptide derivatives[186,187] were synthesized. The blood compatibilities of these copolymers and derivatives were reportedly higher than those of each homopolymer and the original polypeptide, respectively.

7. Polyether-Polysiloxane

Pekala et al.[188,189] developed a new method and material for connecting polyether chains to form blood-compatible materials. This method involves reaction of the hydroxyl end groups of polypropylene glycol (PPG) with the epoxy groups of polyglycidoxy propyl methyl siloxane (PGPMS). The hydroxy/epoxy reaction resulted in the formation of covalently cross-linked networks. The PPG/PGPMS networks exhibited low fibrinogen adsorption and low platelet activation.

E. CARBON COATING

Carbon-carbon materials have been discussed as biomaterials by Adams and Williams.[190] The general structure of carbon-carbon composites is characterized by the presence of two phases: one is made of carbon fibers and plays the role of reinforcing agent (stiffener), corresponding to various arrangements of fibers and meshes; the other is made of pyrolytic carbon and is called the matrix, more or less filling up the gaps between the fibers of the preform and resulting from the chemical vapor deposition process. The hemocompatibility was investigated by Baquey et al.[191]

Carbon-coated vascular prostheses were evaluated by Boyd et al.,[192] who reported the relative nonthrombogenicity of ultralow temperature, isotropically (ULTI) carbon-coated, small-diameter vascular prostheses, compared with polytetrafluoroethylene grafts of the same size. They considered that in addition to their inherent property of biocompatibility, carbon-coated graft surfaces might provide an excellent substrate for attachment of seeded endothelial cells.

Chignier et al.[193] developed two new classes of carbonaceous composites for cardiovascular devices. One is a carbon/carbon composite and the other is a carbon/silicone carbide composite. The former was obtained by chemical vapor infiltration of a binder made of pyrolytic carbon into the pores of the initial texture, up to a density between 1.7 and 1.9. The latter was obtained by two phases of chemical vapor infiltration into the pores of the initial texture: the first phase was carried out following the same process as that described for the carbon/carbon composites. Afterward, a second phase of chemical vapor infiltration was performed using a binder of silicone carbide at 1000°C. They observed that tissue ingrowth was enhanced by the carbon treatment.

III. UTILIZATION OF BIOSPECIFIC ANTICOAGULANTS

The techniques of immobilization of bioactive reagents to synthesize hemocompatible polymers have been extensively reviewed by Wilson[194] as follows: (1) heparinization; (2) heparin plus adjuvant (heparin plus an antibiotic, heparin plus urokinase, heparin plus albumin, and heparin plus chitosan); (3) bioactive agents used without heparin (urokinase and trypsin immobilization, prostaglandin-releasing polymers, brinolase immobilization, polysilicone plus aspirin and theophylline, use of Evan's blue, and Biomer grafted with monomeric prostaglandin analogues); and (4) techniques not involving the use of specific bioactive agents (polymers designed to have intrinsic hemocompatibility, carbon coatings, albumin coating, biolized surfaces, integrally textured surfaces, and radiation grafting on polyetherurethanes). Tanzawa[195] summarized the anticoagulant reagents used for synthesizing blood-compatible materials as shown in Table 3. Most researchers currently consider that an applicable blood-compatible material will be accomplished by combination with biological components.

A. COAGULATION INACTIVATION
1. Heparin
The anticoagulant activity of heparin has been well characterized and occurs by accelerating the inhibitory effect of antithrombin III on factors XIIIa, XIa, Xa, IXa, and thrombin.[197] Heparinization is the most popular method to improve the blood compatibility of medical devices. Heparinization methods are divided mainly into three categories: (1) heparin modification of material coatings; (2) ionically binding of heparin; (3) covalently binding of heparin, as shown in Figure 19. Some reviews in this field have been published.[197] In this section, recent progress will be described.

TABLE 3
Bioactive Reagents for Synthesis of Antithrombogenic Materials

	Inhibition of platelet adhesion	Inhibition of coagulation system	Activation of fibrinolysis system	Inhibition of complement system
Natural product	Prostaglandin Thrombomodulin Albumin Adenyl cyclase Kinethrombin Ditazol	Heparin Antithrombin III Protein C Hirudin	Urokinase Streptokinase Thrombomodulin	
Synthesized product	Aspirin Chicagoic acid	MD-805		FUT-175

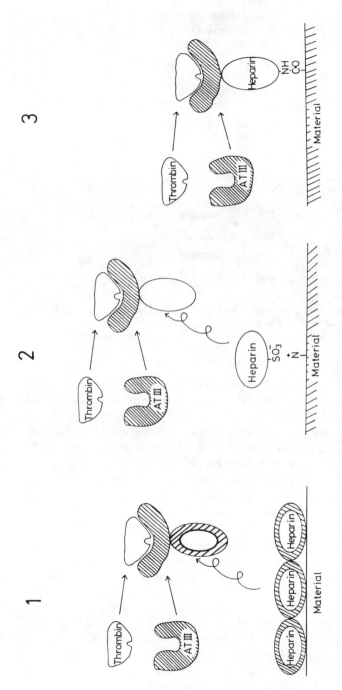

FIGURE 19. Heparinization methods. (1) Coating, (2) ionic binding, and (3) covalent binding.

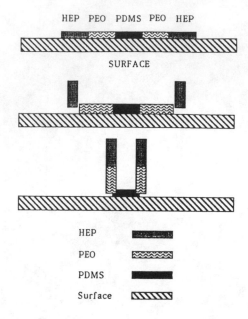

FIGURE 20.   Schematic representation of the possible surface orientations of PDMS-[PEO-HEP]2 pentablock copolymers. (From Piao, A. Z., Nojiri, C., Park, K. D., Jacobs, H., Feijen, J., and Kim, S. W., *J. Biomater. Sci., Polym. Ed.*, 1, 299, 1990. With permission.)

a. Coating

First, heparin was coated by complexing with an ammonium salt, such as benzalkonium chloride. However, the release rate was too high to maintain long-term antithrombogenicity. Moreover, the released cationic reagents were toxic for humans. To overcome these problems, a heparin-PVA compound has been synthesized and investigated by Sefton's group extensively.[198-201]

Recently, as a coating for heparin-bound polymer, Grainger et al.[202-204] synthesized a new amphiphilic triblock copolymer of polydimethylsiloxane (PDMS), POE, and heparin (Hep). Films of this material cast onto glass substrates demonstrated a significant ability to interact with clotting factors, especially antithrombin III and factor Xa, *in vitro* in plasma preparations. Various assays demonstrated measurable quantities of heparin on the surface of these films after exposure to plasma or aqueous media. Because the low-energy PDMS phase dominates surfaces of this copolymer when solvent-cast under air or vacuum conditions, attempts were made to explain the surface restructuring and rearrangements induced in hydrated or aqueous environments that permit surface accessibility and bioactivity of the heparin moieties. Compositional models based on ADESCA, SIMS, and DSC data have been presented for desiccated and hydrated copolymer surfaces as shown in Figure 20. On the other hand, platelet adhesion and release data have suggested that PDMS-POE-Hep and its homopolymer constituents react minimally with platelets and seemed to pacify surface activation compared to Biomer. The group synthesized a polydimethylsiloxane-polyoxyethylene-heparin CBABC-type block copolymer PDMS-[POE-Hep]$_2$ for further improving blood compatibility.

b. Ionic Binding

Heparin has been ionically bound by many researchers. Shiomi et al.[206] immobilized heparin ionically onto the surface of an ethylene-vinyl alcohol copolymer (EVAL) membrane which was derivatized by amino acetalization to produce cationic surface charges. EVAL was suitable as a blood-compatible polymeric material. Its fabrication as a hollow fiber membrane had been applied in hemodialysis; however, the systemic administration of heparin was required in clinical use to prevent thrombus deposition. The binding of heparin to the membrane achieved a nonthrombogenic surface, and prevented excessive heparin administration.

Recently heparinization of polyetherurethaneurea, which is mechanically stable *in vivo,* has been the subject of several investigations. Ito et al. have synthesized derivatives of polyetherurethaneurea capable of heparinization without losing mechanical stability.[42,197] The heparinizable polyetherurethaneureas were designed, synthesized, and investigated for *in vitro* antithrombogenicity, adsorption of plasma proteins, interactions with platelets and cells, and *in vivo* antithrombogenicity.[207,208] The blood-compatibility mechanism of these polyurethanes was recently discussed by Wilson.[210]

Barbucci's group[211-215] and Tanzi and Levi[216] have synthesized a new polymer. This material had some particular properties: ease of synthesis, heparin-binding capacity, and adequate mechanical properties relative to polyurethane.

Heparinized polyurethane for coating was also synthesized by Iida et al.[217] They synthesized several polymerizable cationic surfactants with a methacrylate group. These monomers yielded complexes soluble in some organic solvents to result in ionically equivalent bonding with heparin. Using one of these complexes derived from quaternary vinyl monomers and heparin, radical polymerization with hydrophilic co-monomer was performed in dimethylformamide in the presence of polyurethane. On the other hand, Nagaoka et al.[218] used Anthron® as a coating material for the sensor.

c. Covalent Binding

Covalent binding of heparin had not been believed suitable for antithrombogenicity. However, recent developments have revealed that covalently bound heparin is active by pertinent effective binding and spacer introduction.[197]

Arnander et al.[219,220] immobilized heparin onto polyethylene tubing. The method implies repeated treatments of tubing with polyethyleneimine together with a cross-linking agent (glutaraldehyde) followed by incubation with dextransulfate. Finally, the tubing is treated with polyethyleneimine without the cross-linking agent and partially nitrous degraded heparin (ca. M_r 8000 Da) was covalently bound, by reductive amination with sodium cyanoborohydride, to the polyethyleneimine layer. The heparinized surface caused less intimal hyperplasia and thrombi than the nonheparinized surface. They also found that the function of the heparin surface was dependent on the blood flow rate. Larsson et al.[221] shortly reviewed the end point-attached heparin. This method was applied to poly(methyl methacrylate) surface modification.[222] Chandler et al.[223] discovered a method to estimate the amount of active heparin bound to a surface.

Introduction of POE as a spacer chain has been carried out by some researchers. Ebert and Kim[224] reported that heparin-immobilized surfaces with alkyl spacers prevented fibrin net formation due to the heparin bioactivity which increased with increasing spacer length to n = 12. These surfaces, however, activated platelets due to hydrophobicity of the alkyl spacers and perhaps due to heparin itself resulting in platelet aggregation.[225,226]

The method of immobilization and the introduction of a spacer chain improved the activity of covalently immobilized heparin (Figure 21).

From this perspective, heparin-immobilized segmented polyurethaneurea surfaces using hydrophilic poly(ethylene oxide) chains as spacers were developed.[227] Han et al.[228] used hexamethylene diisocyanate to introduce functional groups onto a polyurethane surface, and produced POE-grafted and heparinized polyurethane surfaces via attached free isocyanate groups; and Barbucci[229] used it to graft amino-terminated poly(amido-amine) onto polyurethane surfaces. The heparinized POE-grafted polyurethane surfaces displayed enhanced blood compatibility due to the synergistic effects of POE and heparin.[230] Nagaoka et al.[231] also immobilized heparin on BrCN-activated Sepharose gel via long POE spaces. They found that the activity of immobilized heparin increased with elongation of the spacer chain and the highest activity was obtained when the chain length of the spacer was between n = 10 to 20.

Tay et al.[232] immobilized commercially obtained (Diosynth) heparin onto poly(vinyl alcohol) (PVA) hydrogels and onto POE hydrogels activated by tresyl chloride. They found that as tresyl chloride activation of PVA increased, the specific activity of the bound heparin toward thrombin and antithrombin decreased by nearly a factor of ten and that commercial heparin bound to POE had nearly tenfold greater activity than when bound to PVA at a comparable concentration. These findings suggested that the long ''leash'' provided by POE hydrogels may give heparin more access to thrombin-antithrombin pairs than tightly bound to PVA and that crowding of heparin units on a surface limits access to thrombin-antithrombin pairs.

Other kinds of spacer arms, such as polyallylamine were used by Liu et al.[233] for ionic and covalent bonding of heparin as shown in Figure 21 B and C.

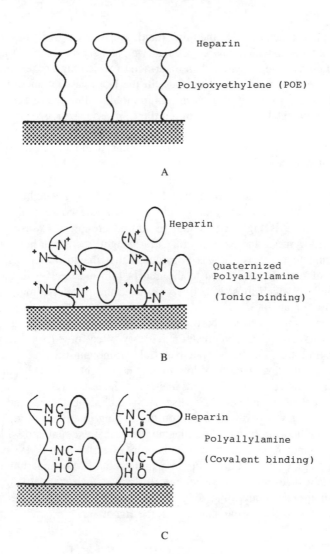

FIGURE 21.   Introduction of several kinds of spacer chains between the material surface and the immobilized heparin.

d. *Heterogenicity of Heparin*

Regarding the effect of heparin on platelets, there has been a diversity of experimental results.[234-236] Heparin interacts with platelets and has been reported to both induce and inhibit platelet aggregation. Salzman and associates[237] reported that the interaction between heparin and platelets was less for the low molecular weight fraction, particularly if it had high affinity for antithrombin III. Kanamori et al.,[238] Holmer et al.,[239] and Caranobe et al.[240] have reported that the low molecular weight fraction of heparin suppresses platelet activity. On the other hand, Grevelink et al.[241] have reported that a surface to which low molecular weight heparins were immobilized adhered platelets more easily than a surface to which high molecular weight heparins were immobilized. However, in the latter case, it should be noted that only a small amount of heparin was immobilized on the surface. It might be argued that the interactions between the film surface uncovered by heparin with plasma proteins or platelets predominate over those between heparin and platelets. Liu et al.[242] reported that the amount of immobilized heparin was so large that the low molecular weight fractions activated AT III more strongly, which probably led to the higher *in vitro* antithrombogenicity. They examined the contents of functional groups in different molecular weight fractions of heparin. In the low molecular weight fraction, the content of negatively charged groups is highest and that of amino groups is lowest. This was considered to mean that the low molecular weight fraction of heparin is least stimulating to platelets. It was considered that the low molecular weight fraction of heparin was bound to the matrix film most easily in the case of ionic binding which most effectively reduced the platelet-stimulating properties of the matrix film.

When compared to standard heparin, the low molecular weight heparin fractions have a relatively greater inhibitory effect on factor Xa activated clotting time than on the activated partial thromboplastin time.[242] Over the past few years, development of low molecular weight (LMW) heparin derivatives has resulted in much current interest in the use of LMW heparin in the management of thromboembolic disease. The clinical interest and rationale for use of LMW heparin are based on the finding in animals that it delivers antithrombotic activity with less bleeding than that encountered with regular heparin. Standard heparin, however, is more effective than LMW heparin in inhibiting aggregation of thrombin on platelets, but LMW heparin has not yet been established clinically to be more effective than standard heparin for minimizing postoperative bleeding. In the future, selection of the nature of heparin for immobilization should become important.

2. Heparinoid

Sulfonate groups were introduced to synthetic polymers to enhance the blood compatibility of materials. For example, van der Does et al.,[243] Sederal et al.,[244] Crassous et al.,[245] and Gebelein and Murphy[246] modified unsaturated polymers with *N*-chlorosulfonyl isocyanate (CSI). Muzzarelli et al.[247] synthesized sulfated *N*-(carboxymethyl)chitosans as anticoagulants. Kanmangne et al.[248-256] and Mauzac and Jozefonvicz[257-260] investigated heparin-like activity of insoluble sulfonated polystyrene and dextran, respectively. Grasel and Cooper[261] and McCoy et al.[262] evaluated the blood-contacting properties of polyurethanes grafted with varying amounts of propyl sulfonate groups and polyetherurethanes based on sulfonic acid-containing diol. Ito et al.[263] introduced sulfanate and carboxylate groups in unsaturated polyurethaneureas.

The introduction of sulfonate groups onto the polymer to enhance the blood compatibility was generalized by using sodium vinyl sulfonate (VS).[264,265] The method is to graft-polymerize VS onto the film which was glow discharged in advance. This technique rendered any material an anticoagulant. The mechanism of antithrombogenicity of this PVS-grafted material is shown in Figure 22.

3. Thrombin Inhibitor

Thrombin inhibitor has been used in order to improve blood compatibility of materials without the inactivation of antithrombin III. Matsuda et al.[266] blended MD-805, synthesized by Kikumoto et al.,[267] with a polymer matrix for slow release. Ito et al.[268-270] synthesized polymerizable MD-805 by coupling with *N*-(2-aminoethyl)acrylamide and graft-polymerized onto glow-discharged polyurethane. Thrombi were not formed on the grafted material. The antithrombogenic mechanism is shown in Figure 23.

Furthermore, Ito et al.[271] immobilized thrombin substrate analogue peptides onto the polyurethane film. In this case, the thrombin reacts with the thrombin substrate analogue faster than with fibrinogen as demonstrated in Figure 24. They demonstrated the mechanism by immobilization of fluorescent synthetic thrombin substrate.[272]

B. FIBRINOLYSIS ACTIVATION

Several components of the fibrinolytic system have been successfully immobilized on graft materials. Sugitachi et al.[273] immobilized urokinase on nylon. Watanabe et al.[274] also immobilized urokinase on collagen. Senatore[275,276] immobilized urokinase on fibrinocollagenous tubes. Rimon and Rimon[277] immobilized streptokinase and plasminogen on two polymeric surfaces. Shankar et al.[278] immobilized plasmin on collagen beads. The plasmin inhibitor α_2-PI did not inhibit the ability of immobilized plasmin to lyse fibrin.

On the other hand, Capet-Antonini and Tamenasse[279] connected urokinase by covalent bonding to the surface of a poly(acrylamide) film. Kaetsu et al.[280] coated the surface of a poly(vinyl chloride) tube with urokinase encapsulated in a 2-hydroxyethyl methacrylate (HEMA)/tetraethylene glycol diacrylate copolymer network by radiation-induced low-temperature polymerization.

In the utilization of urokinase, however, instability against plasma protein inhibitors such as antithrombin III and α_2-macroglobulin is a problem. Remy and Poznansky[281] modified urokinase with serum albumin or POE and found that the stability of the modified urokinase against protease inhibitors was increased accompanying a simultaneous depression of the enzymatic activity. Liu et al.[282] immobilized urokinase on HEMA by encapsulation in a HEMA/*N,N'*-methylenebisacrylamide copolymer network, and they investigated the conditions for the efficient protection of urokinase from attack by protease inhibitors without losing enzymatic activity.

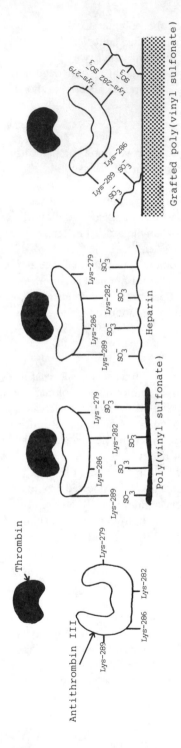

FIGURE 22.   Schematic representation of interactions of thrombin with antithrombin III in the presence of heparin, poly(vinyl sulfonate) (PVS), or immobilized PVS.

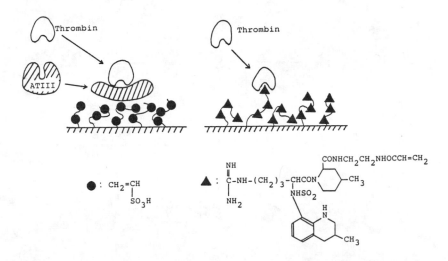

FIGURE 23. Schematic representation of the antithrombogenic mechanism of materials grafted with poly(vinyl sulfonate) or polymer containing thrombin inhibitor (MD-805) in the side chains. (From Ito, Y., Liu, L. S., and Imanishi, Y., *Artif. Organs,* 14, 132, 1990. With permission.)

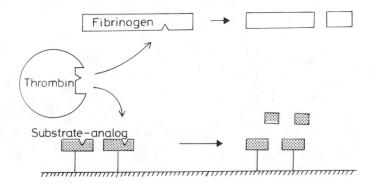

FIGURE 24.. Schematic representation of antithrombogenic mechanism of an immobilized thrombin substrate analogue. The analogue is cleaved instead of fibrinogen cleavage for fibrin formation.

C. PLATELET INACTIVATION

Antiplatelet reagents were released in order to give material antithrombogenicity.[283,284]

On the other hand, Grode et al.[285] demonstrated that the covalent immobilization of prostaglandin E_1 (PGE_1) on glass reduced platelet adhesion to that surface. Ebert et al.[286] immobilized prostacyclin (PGI_2) onto polystyrene. In their study, prostaglandin $F_{2\alpha}$ ($PGF_{2\alpha}$) was immobilized onto the surface followed by chemical conversion to PGI_2 as shown in Figure 25. Limited platelet activity studies on these surfaces showed greatly reduced

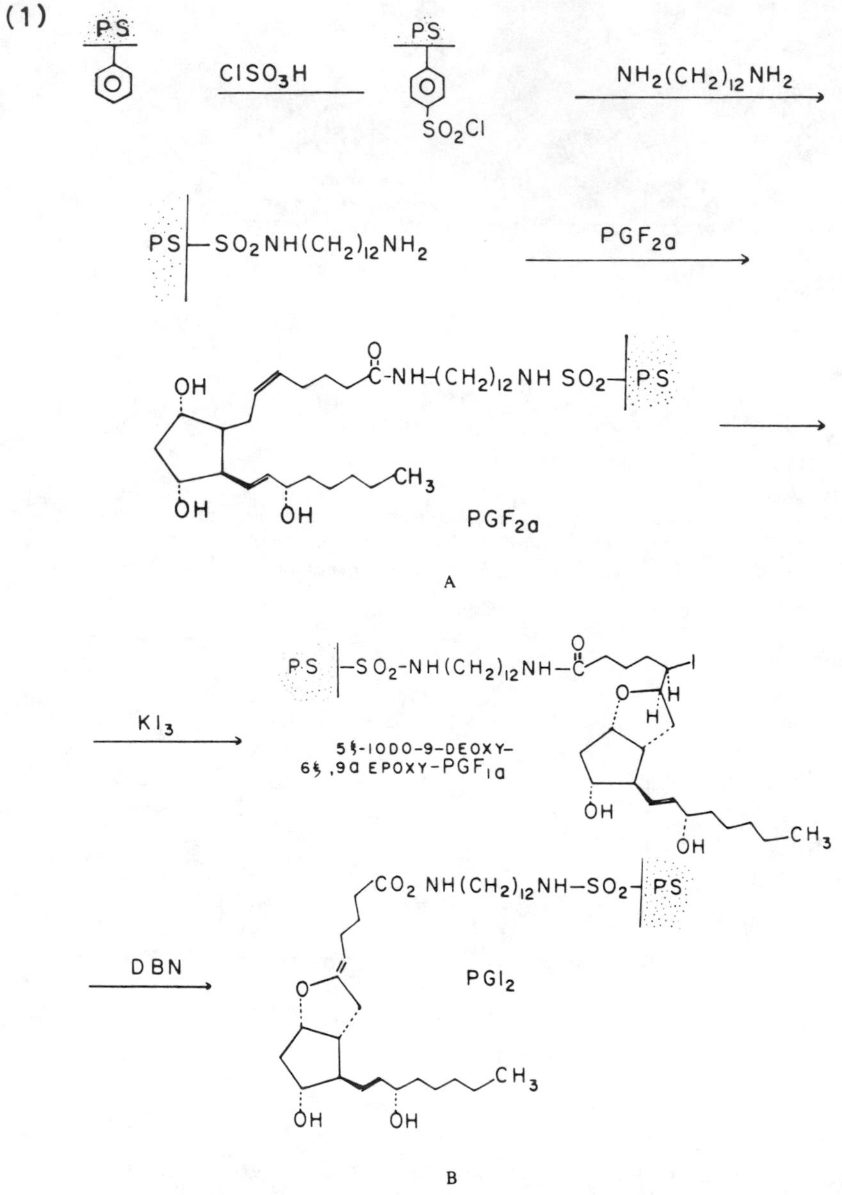

(1)

PS

$ClSO_3H$

PS

SO_2Cl

$NH_2(CH_2)_{12}NH_2$

$PS{-}SO_2NH(CH_2)_{12}NH_2$

$PGF_{2\alpha}$

OH

$\overset{O}{\underset{\|}{C}}{-}NH{-}(CH_2)_{12}NH\ SO_2{-}PS$

CH₃

OH   OH

$PGF_{2\alpha}$

A

$PS{-}SO_2{-}NH(CH_2)_{12}NH{-}\overset{O}{\underset{\|}{C}}$

KI_3

5ξ-IODO-9-DEOXY-
6ξ,9α EPOXY-PGF₁α

OH

OH   CH₃

$CO_2\ NH(CH_2)_{12}NH{-}SO_2{-}PS$

DBN

PGI_2

CH₃

OH   OH

B

FIGURE 25.   Immobilization of prostaglandin derivatives on polymeric materials. (1) (A) Reaction scheme for the immobilization of $PGF_{2\alpha}$ including chlorosulfonation and diaminododecane
derivatization of polystyrene. (B) Reaction scheme for the conversion of immobilized $PGF_{2\alpha}$ to
immobilized prostacyclin. DBN = 1,5-diazabicyclo-5-nonene. (From Ebert, C. D., Lee, E. S.,
and Kim, S. W., *J. Biomed. Mater. Res.*, 16, 629, 1982. With permission.) (2) (a) Formation
of polylysine adduct, (b) immobilization of adduct. (From Llanos, G. and Sefton, M. V.,
Biomaterials, 9, 429, 1988. With permission.)

(2)

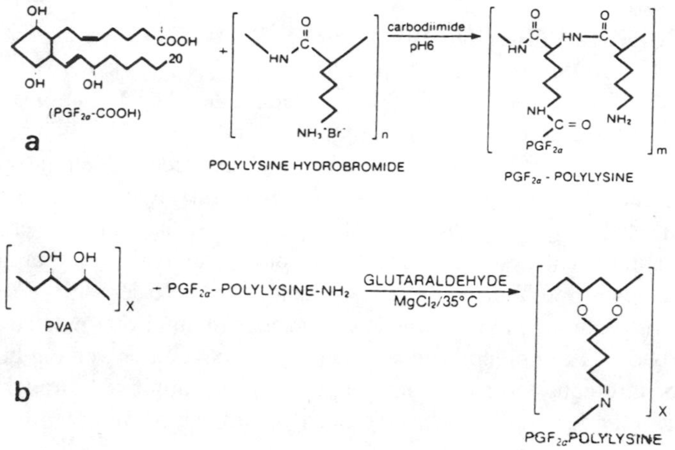

a (PGF$_{2a}$-COOH) + POLYLYSINE HYDROBROMIDE → PGF$_{2a}$ - POLYLYSINE

b PVA - PGF$_{2a}$- POLYLYSINE-NH$_2$ →(GLUTARALDEHYDE / MgCl$_2$/35°C)→ PGF$_{2a}$ POLYLYSINE

FIGURE 25. (continued)

activity compared to control polystyrene surfaces. Bamford et al.[287] reported the synthesis of a hydroxymethacrylate monoester of BW245C (prostacyclin analogue) and subsequent immobilization on polyetherurethane, but gave insufficient data to determine whether platelets were compatible with the surface.

Llanos and Sefton[288] reported upon the chemical characterization of a prostaglandin F$_{2\alpha}$-polylysine adduct which was immobilized to the poly(vinyl alcohol) gel using glutaraldehyde as shown in Figure 25b.

On the other hand, Miura et al.[289] immobilized thrombomodulin (I-TM), which prolonged clotting time, and inhibited thrombin-induced activities, such as aggregation and release of serotonin in platelets. The ability of thrombin to activate protein C was enhanced by I-TM. The anticoagulant activity was considered to be mainly due to the proteolytic activity of activated protein C on coagulation factors Va and VIIIa.

D. COMPLEMENT INACTIVATION

Matsuda et al.[283] blended FUT, which was developed by Torii Co. Ltd., Japan, for controlled release from the matrix to make the material anticoagulant. They described that *ex vivo* results correlated well with those from an *in vitro* hematological evaluation.

E. OTHER BIOLOGICAL AGENTS IMMOBILIZATION

Grafts impregnated with various bioerodible coatings were proposed. They include coating of albumin,[290-298] gelatin,[299-301] and elastin derivatives.[302] Nose et al.[303] called the gelatin-coated surface "biolized surface". These

immobilizations were for preventing platelet adhesion and for promoting endothelialization.

F. HYBRIDIZATION OF BIOACTIVE REAGENTS
1. Prostaglandin-Heparin Conjugate

A dual acting biomolecule, a synthetic PGE_1-heparin conjugate, was
investigated to control both aspects of the clotting system.[304-306] The conjugate
was synthesized by forming an amide bond between the amine group on
heparin and the carboxylic group on PGE_1 via a mixed anhydride procedure.
The released conjugate maintained the patency of the shunts over controls
and did not elicit systemic side effects on the animals. Furthermore, the
conjugate was immobilized on an imidazole carbamate derivatized sepharose
bead surface through hydrophilic spacer groups (diamino-terminated polyethylene oxide). The immobilized conjugate was active in preventing both pathways of thrombus formation and the efficacy was improved through the use
of long-chain hydrophilic spacer groups. Coimmobilization of these agents
was studied by Akashi et al.[307]

2. Albumin-Heparin Conjugate

The development of a dual pharmacological drug capable of inhibiting
fibrin formation and platelet aggregation was studied by Hennink et al.[308-310]
They covalently attached heparin to albumin, incorporating the platelet pas-
sivating nature of albumin with the anticoagulant properties of heparin.

3. Antithrombin III-Heparin Conjugate

Coimmobilization of antithrombin III and heparin was performed by Miura
et al.[311]

4. Fibrolytic System-Heparin Conjugate

A urokinase-heparin bonded synthetic surface was investigated by Kusserow et al.[312] Aoshima et al.[313] immobilized urokinase on a heparinoid,
sulfonated poly(vinylidene fluoride).

Shankar et al.[314] coimmobilized heparin and plasmin on collageno-elastic
tubes in order to develop a thromboresistant and fibrinolytic vascular prosthesis. Soluble heparin exerted a positive synergistic effect on soluble plasmin.
Immobilized heparin enhanced plasmin loading on the tubes as compared to
a heparin-free graft. The degree of enhancement was evaluated kinetically.

5. Collagen-Heparin Conjugate

Immobilization of heparin on a collagenous surface was studied by several
researchers.[315-320] The collagenous material included a bovine arterial heter-
ograft, autogenous fibrocollagenous tubes, and human umbilical vein grafts.
Senatore et al.[320] reported that although carotid arterial interpositions yielded

enhanced patency of heparinized grafts over control grafts, the patency duration was independent of heparin loading for the immobilization conditions examined.

6. Albumin-Immunoglobulin G Conjugate

Mohammad and Olsen[321,322] noted that preexposure of a number of polymers to serum made them less attractive to platelets; they also found that the component in serum that affects platelet adhesion was a protein-protein complex of albumin and immunoglobulin G. Therefore, albumin was covalently bound to the hinge region of immunoglobulin G via disulfide bond(s). Platelet adhesion and activation of coagulation factors were reduced on polyurethane treated with the complex.

7. Albumin-Prostaglandin Conjugate

The coimmobilization of albumin and prostaglandin E_1 was performed by Chandy and Sharma.[323] They observed a reduction of adherent platelets in conjunction with an increase in plasma recalcification times of these surfaces.

G. BIOMIMETIC SURFACES

Biomembranes have many functions such as construction of cell walls, permeation of indispensable substances, rejection of toxins, and, most importantly, excellent biocompatibility. The surface of a biomembrane only mildly interacts with blood proteins and cells and does not adsorb and activate these biological molecules. The biomembrane is a hybrid consisting mainly of two chemical classes, phospholipids and proteins. Therefore, some attempts have been performed to improve biocompatibility of the artificial surfaces by using phospholipids or the derivatives.

Kono et al.[324] coated a nylon microcapsule with a phospholipid bilayer. Platelet adhesion onto the microcapsules was significantly suppressed by the lipid coating. In their report, it was concluded that platelet adhesion onto the polyamide microcapsule was controlled by the nature of the lipid membrane and the protein adsorbed onto the microcapsules.

Some researchers[325-330] assumed that if an artificial surface similar to a biomembrane was constituted by an arrangement of phospholipids from blood on the surface, the surface would show excellent nonthrombogenicity. In arranging the phospholipids, a polymer with phospholipid polar groups was expected to possess affinity for natural phospholipid molecules.

Therefore, polymeric phospholipid and phosphorylcholine were bound to the surface by Hayward and Chapman and colleages.[325-329] Ishihara et al.[330] synthesized a new monomer with a phosphoryl choline moiety, 2-methyacryloyloxyethyl phosphorylcholine (MPC). They reported that the number of adherent platelets decreased and the deformation and aggregation were suppressed with increasing MPC composition in the polymer. Their chemical structures are shown in Figure 26.

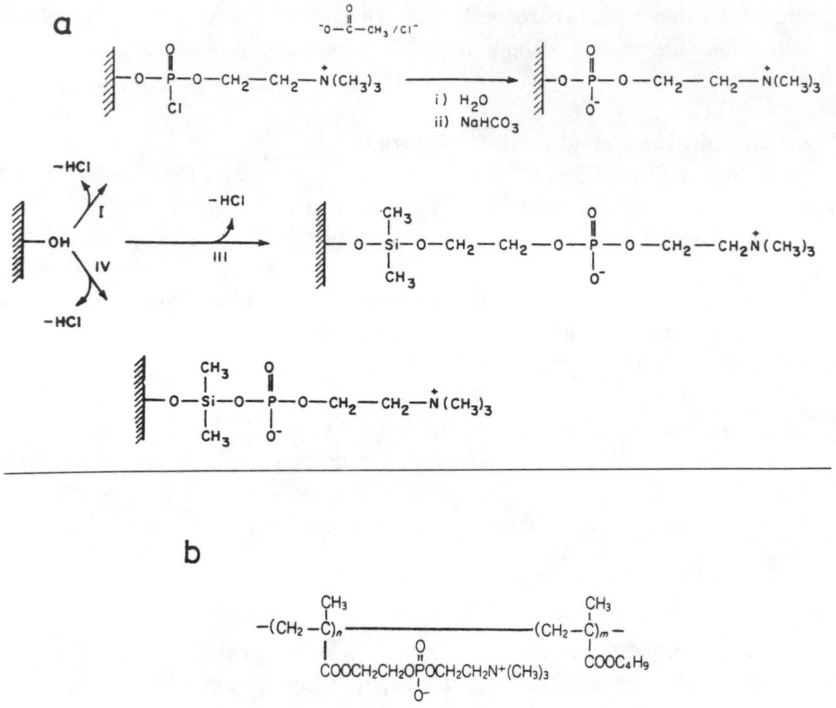

FIGURE 26.   Chemical structures of phosphorylcholine derivatives used by Hayward et al. (a) and Ishihara et al. (b).

IV. HYBRIDIZATION WITH BLOOD ENDOTHELIAL CELLS

Hybridization with endothelial cells has been carried out by many researchers.[331] The first seeding of endothelial cells onto a knitted Dacron tube with a 6-mm diameter was performed by Herring et al.[332] The tube was completely covered with endothelial cells for 2 weeks after implantation. The patency ratio of the seeded graft was 76.4%, whereas that of the nontreated graft was 22.4%. Graham et al.[333] also seeded endothelial cells, which were harvested using trypsin and collagenase, onto a double velour knitted Dacron by the preclotting method.

However, there were some disadvantages in the preclotting method. First, since the treated tube was washed by physiological saline containing heparin to remove excessive clots from the tube, very few cells remained on the tube. Rosenman[334] reported that only 4.4% of feeder cells remained on the tube after washing. Second, the long-term survival of the formed endothelium must be taken into consideration.

In order to overcome these difficulties, three approaches have been adopted:

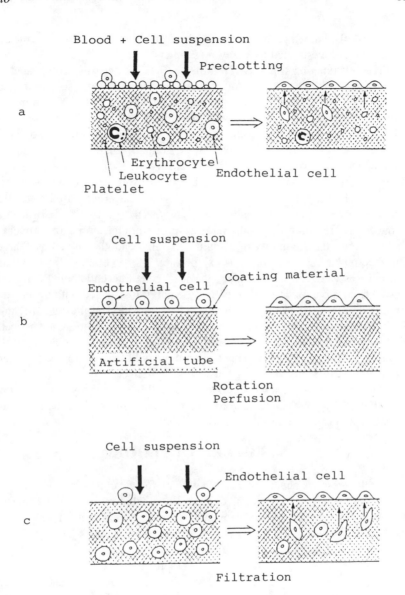

FIGURE 27.   Endothelialization by (a) preclotting, (b) rotation or perfusion, and (c) filtration.

first, the endothelial cells lined by rotation,[335-337] rotation combined with perfusion,[338] perfusion,[339] a two-step method (preclotting and rotation),[340] and filtration culture[341] as illustrated in Figure 27.

Second, adhesive proteins, collagen, fibronectin, laminin, gelatin, and fibrin were coated to enhance cell attachment.[342-349] Greisler et al.[350,351] pretreated tubes with an endothelial cell growth factor to enhance cell growth. As mentioned in Chapter 5.2, insulin immobilization was performed to enhance endothelial cell proliferation.[352] Third, the porosity or fibrous structure of tubes was investigated by many researchers.[355-358]

The cell function which covered the artificial vessel was investigated by various methods. The lifetime of platelets and erythrocytes, secretion of 6-keto-PGF$_{1\alpha}$ and PGI$_2$, and presentation of coagulation factor VIII were examined using animals implanted with endothelial cell-seeded grafts by ^{111}In-labeling, ^{99m}Tc-labeling, radioimmuno assay, and fluorescentimmuno assay, respectively.[359-362] Effects of blood flow,[363] upon leukocytes,[364,365] and of antiplatelet or anticoagulant drugs[364-369] were also investigated for clinical use. The implant conditions have been investigated using several animals.[372-382] Mesothelial cells[383] and smooth muscle cells[384,385] have been used instead of endothelial cells. Several clinical uses have been reported.[386-392] However, some clinical results were not the same as those from the animal experiments.

Recently, the possibility of using the vascular endothelial cell as a target for gene replacement therapy was explored by Wilson et al.[393] Recombinant retroviruses were used to transduce the lacZ gene into endothelial cells harvested from mongrel dogs. Prosthetic vascular grafts seeded with genetically modified cells were implanted as carotid interposition grafts into the dogs from which the original cells were harvested. Analysis of the graft 5 weeks after implantation revealed genetically modified endothelial cells lining the luminal surface of the graft. This technology could be used in the treatment of atherosclerosis and the design of new drug delivery systems.

The quantitative analysis or endothelial cell behavior was carried out by Niu et al.[394] They measured adhesion, spreading, proliferation, and migration (velocity and directionality) of the cells *in vitro*.

V. FUTURE OUTLOOK

At the present stage, hybridization with biological components is the most promising strategy for material design. Endothelialization will be studied further by several researchers. However, there are some disadvantages, such as the difficulties of urgent use and industrial production.

Therefore, combination with biological macromolecules or anticoagulant drugs is very promising. The recent development of blood science has provided excellent anticoagulant reagents and information about the anticoagulant mechanism. For example, Waxman et al.[395] purified a low molecular weight serine protease inhibitor (TAP) from extracts of the soft tick, *Ornithodoros moubata*. The peptide is a slow, tight-binding inhibitor, specific for factor Xa. The most potent natural inhibitor, hirudin, has been thoroughly investigated by using a recombinant type.[396] This information will contribute significantly to the development of blood-compatible materials.

REFERENCES

1. **Williams, D. F., Ed.,** *Blood Compatibility,* Vol. 1, CRC Press, Boca Raton, FL, 1987.
2. **Lelah, M. D. and Cooper, S. L.,** *Polyurethanes in Medicine,* CRC Press, Boca Raton, FL, 1986.
3. **Spaet, T. H., Leonard, E. F., Turritto, V. T., and Vroman, L., Eds.,** Blood in contact with natural and artificial surfaces, *Ann. N.Y. Acad. Sci.,* 1987.
4. **Sevastianov, V. I.,** Role of protein adsorption in blood compatibility of polymers, *CRC Crit. Rev. Biocompatibility,* 4, 109, 1988.
5. **Chuang, H. Y. K.,** Interaction of plasma proteins with artificial surfaces, in *Blood Compatibility,* Williams, D. F., Ed., CRC Press, Boca Raton, FL, 1987.
6. **Andrade, J. D. and Hlady, V.,** Plasma protein adsorption: the big twelve, *Ann. N.Y. Acad. Sci.,* 516, 158, 1987.
7. **Jayalumari, N., Chitra, M., and Iyer, K. S.,** Interaction of human serum lipoproteins with biomaterials, *J. Biomed. Mater. Res.,* 23, 1261, 1989.
8. **Mulzer, S. R. and Brash, J. L.,** Identification of plasma proteins adsorbed to hemo-dialyzers during clinical use, *J. Biomed. Mater. Res.,* 23, 1483, 1989.
9. **Ito, Y., Sisidio, M., and Imanishi, Y.,** Adsorption of plasma proteins to the derivatives of polyetherurethaneurea carrying tertiary amino groups in the side chains, *J. Biomed. Mater. Res.,* 20, 1139, 1986.
10. **Sanada, T., Ito, Y., Sisido, M., and Imanishi, Y.,** Adsorption of plasma proteins to the derivatives of polyetherurethaneurea: the effect of hydrogen-bonding property of the material surface, *J. Biomed. Mater. Res.,* 20, 1179, 1986.
11. **Leininger, R. I., Hutson, T. B., and Jakobsen, R. J.,** Spectroscopic approaches to the investigation of interactions between artificial surfaces and proteins, *Ann. N.Y. Acad. Sci.,* 516, 173, 1987.
12. **Lenk, T. J., Ratner, B. D., Gendreau, R. M., and Chittur, K. K.,** IR spectral changes of bovine serum albumin upon surface adsorption, *J. Biomed. Mater. Res.,* 23, 549, 1989.
13. **Lu, D. R. and Park, K.,** Protein adsorption on polymer surfaces: calculation of adsorption energies, *J. Biomat. Sci., Polym. Ed.,* 4, 243, 1990.
14. **Horbett, T. A., Cheng, C. M., Ratner, B. D., Hoffman, A. S., and Hanson, S. R.,** The kinetics of baboon fibrinogen adsorption to polymers: *in vitro* and *in vivo* studies, *J. Biomed. Mater. Res.,* 20, 739, 1986.
15. **Salzman, E. W., Lindon, J., McManama, G., and Ware, J. A.,** Role of fibrinogen in activation of platelets by artificial surfaces, *Ann. N.Y. Acad. Sci.,* 516, 184, 1987.
16. **Brash, J. L.,** The fate of fibrinogen following adsorption at the blood-biomaterial interface, *Ann. N.Y. Acad. Sci.,* 516, 206, 1987.
17. **Slack, S. M., Bohnert, and Horbett, T. A.,** The effect of surface chemistry and coagulation factors on fibrinogen adsorption from plasma, *Ann. N.Y. Acad. Sci.,* 518, 223, 1987.
18. **Cuypers, P. A., Willems, G. M., Hemker, H. C., and Hermens, W. H.,** Adsorption kinetics of protein mixtures. A tentative explanation of the Vroman effect, *Ann. N.Y. Acad. Sci.,* 518, 244, 1987.
19. **Colman, R. W., Scott, C. F., Schmaier, A. H., Wachtfogel, Y. T., Pixley, R. A., and Edmunds, L. H., Jr.,** Initiation of blood coagulation at artificial surfaces, *Ann. N.Y. Acad. Sci.,* 518, 253, 1987.
20. **Pitt, W. G. and Cooper, S. L.,** FTIR-ATR studies of the effect of shear rate upon albumin adsorption onto polyurethaneurea, *Biomaterials,* 7, 340, 1986.
21. **Grinnell, F.,** The role of fibronectin in the bioreactivity of material surfaces, in *Bio-compatible Polymers, Metals, and Composites,* Szycher, M., Ed., Technomic, Lancaster, PA, 1983, 674.

22. **Ito, Y., Sisido, M., and Imanishi, Y.**, Adsorption of plasma proteins and adhesion of platelets onto novel polyetherurethaneureas — Relationship between denaturation of adsorbed proteins and platelet adhesion, *J. Biomed. Mater. Res.*, 24, 229, 1990.
23. **Hirsch, J., Buchanan, M. R., Ofosu, F. A., and Weitz, J.**, Evolution of thrombosis, *Ann. N.Y. Acad. Sci.*, 516, 586, 1987.
24. **Silverberg, M. and Diehl, S. V.**, The activation of the contact system of human plasma by polysaccharide sulfates, *Ann. N.Y. Acad. Sci.*, 518, 268, 1987.
25. **Coller, B. S.**, Blood elements at surfaces: platelets, *Ann. N.Y. Acad. Sci.*, 516, 362, 1987.
26. **Grasel, T. G. and Cooper, S. L.**, Surface properties and blood compatibility of polyurethaneureas, *Biomaterials*, 7, 315, 1986.
27. **Pelzer, H. and Heimburger, N.**, Evaluation of the *in vitro* and *ex vivo* blood compatibility of primary reference materials, *J. Biomed. Mater. Res.*, 20, 1401, 1986.
28. **Engbergs, G. H. M., Dost, L., Hennink, W. E., Aarts, P. A. A. M., Sixma, J. J., and Feijen, J.**, An *in vitro* study of the adhesion of blood platelets onto vascular catheters. I, *J. Biomed. Mater. Res.*, 21, 613, 1987.
29. **Ito, Y., Sisido, M., and Imanishi, Y.**, Platelet adhesion onto polyetherurethaneurea derivatives — Effect of cytoskeleton proteins of platelet, *Biomaterials*, 8, 458, 1987.
30. **Goodman, S. L., Grasel, T. G., Cooper, S. L., and Albrecht, R. M.**, Platelet shape change and cytoskeletal reorganization on polyurethaneureas, *J. Biomed. Mater. Res.*, 23, 105, 1989.
31. **Kang, I.-K., Ito, Y., Sisido, M., and Imanishi, Y.**, Serotonin and β-thromboglobulin release reaction from platelet as triggered by interaction with polypeptide derivatives, *J. Biomed. Mater. Res.*, 22, 595, 1988.
32. **Vankampen, C. L., Gibbons, D. F., and Jones, R. D.**, Effect of implant surface chemistry upon arterial thrombosis, *J. Biomed. Mater. Res.*, 13, 517, 1979.
33. **Herzlinger, G. A. and Cumming, R. D.**, Role of complement activation in cell adhesion to polymer blood contact surfaces, *Trans. Am. Soc. Artif. Intern. Organs*, 26, 165, 1980.
34. **Kottke-Marchant, K., Anderson, J. M., Miller, K. M., Marchant, R. E., and Lazarus, H.**, Vascular graft-associated complement activation and leukocyte adhesion in an artificial circulation, *J. Biomed. Mater. Res.*, 21, 379, 1987.
35. **Kottke-Marchant, K., Anderson, J. M., Miller, K. M., Marchant, R. E., and Lazarus, H.**, Vascular graft-associated complement activation and leukocyte in an artificial circulation, *J. Biomed. Mater. Res.*, 21, 379, 1987.
36. **Brunstedt, M. R., Anderson, J. M., Spilizewski, K. L., Marchant, R. E., and Hiltner, A.**, *In vivo* leucocyte interactions on Pellethane R surfaces, *Biomaterials*, 11, 370, 1990.
37. **Chenoweth, D. E.**, Complement activation in extracorporeal circuits, *Ann. N.Y. Acad. Sci.*, 518, 306, 1987.
38. **Fukumura, H., Hayashi, K., Yoshikawa, S., Miya, M., Yamamoto, N., and Yamashita, I.**, Complement-induced thrombus formation on the surface of poly(N-vinylpyrrolidone)-grafted polyethylene, *Biomaterials*, 8, 74, 1987.
39. **Hayashi, K., Fukumura, H., and Yamamoto, N.**, *In vivo* thrombus formation induced by complement activation on polymer surfaces, *J. Biomed. Mater. Res.*, 24, 1385, 1990.
40. **Payne, M. S. and Horbett, T. A.**, Complement activation by hydroxyethylmethacrylate-ethylmethacrylate copolymers, *J. Biomed. Mater. Res.*, 21, 843, 1987.
41. **Gimbrone, M. A., Jr.**, Vascular endothelium: nature's blood compatible container, *Ann. N.Y. Acad. Sci.*, 516, 5, 1987.
42. **Ito, Y. and Imanishi, Y.**, Blood-compatibility of polyurethanes, *CRC Crit. Rev. Biocompatibility*, 5, 45, 1989.
43. **Fiala, V., Sotolova, O., Trbusek, V., Vrbova, M., Vasku, A., Urbanek, P., Vasku, J., Lucas, J., Houska, M., and Tyrackova, V.**, Haemocompatibility of segmented polyurethanes investigated *in vivo*, *Biomaterials*, 8, 259, 1987.

44. **Grasel, T. G., Pitt, W. G., Murthy, K. D., McCoy, T. J., and Cooper, S. L.,** Properties of extruded poly(tetramethylene oxide)-polyurethane block copolymers for blood-contacting applications, *Biomaterials,* 8, 329, 1987.
45. **Brothers, T. E., Stanley, J. C., Burkel, W. E., and Graham, L. M.,** Small-caliber polyurethane and polytetrafluoroethylene grafts: a comparative study in a canine aortiliac model, *J. Biomed. Mater. Res.,* 24, 761, 1990.
46. **Bakker, D., Blitterswijk, C. A. V., Daems, W. T., and Grote, J. J.,** Biocompatibility of six elastomers *in vitro, J. Biomed. Mater. Res.,* 22, 423, 1988.
47. **Teijeira, F. J., Lamoureux, G., Tetreault, J.-P., Bauset, R., Guidoin, R., Marois, Y., Paynter, R., and Assayed, F.,** Hydrophilic polyurethane versus autologous femoral vein as substitutes in the femoral arteries of dogs: quantification of platelets and fibrin deposits, *Biomaterials,* 10, 80, 1989.
48. **Therrien, M., Guidoin, R., Adnot, A., and Paynter, R.,** Hydrophobic and fibrillar microporous polyetherurethane urea prosthesis: an ESCA study on the internal and external surfaces of expanded grafts, *Biomaterials,* 10, 517, 1989.
49. **Marois, Y., Guidoin, R., Boyer, D., Assayed, F., Dollin, C. J., Paynter, R., and Marois, M.,** *In vivo* evaluation of hydrophobic and fibrillar microporous polyetherurethane urea graft, *Biomaterials,* 10, 521, 1989.
50. **Martz, H., Paynter, R., Slimane, S. B., Beaudoin, G., Borzone, J., Simhon, H. B., Stain, R., and Sheiner, N.,** Hydrophilic microporous polyurethane versus expanded PTFE grafts as substitutes in the carotid arteries of dogs. A limited study, *J. Biomed. Mater. Res.,* 22, 63, 1988.
51. **Ratner, B. D.,** Surface characterization of biomaterials by electron spectroscopy for chemical analysis, *Ann. Biomed. Eng.,* 11, 313, 1983.
52. **Grasel, T. G., Lee, D. C., Okkema, A. Z., Slowinski, T. J., and Cooper, S. L.,** Extraction of polyurethane block copolymers: effects on bulk and surface properties and biocompatibility, *Biomaterials,* 9, 383, 1988.
53. **Ratner, B. D., Johnson, A. B., and Lenk, T. J.,** Biomaterial surfaces, *J. Biomed. Mater. Res., Appl. Biomat.,* 21, A1, 59, 1987.
54. **Castner, D. G., Ratner, B. D., and Hoffman, A. S.,** Surface characterization of a series of polyurethanes by X-ray photoelectron spectroscopy and contact angle methods, *J. Biomater. Sci., Polym. Ed.,* 1, 191, 1990.
55. **Kasemo, B. and Lausmaa, J.,** Biomaterial and implant surfaces: on the role of cleanliness, contamination, and preparation procedures, *J. Biomed. Mater. Res.,* 22, A2, 145, 1988.
56. **Takahara, A., Jo, N. J., and Kajiyama, T.,** Surface molecular mobility and platelet reactivity of segmented poly(etherurethaneureas) with hydrophilic and hydrophobic soft segment components, *J. Biomater. Sci., Polym. Ed.,* 1, 17, 1989.
57. **Tingey, K. G., Andrade, J. D., Zdrahala, R. J., Chittur, K. K., and Gendreau, R. M.,** Surface analysis of polyether and polysiloxane soft segment polyurethanes, in *Surface Characterization of Biomaterials,* Ratner, B. D., Ed., Elsevier, Amsterdam, 1988, 10.
58. **Kottke-Marchant, K., Anderson, J. M., and Rabinovitch, A.,** The platelet reactivity of vascular graft prostheses: an *in vitro* model to test the effect of preclotting, *Biomaterials,* 7, 441, 1986.
59. **Hodgkin, J. H., Heath, G. R., Norris, W. D., Donald, G. S., and Johnson, G.,** Endothelial cell response to polyvinyl chloride-packaged GORETEX: effect of surface contamination, *Biomaterials,* 11, 9, 1990.
60. **Haubold, A. D., Shim, H. S., and Bokros, J. C.,** Carbon in medical devices, in *Biocompatibility of Clinical Implant Materials,* Vol. 2, Williams, D. F., Ed., CRC Press, Boca Raton, FL, 1982, 3.
61. **Spencer, C. P., Schmidt, B., Samtleben, W., Bosch, T., and Gurland, H. J.,** *Ex vivo* model of hemodialysis membrane biocompatibility, *Trans. Am. Soc. Artif. Intern. Organs,* 31, 495, 1985.

62. **Akizawa, T., Kitao, T., Watanabe, T., Imamura, K., Tsurumi, T., Suma, Y., and Eiga, S.,** Development of noncomplement activating membrane of regenerated cellulose for hemodialysis, *Jpn. J. Artif. Organs,* 16, 822, 1987.
63. **Gissinger, D. and Stamm, A.,** A comparative study of cross-linked carboxymetylcellulose as tablet disintegrant, *Pharm. Ind.,* 42, 189, 1980.
64. **Nakano, Itoh, M., Juni, K., Sekikawa, H., and Arita, T.,** Sustained urinary excretion of sulfamethizole following oral administration of enteric coated microcapsules in humans, *Int. J. Pharm.,* 4, 291, 1980.
65. **Franz, G.,** Polysaccharides in pharmacy, *Adv. Polym. Sci.,* 76, 1, 1986.
66. **Okamoto, Y., Kawashima, M., Yamamoto, K., and Hatada, K.,** Useful kiral packing materials for chromatographic resolution, *Chem. Lett.,* 739, 1984.
67. **Kamide, K., Okajima, K., Matsui, T., Ohnishi, M., and Kobayashi, H.,** Roles of molecular characteristics in blood anticoagulant activity and acute toxicity of sodium cellulose sulfate, *Polym. J.,* 15, 309, 1983.
68. **Ito, H., Shibata, T., Miyamoto, H., Inagaki, H., and Noishiki, Y.,** Formation of polyelectrolyte complexes between cellulose derivatives and their blood compatibility, *J. Appl. Polym. Sci.,* 31, 2491, 1986.
69. **Ito, H., Miyamoto, T., Inagaki, H., Noishiki, Y., Iwata, H., and Matsuda, T.,** *In vivo* and *in vitro* blood compatibility of polyelectrolyte complexes formed between cellulose derivatives, *J. Appl. Polym. Sci.,* 32, 3413, 1986.
70. **Miyamoto, T., Takahashi, S., Tsuji, S., Ito, H., and Inagaki, H.,** Preparation of cellulose graft copolymers having polypeptide side chains and their blood compatibility, *J. Appl. Polym. Sci.,* 31, 2302, 1986.
71. **Woffindin, C. and Hoenich, N. A.,** Blood-membrane interactions during haemodialysis with cellulose and synthetic membranes, *Biomaterials,* 9, 53, 1988.
72. **Miyamoto, T., Takahashi, S., Ito, H., Inagaki, H., and Noishiki, Y.,** Tissue biocompatibility of cellulose and its derivatives, *J. Biomed. Mater. Res.,* 23, 125, 1989.
73. **Matsuzaki, K., Yamamoto, I., Sato, T., and Oshima, R.,** Synthesis of water-soluble branched polysaccharides and their antitumor activity. I, *Macromol. Chem.,* 186, 449, 1985.
74. **Hayashi, K.,** Mechanical properties of biomaterials: relationship to clinical applications, in *Contemporary Biomaterials: Material and Host Response, Clinical Applications, New Technology and Legal Aspects,* Boretos, J. W. and Eden, M., Eds., Noyes Publ., Park Ridge, NJ, 1984, 46.
75. **Szlagyi, D. E., Smith, R. F., Elliot, J. P., Hageman, J. H., and Dall'Olmo, C. A.,** Anastomic aneurysms after vascular reconstruction: problems of incidence, etiology, and treatment, *Surgery,* 78, 800, 1975.
76. **Hayashi, K., Takano, H., and Umezu, M.,** Significance of biomechanics in the design of artificial organs and biomaterials, *Jpn. J. Artif. Organs,* 9, 629, 1980.
77. **Abbot, W. M. and Bouchier-Hayes, D. J.,** Role of mechanical properties in graft design, in *Graft Materials in Vascular Surgery,* Dardik, H., Ed., Year Book Med. Pub., Chicago, 1978, 59.
78. **Paasche, P. E., Kinley, C. E., Dolan, F. G., Gozna, E. R., and Marble, A. E.,** Consideration of suture line stresses in the selection of synthetic grafts for implanatation, *J. Biomechanics,* 6, 253, 1973.
79. **Kinley, C. E. and Marble, A. E.,** Compliance: a continuing problem with vascular grafts, *J. Cardiovasc. Surg.,* 21, 163, 1980.
80. **DeWeese, J. A.,** Anastomotic intimal hyperplasia, in *Vascular Grafts,* Sawyer, P. N. and Kaplitt, M. J., Eds., Appleton-Century-Crofts, New York, 1978, 147.
81. **Ghista, D. N. and Moskal, T.,** Biomechanical estimate for optimal coronary bypass graft diameter for maximal coronary perfusion, 1983 Biomechanics Symposium, Woo, S. L.-Y. and Mates, R. E., Eds., Am. Soc. Mech. Eng., New York, 1983, 181.
82. **How, T. V. and Annis, D.,** Viscoelastic behavior of polyurethane vascular prostheses, *J. Biomed. Mater. Res.,* 21, 1093, 1987.

83. **Kidson, I. G. and Abbot, W. M.**, Low compliance and arterial graft occlusion, *Circulation*, 58, 1, 1978.
84. **Walden, R. G., L'Italien, G. J., Megerman, J., and Abbott, W. M.**, Matched elastic properties and successful arterial grafting, *Arch. Surg.*, 115, 1166, 1980.
85. **Lyman, D. J., Lyman, E. C., and Eichwald, E.**, The effect of graft compliance on the patency of small-diameter polyurethane vascular grafts, Trans. 3rd World Biomat. Cong., Kyoto, Apr. 21—25, 1988, 55.
86. **Pordeyhimi, B. and Wagner, D.**, On the correlation between the failure of vascular grafts and their structural and material properties: a critical analysis, *J. Biomed. Mater. Res.*, 20, 375, 1986.
87. **Andrade, J. D.**, Interface phenomena and biomaterials, *Med. Instrum. (Baltimore)*, 7, 110, 1973.
88. **Baier, R. E.**, The role of surface energy in thrombogenesis, *Bull. N.Y. Acad. Med.*, 48, 257, 1972.
89. **Costello, M., Stanczewski, B., Vriesman, P., Lucas, S., Srinivasan, S., and Sawyer, P. N.**, Correlations between electrochemical and antithrombogenic characteristics of polyelectrolyte materials, *Trans. Am. Soc. Artif. Intern. Organs*, 16, 1, 1970.
90. **Ikada, Y.**, Blood-compatible polymers, *Adv. Polym. Sci.*, 57, 103, 1984.
91. **Baier, R. E.**, Selected methods of investigation for blood-contact surfaces, *Ann. N.Y. Acad. Sci.*, 516, 68, 1991.
92. **Ruckenstein, E. and Gourisankar, S. V.**, A surface energetic criterion of blood compatibility of foreign surfaces, *J. Colloid Interface Sci.*, 101, 436, 1984.
93. **Ruckenstein, E. and Gourisankar, S. V.**, Preparation and characterization of thin film surface coatings for biological environments, *Biomaterials*, 7, 403, 1986.
94. **Lyman, D. J., Muir, W. M., and Lee, I. J.**, The effect of chemical structure and surface properties of polymers on coagulation of blood. I. Surface free energy effects, *Trans. Am. Soc. Artif. Intern. Organs*, 11, 301, 1965.
95. **Nylias, E., Morton, W. A., Lederman, D. M., Chiu, T.-H., and Cumming, R. D.**, Interdependence of hemodynamic and surface parameters in thrombosis, *Trans. Am. Soc. Artif. Intern. Organs*, 21, 55, 1975.
96. **Kaelble, D. H. and Moacanin, J.**, Surface energy analysis of bioadhesion, *Polymer*, 18, 475, 1977.
97. **Ratner, B. D., Hoffman, A. S., Hanson, S. R., Harker, L. A., and Whiffen, J. D.**, Blood compatibility — water content relationship for radiation grafted hydrogels, *J. Polym. Sci., Polym. Symp.*, 66, 363, 1979.
98. **Baszkin, A. and Lyman, D. J.**, The interaction of plasma proteins with polymers. Relationship between polymer surface energy and protein adsorption/desorption, *J. Biomed. Mater. Res.*, 14, 393, 1980.
99. **Okano, T., Nishiyama, S., Shinohara, I., Akaike, T., Sakurai, Y., Kataoka, K., and Tsuruta, T.**, Effect of hydrophilic and hydrophobic microdomains on mode of interaction between block polymer and blood platelets, *J. Biomed. Mater. Res.*, 15, 393, 1981.
100. **Okano, T., Aoyagi, T., Kataoka, K., Abe, K., Sakurai, Y., Shimada, M., and Shinohara, I.**, Hydrophilic-hydrophobic microdomain surfaces having an ability to suppress platelet aggregation and their *in vivo* antithrombogenicity, *J. Biomed. Mater. Res.*, 20, 919, 1986.
101. **Okano, T., Uruno, M., Sugiyama, N., Shimada, M., Shinohara, I., Kataoka, K., and Sakurai, Y.**, Suppression of platelet activity on microdomain surfaces of 2-hydroxyethyl methacrylate-polyether block copolymers, *J. Biomed. Mater. Res.*, 20, 1035, 1986.
102. **Kataoka, K.**, Polymers for cell separation, *CRC Crit. Rev. Biocompatibility*, 4, 341, 1988.
103. **Maruyama, A., Tsuruta, T., Kataoka, K., and Sakurai, Y.**, Elimination of cellular active adhesion on microdomain-structured surface of graft-polyamine copolymers, *Biomaterials*, 10, 291, 1989.

104. **Maruyama, A., Tsuruta, T., Kataoka, K., and Sakurai, Y.,** Separation of B and T lymphocytes by cellular adsorption chromatography with polyamine graft copolymers as column matrices. II. Recovery of adsorbed B cell enriched populations from the column, *Biomaterials,* 10, 393, 1989.

105. **Nojiri, C., Okano, T., Jakobs, H. A., Park, K. D., Mohammad, S. F., Olsen, D. B., and Kim, S. W.,** Blood compatibility of PEO grafted polyurethane and HEMA/ styrene block copolymer surfaces, *J. Biomed. Mater. Res.,* 24, 1151, 1990.

106. **Hasegawa, H. and Hashimoto, T.,** Morphology of block polymers near a free surface, *Macromolecules,* 18, 589, 1985.

107. **Hasegawa, H. and Hashimoto, T.,** Bicontinuous microdomain morphology of block copolymers. I. Tetrapod-network structure of polystyrene-polyisoprene diblock polymers, *Macromolecules,* 20, 1651, 1987.

108. **Ishizu, K.,** Surface changes on block copolymers by crosslinking of spherical microdomains, *Polymer,* 30, 793, 1989.

109. **Suzuki, K., Okano, T., Sakurai, Y., Nakahama, S., and Nagaoka, T.,** Activation of platelets on microdomain structure, *Polym. Prepr., Jpn.,* 38, 3090, 1989.

110. **Cho, C. S., Takayama, T., Kunou, M., and Akaike, T.,** Platelet adhesion onto Langmuir-Blodgett film of poly(γ-benzyl L-glutamate)-poly(ethylene oxide)-poly(γ-benzyl L-glutamate) block copolymer, *J. Biomed. Mater. Res.,* 24, 1369, 1990.

111. **Yui, N., Sanui, K., Ogata, N., Kataoka, K., Okano, T., and Sakurai, Y.,** Effect of microstructure of poly(propylene oxide)-segmented polyamides on platelet adhesion, *J. Biomed. Mater. Res.,* 20, 929, 1986.

112. **Yui, N., Kataoka, K., Sakura, Y., Aoki, T., Sanui, K., and Ogata, N.,** *In vitro* and *in vivo* studies on antithrombogenicity of poly(propylene oxide) segmented nylon 610 in relation to its crystalline-amorphous microstructure, *Biomaterials,* 9, 225, 1988.

113. **Matsuda, T.,** *In vitro* evaluation of antithrombogenicity, in *Polymers and Medicine,* Takemoto, K., Sunamoto, J. and Akashi, M., Eds., Mita Publ., Tokyo, 1989, 52.

114. **Merrill, E. W.,** Distinctions and correspondences among surface contacting blood, *Ann. N.Y. Acad. Sci.,* 516, 196, 1987.

115. **Hoffman, A. S.,** Letters to the editor: a general classification scheme for "hydrophilic" and "hydrophobic" biomaterial surfaces, *J. Biomed. Mater. Res.,* 20, ix, 1986.

116. **Gombotz, W. R. and Hoffman, A. S.,** Functionalization of polymeric films by plasma polymerization of allyl alcohol and allyl amine, *J. Appl. Polym. Sci., Appl. Polym. Symp.,* 42, 285, 1988.

117. **Marchant, R. E., Johnson, S. D., Schneider, B. H., Agger, M. P., and Andrson, J. M.,** A hydrophilic plasma polymerized film composite with potential application as an interface for biomaterials, *J. Biomed. Mater. Res.,* 24, 1521, 1990.

118. **Paul, J. F., Peter, P., and Karlheinz, B.,** Methods for preparation and characterization of plastics with improved blood compatibility, *Angew. Makromol. Chemie,* 105, 131, 1982.

119. **Wichterle, O. and Lim, D.,** Hydrophilic gels for biological use, *Nature (London),* 185, 117, 1960.

120. **Hsiue, G.-H., Yang, J.-M., and Wu, R.-L.,** Preparation and properties of a biomaterial: HEMA grafted SBS by X-ray irradiation, *J. Biomed. Mater. Res.,* 22, 405, 1988.

121. **Nagaoka, S. and Akashi, R.,** Low-friction hydrophilic surface for medical devices, *Biomaterials,* 11, 419, 1990.

122. **Nagaoka, S. and Akashi, R.,** Low friction hydrophilic surfaces for medical devices, *J. Bioactive Comp. Polym.,* 5, 212, 1990.

123. **Ratner, B. D.,** Blood-compatibility-water-content relationships for radiation-grafted hydrogels, *J. Polym. Sci., Polym. Symp.,* 66, 363, 1979.

124. **Merrill, E. W. and Salzman, E. W.,** Polyethylene oxide as a biomaterial, *ASAIO, J.,* 6, 60, 1983.

125. **Andrade, J. D., Nagaoka, S., Cooper, S., Okano, T., and Kim, S. W.,** Surfaces and blood compatibility: current hypotheses, *ASAIO J.,* 10, 75, 1987.

126. **Merrill, E. W., Salzman, E. W., Wan, S., Mahmud, N., Kushner, L., Lindon, J. N., and Curme, J.,** Platelet-compatible hydrophilic segmented polyurethanes from polyethylene glycols and cyclohexane diisocyanate, *Trans. Am. Soc. Artif. Intern. Organs,* 28, 482, 1982.

127. **Hiatt, C. W., Shelokov, A., Rosenthal, E. J., and Galimore, J. M.,** Treatment of controlled pore glass with poly(ethylene oxide) to prevent adsorption of rabies virus, *J. Chromatogr.,* 56, 362, 1971.

128. **Hawk, G. L., Cameron, J. A., and Dufault, L. B.,** Chromatography of biological materials on polyethylene glycol treated controlled-pore glass, *Prep. Biochem.,* 2, 193, 1972.

129. **Nagaoka, S. and Nakao, A.,** Clinical application of antithrombogenic hydrogel with long poly(ethylene oxide) chains, *Biomaterials,* 11, 119, 1990.

130. **Okkema, A. Z., Grasel, T. G., Zdrahala, R. J., Solomon, D. D., and Cooper, S. L.,** Bulk, surface, and blood-contacting properties of polyetherurethanes modified with polyethylene oxide, *J. Biomat. Sci., Polym. Ed.,* 1, 43, 1989.

131. **Brinkman, E., Poot, A., Beugeling, T., van der Does, L., and Bantjes, A.,** Surface modification of copolyether-urethane catheters with poly(ethylene oxide), *Int. J. Artif. Organs,* 12, 390, 1989.

132. **Feng, X. D., Sun, Y. H., and Qiu, K. Y.,** Selective grafting of hydrogels onto multiphase block copolymers, *Makromol. Chem.,* 186, 1533, 1985.

133. **Poot, A., Beugeling, T., Cazenave, J. P., Bantjes, A., and van Aken, W. G.,** Platelet deposition in a capillary perfusion model: quantitative and morphological aspects, *Biomaterials,* 9, 126, 1988.

134. **Brinkman, E., Poot, A., van der Does, L., and Bantjes, A.,** Platelet deposition studies on copolyether urethanes modified with poly(ethylene oxide), *Biomaterials,* 11, 200, 1990.

135. **Liu, S. Q., Ito, Y., and Imanishi, Y.,** Synthesis and nonthrombogenicity of polyurethanes with poly(oxyethylene) side chains in soft segment regions, *J. Biomat. Sci.,* 1, 111, 1989.

136. **Maechling-Strasser, C., Dejardin, Ph., Galin, J. C., and Schmitt, A., Housse-Ferrari, V., Sebille, B., Mulvihill, J. N., and Cazenave,** Synthesis and adsorption of a poly(*N*-acetylethyleneimine)-polyethyleneoxide-poly(*N*-acetylethyleneimine) triblock-copolymer at a silica/solution interface. Influence of its preadsorption on platelet adhesion and fibrinogen adsorption, *J. Biomed. Mater. Res.,* 23, 1395, 1989.

137. **Lee, J. H., Kopecek, J., and Andrade, J. D.,** Protein-resistant surfaces prepared by PEO-containing block copolymer surfactants, *J. Biomed. Mater. Res.,* 23, 351, 1989.

138. **Lee, J. H., Kopeckova, P., Kopecek, J., and Andrade, J. D.,** Surface properties of copolymers of alkyl methacrylates with methoxy (polyethylene oxide) methacrylates and their application as protein-resistant coatings, *Biomaterials,* 11, 455, 1990.

139. **Graiger, D. W., Nojiri, C., Okano, T., and Kim, S. W.,** *In vitro* and *ex vivo* platelet interactions with hydrophilic-hydrophobic poly(ethylene oxide)-polystyrene multiblock copolymers, *J. Biomed. Mater. Res.,* 23, 979, 1989.

140. **Bots, J. G. F., van der Does, L., and Bantjes, A.,** Small diameter blood vessel prostheses from blends of poly(ethylene oxide) and poly(propylene oxide), *Biomaterials,* 7, 393, 1986.

141. **Bots, J. G. F., van der Does, L., Bantjes, A., and Lutz, B.,** Polyethers for biomedical applications: bulk and surface characterization of cross-linked blends of poly(ethylene oxide) and poly(propylene oxide), *Br. Polym. J.,* 19, 527, 1987.

142. **Chawla, A. S.,** Evaluation of plasma polymerized hexamethylcyclotrisiloxane biomaterials towards adhesion of canine platelets and leukocytes, *Biomaterials,* 2, 83, 1981.

143. **Hasirci, N. and Akovali, G.,** Polymer coating for hemoperfusion over activated charcoal, *J. Biomed. Mater. Res.,* 20, 963, 1986.

144. **Ho, C. P. and Yasuda, H.,** Ultrathin coating of plasma polymer of methane applied on the surface of silicone contact lenses, *J. Biomed. Mater. Res.,* 22, 919, 1988.

145. **Cunningham, J. C., Kitinoja, D., and Kamel, I. L.,** RF induced modification of surfaces for reduced blood plasma contact activation, *Trans. Soc. Biomater.,* 15, 45, 1989.
146. **Hahn, A. W., Nichols, M. F., Sharma, A. K., and Hellmuth, E. W.,** *Biomedical and Dental Applications of Polymers,* Gebelein, C. G. and Koblitz, F. F., Eds., Plenum Press, New York, 1981, 85.
147. **deGroot, K., Geesink, R., Klein, C. P. A. T., and Serekian, P.,** Plasma-sprayed coating of hydroxylapatite, *J. Biomed. Mater. Res.,* 21, 1375, 1987.
148. **Ducheyne, P. and Healy, K. E.,** The effect of plasma-sprayed calcium phosphate ceramic coatings on the metal ion release from porous titanium and cobalt-chromium alloys, *J. Biomed. Mater. Res.,* 21, 1137, 1988.
149. **Garfinkle, A. M., Hoffman, A. S., Ratner, B. D., Reynolds, L. O., and Hanson, S. R.,** Effects of tetrafluoroethylene glow discharge on patency of small diameter Dacron vascular grafts, *Trans. Am. Soc. Artif. Intern. Organs,* 30, 432, 1984.
150. **Bohnert, J. L., Fowler, B. C., Horbett, T. A., and Hoffman, A. S.,** Plasma gas discharge deposited fluorocarbon polymers exhibit reduced elutability of adsorbed albumin and fibrinogen, *J. Biomater. Sci., Polym. Ed.,* 1, 279, 1990.
151. **Yeh, Y. S., Iriyama, Y., Matsuzawa, Y., Hanson, S. R., and Yasuda, H.,** Blood compatibility of surfaces modified by plasma polymerization, *J. Biomed. Mater. Res.,* 22, 7905, 1988.
152. **Munro, M. S., Quattrone, A. J., Ellsworth, S. R., Kulkarni, P., and Eberhart, R. C.,** Alkyl-substituted polymers with enhanced albumin affinity, *Trans. Am. Soc. Artif. Intern. Organs,* 27, 499, 1981.
153. **Spector, A. A.,** Fatty acid binding to plasma albumin, *J. Lipid Res.,* 16, 165, 1975.
154. **Goodman, D. S.,** The interaction of human serum albumin with long-chain fatty acid anions, *J. Am. Chem. Soc.,* 80, 3892, 1958.
155. **Uzhinova, L. D., Gracheva, N. A., and Plate, N. A.,** A study of the binding of serum albumin with cetyl methacrylate in solution by the fluorescent probe method, *Soviet J. Bioorganic Chem.,* 5, 877, 1980.
156. **Grasel, T. G., Pierce, J. A., and Cooper, S. L.,** Effects of alkyl grafting on surface properties and blood compatibility of polyurethane block copolymers, *J. Biomed. Mater. Res.,* 21, 815, 1987.
157. **McCoy, T. J., Wabers, H. D., and Cooper, S. L.,** Series shunt evaluation of polyurethane vascular graft materials in chronically AV-shunted canines, *J. Biomed. Mater. Res.,* 24, 107, 1990.
158. **Grasel, T. C., Castner, D. G., Ratner, B. D., and Cooper, S. L.,** Characterization of alkyl grafted polyurethane block copolymers by variable takeoff angle X-ray photoelectron spectroscopy, *J. Biomed. Mater. Res.,* 24, 605, 1990.
159. **Pitt, W. G. and Cooper, S. L.,** Albumin adsorption on alkyl chain derivatized polyurethanes. I. The effect of C-18 alkylation, *J. Biomed. Mater. Res.,* 22, 359, 1988.
160. **Pitt, W. G., Grasel, T. G., and Cooper, S. L.,** Albumin adsorption on alkyl chain derivatized polyurethanes. II. The effect of alkyl chain length, *Biomaterials,* 9, 40, 1988.
161. **Eberhart, R. C., Munro, M. S., Frautschi, J. R., Lubin, M., Clubb, F. J., Jr., Miller, C. W., and Sevastianov, V. I.,** Influence of endogenous albumin binding on blood-material interactions, *Ann. N.Y. Acad. Sci.,* 516, 78, 1987.
162. **Yamauchi, J.,** Anticoagulant coating (65493 CA 1976), Japan Kokai Patent, 75-38790, 4,110,175.
163. **Singh, J., Ray, A. R., Singhal, J. P., and Singh, H.,** Radiation-induced graft copolymerization of methacrylic acid on to poly(vinyl chloride) films and their thrombogenicity, *Biomaterials,* 11, 473, 1990.
164. **Nigretto, J. M., Corretge, and Jozefowicz, M.,** Contributions of negatively charged chemical groups to the surface-dependent activation of human plasma by soluble dextran derivatives, *Biomaterials,* 10, 449, 1989.

165. **Criep, J. J., Fujikawa, K., and Nelsestuen, G. L.,** Possible basis for the apparent surface selectivity of the contact activation of human blood coagulation factor XII, *Biochemistry,* 25, 6688, 1986.
166. **Miyamoto, M., Sasakawa, S., Ozawa, T., Kawaguchi, H., and Ohtsuka, Y.,** Mechanisms of blood coagulation induced by latex particles and the roles of blood cells, *Biomaterials,* 11, 385, 1990.
167. **Grasel, T. G. and Cooper, S. L.,** Surface properties and blood compatibility of polyurethaneureas, *Biomaterials,* 7, 315, 1986.
168. **Huang, S. J. and Roby, M. S.,** Biodegradable polymers poly(amide-urathanes). I, *J. Bioact. Comp. Polym.,* 1, 61, 1986.
169. **Okkema, A. Z., Fabrizius, D. J., Grasel, T. G., Cooper, S. L., and Zdrahala, R. J.,** Bulk, aurface and blood-contacting properties of polyether polyurethanes modified with polydimethylsiloxane macroglycols, *Biomaterials,* 10, 23, 1989.
170. **Pizzoferrato, A., D'Addato, M., Cenni, E., Cavedagna, D., Tarabusi, C., and Curti, T.,** Release of β-thromboglobulin and platelet factor 4 induced by alloplastic materials, in *Biomaterials and Biomechanics, 1983,* Ducheyne, P., Van der Perre, G., and Albert, A. E., Eds., Elsevier Science Publ., Amsterdam, 1984, 337.
171. **Deaves, J., Smith, R. C., and Pepper, D. S.,** The release, distribution and clearance of human β thromboglobulin and platelet factor 4, *Thromb. Res.,* 12, 85, 1978.
172. **Ludlan, C. A., Moore, S., Bolton, A. E., Pepper, D. S., and Cash, J. D.,** The release of human platelet specific protein measured by a radioimmunoassay, *Thromb. Res.,* 6, 543, 1975.
173. **Leake, D. L., Cenni, E., Cavedagna, D., Stea, S., Ciapetti, G., and Pizzoferrato, A.,** Comparative study of the thromboresistance of Dacron combined with various polyurethanes, *Biomaterials,* 10, 441, 1989.
174. **Ota, K., Sakaki, Y., Nakagawa, Y., and Teraoka, S.,** A completely new poly(etherurethane) graft ideal for hemodialysis blood access, *Trans. Am. Soc. Artif. Intern. Organs,* 33, 129, 1987.
175. **Nakagawa, Y., Ota, K., Osima, T., Osaki, S., Teraoka, S., and Agishi, T.,** Improved polyester-polyurethane graft applicable to blood access for hemodialysis, *Jpn. J. Artif. Organs,* 20, 570, 1991.
176. **Gogolewski, S. and Pennings, A. J.,** An artificial skin based on biogradable mixtures of polylactides and polyurethanes for full-thickness skin wound covering, *Makromol. Chem. Rapid Commun.,* 4, 675, 1983.
177. **Bruin, P., Jenkman, M. F., Meijer, H. J., and Pennings, A. J.,** A new porous polyetherurethane wound covering, *J. Biomed. Mater. Res.,* 24, 217, 1990.
178. **Kashiwagi, T., Miyake, E., Ito, Y., and Imanishi, Y.,** Synthesis of polyurethane-polypeptide block copolymer, *Prepr. Polym. Symp.,* Kobe, 1986, 86.
179. **Imanishi, Y.,** Polyetherurethanes with specific properties, *Makromol. Chem.,* (Suppl. 12), 83, 1985.
180. **Maeda, M., Kimura, M., Inoue, S., Kataoka, K., Okano, T., and Sakurai, Y.,** Adhesion behavior of rat lymphocyte subpopulations (B cell and T cell) on the surface of polystyrene/polypeptide graft copolymer, *J. Biomed. Mater. Res.,* 20, 25, 1986.
181. **Mori, A., Ito, Y., Sisido, M., and Imanishi, Y.,** Interaction of polystyrene/poly(γ-benzyl L-glutamate) and poly(methyl methacrylate)/poly(γ-benzyl L-glutamate) block copolymer with plasma proteins and platelets, *Biomaterials,* 7, 386, 1986.
182. **Kang, I.-K., Ito, Y., Sisido, M., and Imanishi, Y.,** Adsorption of plasma proteins and platelet adhesion onto polydimethylsiloxane/poly(γ-benzyl L-glutamate) block copolymer films, *Biomaterials,* 9, 138, 1988.
183. **Kang, I.-K., Ito, Y., Sisido, M., and Imanishi, Y.,** Gas permeability of the film of block and graft copolymers of polydimethylsiloxane and poly(γ-benzyl L-glutamate), *Biomaterials,* 9, 349, 1988.
184. **Kang, I.-K., Ito, Y., Sisido, M., and Imanishi, Y.,** Synthesis, blood compatibility and gas permeability of block copolymers consisting of polyoxypropylene and poly(γ-benzyl L-glutamate) block copolymer films, *Polym. J.,* 19, 1329, 1987.

185. **Yokoyama, M., Nakahashi, T., Nishimura, T., Maeda, M., Inoue, S., Kataoka, K., and Sakurai, Y.,** Adhesion behavior of rat lymphocytes to poly(ether)-poly(amino acid) block and graft copolymers, *J. Biomed. Mater. Res.*, 20, 867, 1986.
186. **Kang, I.-K., Ito, Y., Sisido, M., and Imanishi, Y.,** Synthesis, antithrombogenicity and gas permeability of copolypeptide having silyl groups or dimethylsiloxane oligomers substituted in the side chains, *Int. J. Biol. Macromol.*, 10, 169, 1988.
187. **Ito, Y., Iwata, K., Kang, I.-K., Sisido, M., and Imanishi, Y.,** Synthesis, blood compatibility and gas permeability of copolypeptides containing fluoroalkyl side groups, *Int. J. Biol. Macromol.*, 10, 201, 1988.
188. **Pekala, R. W., Rudoltz, M., Lang, E. R., Merrill, E. W., Lindon, J., Kushner, L., McManama, G., and Salzman, E. W.,** Crosslinked polyether/polysiloxane networks for blood-interfacing applications, *Biomaterials*, 7, 372, 1986.
189. **Pekala, R. W., Merrill, E. W., Lindon, J., Kushner, L., and Salzman, E. W.,** Fibrinogen adsorption and platelet adhesion at the surface of modified polypropylene glycol/polysiloxane networks, *Biomaterials*, 7, 379, 1986.
190. **Adams, D. and Williams, D. F.,** The response of bone to carbon-carbon composites, *Biomaterials*, 5, 59, 1984.
191. **Baquey, C., Bordenave, L., More, N., Caix, J., and Basse-Cathalinat, B.,** Biocompatibility of carbon-carbon materials: blood tolerability, *Biomaterials*, 10, 435, 1989.
192. **Boyd, K. L., Schmidt, S. P., Pippert, T. D., and Sharp, W. V.,** Endothelial cell seeding of UTLI carbon-coated small-diameter PTFE vascular grafts, *Trans. Am. Soc. Artif. Intern. Organs*, 33, 631, 1987.
193. **Chignier, E., Monties, J. R., Butazzoni, B., Dureau, G., and Eloy, R.,** Haemocompatibility and biological course of carbonaceous composites for cardiovascular devices, *Biomaterials*, 8, 18, 1987.
194. **Wilson, J. E.,** Hemocompatible polymers: preparation and properties, *Polym. Plast. Technol. Eng.*, 25, 233, 1986.
195. **Tanzawa, H.,** *Antithrombogenic Materials and Artificial Organs*, Kataoka, K., Ed., Shoukabo, Tokyo, 1991, 18.
196. **Rosenberg, R. D.,** Actions and interactions of antithrombin and heparin, *N. Engl. J. Med.*, 292, 146, 1975.
197. **Ito, Y.,** Antithrombogenic heparin-bound polyurethanes, *J. Appl. Biomat.*, 2, 235, 1987.
198. **Goosen, M. F. A. and Sefton, M. V.,** Properties of heparin-polyvinyl alcohol hydrogel coating, *J. Biomed. Mater. Res.*, 17, 359, 1983.
199. **Rollason, G. and Sefton, M. V.,** Factor Xa inactivation by a heparinized hydrogel, *Thromb. Res.*, 44, 517, 1986.
200. **Cholakis, C. H., Zingg, W., and Sefton, M. V.,** *In vitro* platelet interaction with a heparin-polyvinyl alcohol hydrogel, *J. Biomed. Mater. Res.*, 23, 417, 1989.
201. **Ip, W. F. and Sefton, M. V.,** Patency of heparin-PVA coated tubes at low flow rates, *Biomaterials*, 10, 313, 1989.
202. **Grainger, D. W., Kim, S. W., and Feijen, J.,** Poly(dimethylsiloxane)-poly(ethylene oxide)-heparin block copolymers. I. Synthesis and characterization, *J. Biomed. Mater. Res.*, 22, 231, 1988.
203. **Grainger, D. W., Knutson, K., Kim, S. W., and Feijen, J.,** Poly(dimethylsiloxane)-poly(ethylene oxide)-heparin block copolymers. II. Surface characterization and *in vitro* assessments, *J. Biomed. Mater. Res.*, 24, 403, 1990.
204. **Grainger, D. W., Okano, T., Kim, S. W., Castner, D. G., Ratner, B. D., Briggs, D., and Sung, Y. K.,** Poly(dimethylsiloxane)-poly(ethylene oxide)-heparin block copolymers. III. Surface and bulk compositional differences, *J. Biomed. Mater. Res.*, 24, 547, 1990.
205. **Piao, A. Z., Nojiri, C., Park, K. D., Jacobs, H., Feijen, J., and Kim, S. W.,** Synthesis and characterization of poly(dimethylsiloxane)-poly(ethylene)-heparin CBABC type block copolymers, *J. Biomat. Sci., Polym. Ed.*, 1, 299, 1990.

206. **Shiomi, T., Satoh, M., Miya, M., Imai, K., Akasu, H., and Ohtake, K.,** Binding of heparin onto ethylene-vinyl alcohol copolymer membrane, *J. Biomed. Mater. Res., Appl. Biomat.,* 22, A3, 269, 1988.

207. **Ito, Y., Sisido, M., and Imanishi, Y.,** Synthesis and antithrombogenicity of polyurethaneurea containing quaternary ammonium groups in the side chains and of the polymer/heparin complex, *J. Biomed. Mater. Res.,* 20, 1017, 1986.

208. **Ito, Y., Sisido, M., and Imanishi, Y.,** Attachment and proliferation of fibroblast cells on polyetherurethaneurea derivatives, *Biomaterials,* 8, 464, 1987.

209. **Ito, Y., Sisido, M., and Imanishi, Y.,** *In vivo* antithrombogenicity of and platelet adhesion onto ionically or covalently heparinized polyetherurethaneurea, *Biomaterials,* 9, 235, 1988.

210. **Wilson, J. E.,** How surface-bound drugs inhibit thrombus formation, *Drug Dev. Res.,* 21, 79, 1990.

211. **Barbucci, R., Baszkin, A., Benvenuti, M., Costa, M. L., and Ferruti, P.,** Surface characterization of heparin-complexing poly(amido amine) chains grafted on polyurethane and glass surfaces, *J. Biomed. Mater. Res.,* 21, 443, 1987.

212. **Azzuoli, G., Barbucci, R., Benvenuti, M., Ferruti, P., and Necentini, M.,** Chemical and biological evaluation of heparinized poly(amido-amine) grafted polyurethane, *Biomaterials,* 8, 61, 1987.

213. **Barbucci, R.,** Workshop on the characterization of biomaterials in contact with blood, *Biomaterials,* 11, 138, 1990.

214. **Barbucci, R., Benvenuti, M., Maso, G. D., Nocentini, M., Tempesti, F., Losi, M., and Russo, R.,** Synthesis and physicochemical characterization of a nem material (PUPA) based on polyurethane and poly(amido-amine) components capable of strongly adsorbing quantities of heparin, *Biomaterials,* 10, 299, 1989.

215. **Barbucci, R. and Magnani, A.,** Physiochemical characterization and coating of polyurethane with a new heparin-adsorbing material, *Biomaterials,* 10, 429, 1989.

216. **Tanzi, M. C. and Levi, M.,** Heparinizable segmented polyurethanes containing polyamidoamine blocks, *J. Biomed. Mater. Res.,* 23, 863, 1989.

217. **Iida, K., Kaneko, N., and Shimomura, Y.,** A new method for heparinization, *Jpn. J. Artif. Organs,* 20, 488, 1991.

218. **Nagaoka, S., Mikami, M., and Shimizu, Y.,** Antithrombogenic pO_2 sensor for continuous intravascular oxygen monitoring, *Biomaterials,* 11, 414, 1990.

219. **Arnander, C., Bagger-Sjoback, Frebelius, S., Larsson, R., and Swedenborg, J.,** Long-term stability *in vivo* of a thromboresistant heparinized surface, *Biomaterials,* 8, 497, 1987.

220. **Arnander, C., Olsson, P., and Larm, O.,** Influence of blood flow and the effect of protamine on the thromboresistant properties of a covalently bonded heparin surface, *J. Biomed. Mater. Res.,* 22, 859, 1988.

221. **Larsson, R., Larm, O., and Olsson, P.,** The search for thromboresistance using immobilized heparin, *Ann. N.Y. Acad. Sci.,* 516, 102, 1987.

222. **Larsson, R., Selen, G., Bjorklund, H., and Fagerholm, P.,** Intraocular PMMA lenses modified with surface-immobilized heparin: evaluation of biocompatibility *in vitro* and *in vivo, Biomaterials,* 10, 511, 1989.

223. **Chandler, W. L., Solomon, D. D., Hu, C. B., and Schmer, G.,** Estimation of surface-bound heparin activity: a comparison of methods, *J. Biomed. Mater. Res.,* 22, 497, 1988.

224. **Ebert, C. D. and Kim, S. W.,** Immobilized heparin spacer arm effect on biological interactions, *Thromb. Res.,* 26, 43, 1982.

225. **Heyman, P. W., Cho, C. S., McRes, J. C., Olsen, D. B., and Kim, S. W.,** Heparinized polyurethanes: *in vitro* and *in vivo* studies, *J. Biomed. Mater. Res.,* 19, 419, 1985.

226. **Brace, L. D. and Fareed, J.,** An objective assessment of the interactions of heparin and its fractions on human platelets, *Thromb. Haemostas.,* 111, 190, 1985.

227. **Park, K. D., Okano, T., Nojiri, C., and Kim, S. W.,** Heparin immobilization onto segmented polyurethaneurea surfaces — effect of hydrophilic spacers, *J. Biomed. Mater. Res.,* 22, 977, 1988.
228. **Han, D. K., Park, K. D., Ahn, K.-D., Jeong, S. Y., and Kim, Y. H.,** Preparation and surface characterization of PEO-grafted and heparin-immobilized polyurethanes, *J. Biomed. Mater. Res.,* 23, 87, 1989.
229. **Barbucci, R., Benvenuti, M., Casini, G., Ferruti, P., and Necentini, M.,** Heparinizable materials 5: preparation and FTIR characterization of polyurethane surfaces grafted with heparin-complexing poly(amido-amine)chain, *Makromol. Chem.,* 186, 2291, 1985.
230. **Han, D. K., Jeong, S. Y., and Kim, Y. H.,** Evaluation of blood compatibility of PEO grafted and heparin immobilized polyurethanes, *J. Biomed. Mater. Res.,* 23, 211, 1989.
231. **Nagaoka, S., Kurumatani, H., Mori, Y., and Tanzawa, H.,** The activity of immobilized heparin via long poly(ethylene oxide) spacers, *J. Bioact. Comp. Polym.,* 4, 323, 1989.
232. **Tay, S. W., Merrill, E. W., Salzman, E. W., and Lindon, J.,** Activity toward thrombin-antithrombin of heparin immobilized on two hydrogels, *Biomaterials,* 10, 11, 1989.
233. **Liu, L. S., Ito, Y., and Imanishi, Y.,** Synthesis and antithrombogenicity of heparinized polyurethanes with intervening spacer chains of various kinds, *Biomaterials,* in press.
234. **Eika, C.,** The platelet aggregating effect eight commercial heparins, *Scand. J. Haematol.,* 9, 480, 1972.
235. **O'Brien, J. R., Shoobridge, S. M., and Finch, W. J.,** Comparison of the effect of heparin and citrate on platelet aggregation, *J. Clin. Pathol.,* 22, 28, 1973.
236. **Thomson, C., Forbes, C. D., and Prentice, C. R. M.,** The potentiation of platelet aggregation and adhesion by heparin *in vitro* and *in vivo, Clin. Sci. Mol. Med.,* 45, 485, 1973.
237. **Salzman, E. W., Rosenberg, R. D., Smith, M. H., and Lindon, J. N.,** Effects of heparin and heparin fractions of platelet aggregation, *J. Clin. Invest.,* 65, 64, 1980.
238. **Kanamori, N., Nakashima, T., Kinugasa, E., Nakayama, F., Sekiguchi, T., Akizawa, T., Koahikawa, T., and Sugiyama, T.,** Characteristics of low molecular weight heparin as an anticoagulant for hemodialysis — *in vitro* and animal study, *Jpn. J. Artif. Organs,* 17, 86, 1988.
239. **Holmer, E., Soderberg, K., Bergovist, D., and Lindahl, U.,** Heparin and its low molecular weight derivatives: anticoagulant and antithrombotic properties, *Haemostasis,* 16, 1, 1986.
240. **Caranobe, C., Dupouy, D., Gabaig, A. M., and Boneu, B.,** Comparative pharmacokinetic study of standard heparin and of low molecular weight heparin in the rabbit, *Thromb. Haematol.,* 54, 196, 1985.
241. **Grevelink, J. M., Mohammad, S. F., Cho, C. S., Pantalos, G. M., Hughes, S. D., Chiang, B. Y., Taenaka, Y., Kim, S. W., Olsen, D. B., and Kolff, W. J.,** Pulmonary artery to pulmonary artery shunt (PAPAS), *Trans. Am. Soc. Artif. Intern. Organs,* 31, 377, 1985.
242. **Carter, J., Kelton, J. B., Hirsch, J., Cerskus, A., Santos, A. V., and Gent, M.,** The relationship between the hemorrhagic and antithrombotic properties of low molecular weight heparin in rabbits, *Blood,* 59, 1239, 1982.
243. **van der Does, L., Bevgeling, T., Froehling, P. E., and Bantjes, A.,** Synthetic polymers with anticoagulant activity, *J. Polym. Sci., Symp.,* 66, 337, 1979.
244. **Sederal, L. C., van der Does, L., van Duijl, J. F., Beugeling, T., and Bantjes, A.,** Anticoagulant activity of a synthetic heparinoid in relation to molecular weight and N-sulfate content, *J. Biomed. Mater. Res.,* 15, 819, 1981.
245. **Crassous, G., Harjanto, F., Mendjel, H., Sledz, J., and Schue, F.,** A new asymmetric membrane having blood compatibility, *J. Membr. Sci.,* 22, 269, 1985.

246. **Gebelein, C. G. and Murphy, D.**, The synthesis of some potentially blood compatible heparin-like polymeric materials, in *Advances in Biomedical Polymers*, Gebelein, C. G., Ed., Plenum Press, New York, 1987, 277.

247. **Muzzarelli, R. A., Tanfani, F., Emanuelli, M., Pace, D. P., Chiurazzi, E., and Piani, M.**, Sulfated *N*-(carboxymethyl)chitosans: novel blood anticoagulants, *Carbohydr. Res.*, 126, 225, 1984.

248. **Kanmangne, F. M., Serne, H., Labarre, D., and Jozefowicz, M.**, Heparin-like activity of insoluble sulfonated polystyrene resins, *J. Colloid Interface Sci.*, 110, 21, 1986.

249. **Douzon, C., Kanmangne, F. M., Serne, H., Labarre, D., and Jozefowicz, M.**, Heparin-like activity of insoluble sulphonated polystyrene resins. III. Binding of dicarboxylic amino acids, *Biomaterials*, 8, 190, 1987.

250. **Boisson, C., Jozenfonvicz, J., and Brash, J. L.**, Adsorption of thrombin from buffer and modified plasma to polystyrene resins containing sulphonate and sulphamide arginyl methyl ester groups, *Biomaterials*, 9, 47, 1988.

251. **Caix, J., Migonney, V., Baquey, A., Fouhnot, C., Minnot, A. P., Beziade, A., Vuillemin, L., Baquey, C., and Ducassou, D.**, Control and isotopic quantification of affinity of antithrombin III for heparin-like surfaces, *Biomaterials*, 9, 62, 1988.

252. **Migonney, V., Fougnot, C., and Jozefowicz, M.**, Heparin-like tubings. I. Preparation, characterization and biological *in vitro* activity assessment, *Biomaterials*, 9, 145, 1988.

253. **Migonney, V., Fougnot, C., and Jozefowicz, M.**, Heparin-like tubings. II. Mechanism of the thrombin-antithrombin III reaction at the surface, *Biomaterials*, 9, 230, 1988.

254. **Dulos, E., Dufourcq, J., Fougnot, C., and Jozefowicz, M.**, Adsorption of plasma proteins onto anticoagulant polystyrene derivatives: a fluorescence study, *Biomaterials*, 9, 405, 1988.

255. **Mignney, V., Fougnot, C., and Jozefowicz, M.**, Heparin-like tubings. III. Kinetics and mechanism of thrombin, antithrombin III and thrombin-antithrombin complex adsorption under controlled-flow conditions, *Biomaterials*, 9, 413, 1988.

256. **Charef, S., Jozefowicz, M., Labarre, D., Tapon-Bretaudiere, J., Fischer, A. M., and Bros, A.**, Plasmatic antiproteinase activity enhancement by insoluble functionalized polystyrene surface, *Biomaterials*, 11, 425, 1990.

257. **Mauzac, M. and Jozefonvicz, J.**, Anticoagulant activity of dextran derivatives. I. Synthesis and characterization, *Biomaterials*, 5, 301, 1984.

258. **Aubert, N., Mauzac, M., and Jozefonvicz, J.**, Anticoagulant hydrogels derived from crosslinked dextran. I. Synthesis, characterization and antithrombic activity, *Biomaterials*, 8, 24, 1987.

259. **Aubert, N., Mauzac, M., Gulino, D., and Jozefonvicz, J.**, Anticoagulant hydrogels derived from crosslinked dextran. II. Mechanism of thrombin inactivation, *Biomaterials*, 8, 100, 1987.

260. **Jozefonvicz, J. and Jozefowicz, M.**, Interactions of biospecific functional polymers with blood proteins and cells, *J. Biomater. Sci., Polym. Ed.*, 1, 147, 1990.

261. **Grasel, T. G. and Cooper, S. L.**, Properties and biological interactions of polyurethane anionomers: effect of sulfonate incorporation, *J. Biomed. Mater. Res.*, 23, 311, 1989.

262. **McCoy, T. J., Wabers, H. D., and Cooper, S. L.**, Series shunt evaluation of polyurethane vascular graft materials in shronically AV-shunted canines, *J. Biomed. Mater. Res.*, 24, 107, 1990.

263. **Ito, Y., Iguchi, Y., and Imanishi, Y.**, Synthesis and nonthrombogenicity of heparinoid polyurethaneureas, *Biomaterials*, in press.

264. **Ito, Y., Iguchi, Y., Kashiwagi, T., and Imanishi, Y.**, Synthesis and nonthrombogenicity of polyetherurethaneurea film grafted with poly(sodium vinyl sulfonate), *J. Biomed. Mater. Res.*, in press.

265. **Ito, Y., Liu, L. S., and Imanishi, Y.**, Interaction of poly(sodium vinyl sulfonate) and its surface graft with antithrombin III, *J. Biomed. Mater. Res.*, 25, 99, 1991.

266. **Matsuda, T., Iwata, H., Noda, H., Toyosaki, T., Takano, H., and Akutsu, T.,** Antithrombogenic elastomers: novel anticoagulant/complement inhibitor-controlled release systems, *Trans. Am. Soc. Artif. Intern. Organs,* 31, 244, 1985.
267. **Kikumoto, R., Yamao, Y., Tezuka, T., Tonomura, S., Hara, H., Ninomiya, K., Hijikata, A., and Okamoto, S.,** Selective inhibition of thrombin by (2R, 4R)-4-methyl-1-1-[N²-[3-methyl-1,2,3,4-tetrahydro-8-quinolinyl)-sulfonyl]-L-arginyl)]]-2-piperidine-carboxylic acid, *Biochemistry,* 23, 85, 1984.
268. **Ito, Y., Liu, L. S., and Imanishi, Y.,** Blood-compatibility by grafting with anticoagulants, *Artif. Organs,* 14, 132, 1990.
269. **Ito, Y., Liu, L. S., and Imanishi, Y.,** Blood-compatibility by surface-grafting of polymerizable anticoagulants, in *Artif. Heart III,* Akutsu, T., Ed., Springer-Verlag, Tokyo, in press, 1991.
270. **Ito, Y., Liu, L. S., Matsuo, R., and Imanishi, Y.,** Synthesis and nonthrombogenicity of polymer film with polymerizable thrombin-inhibitors graft-polymerized, *J. Biomed. Mater. Res.,* submitted.
271. **Ito, Y., Liu, L. S., and Imanishi, Y.,** Nonthrombogenicity of thrombin-substrate-immobilized polymer film by the inhibition of thrombin activity, *J. Biomater. Sci., Polym. Ed.,* in press.
272. **Ito, Y., Liu, L. S., and Imanishi, Y.,** Interaction of thrombin with fluorescent synthetic thrombin substrate immobilized on polymer film, *Biomaterials,* submitted.
273. **Sugitachi, A., Takagi, K., Inaska, S., and Kosaki, G.,** Immobilization of plasminogen activator, urokinase on nylon, *Thromb. Haemostasis,* 39, 426, 1978.
274. **Watanabe, S., Shimizu, Y., Teramatsu, T., Murachi, T., and Hino, T.,** The *in vitro* and *in vivo* behavior of urokinase immobilized onto collagen polymer composite material, *J. Biomed. Mater. Res.,* 15, 535, 1981.
275. **Senatore, F. F., Bernath, F. R., Meisner, K.,** Clinical study of urokinase-bound fibrinocollagenous tubes, *J. Biomed. Mater. Res.,* 20, 177, 1986.
276. **Senatore, F. F. and Bernath, F. R.,** Parameters affecting optimum activity and stability of urokinase-bound fibrinocollagenous tubes, *Biotechnol. Bioeng.,* 28, 64, 1986.
277. **Rimon, A. and Rimon, S.,** Immobilized components of the plasmin system, in *Insolubilized Enzymes,* Salmona, M., Saronino, C., and Garattini, S., Eds., Raven Press, New York, 1974, 78.
278. **Shankar, H., Senator, F. F., Zuniga, P., and Venkataramani, E.,** Enhanced fibrinolytic activity of immobilized plasmin on collagen beads, *J. Biomed. Mater. Res.,* 21, 897, 1987.
279. **Capet-Antonini, F. C. and Tamenasse, J.,** Kinetic studies on soluble and insoluble urokinases, *Can. J. Biochem.,* 53, 890, 1975.
280. **Kaetsu, I., Kumakura, M., Asano, M., Yamada, A., and Sakurai, Y.,** Immobilization of enzymes for medical uses on plastic surfaces by radiation-induced polymerization at low temperatures, *J. Biomed. Mater. Res.,* 14, 199, 1980.
281. **Remy, M. H. and Poznansky, M. J.,** Immunogenicity and antigenicity of soluble cross-linked enzyme/albumin polymers: advantages for enzyme therapy, *Lancet,* 2, 68, 1978.
282. **Liu, L.-S., Ito, Y., and Imanishi, Y.,** Biological activity of urokinase immobilized to cross-linked poly(2-hydroxyethyl methacrylate), *Biomaterials,* in press.
283. **McRea, J. C., Ebert, C. D., and Kim, S. W.,** Prostaglandin releasing polymers stability and efficacy, *Trans. Am. Soc. Artif. Intern. Organs,* 27, 511, 1981.
284. **Lin, J. Y., Okano, T., Dost, L., Feijen, J., and Kim, S. W.,** Prevention of platelet contact activation by PGE₁ released from polyurethane surfaces, *Trans. Am. Soc. Artif. Intern. Organs,* 31, 468, 1985.
285. **Grode, G. A., Pitman, J., Crowley, J. P., Lieninger, R. I., and Falb, R. D.,** Surface immobilized prostaglandin as a platelet protective agent, *Trans. Am. Soc. Artif. Intern. Organs,* 20, 38, 1974.
286. **Ebert, C. D., Lee, L. S., and Kim, S. W.,** The antiplatelet activity of immobilized prostacyclin, *J. Biomed. Mater. Res.,* 19, 629, 1982.

287. **Bamford, C., Middleton, I., and Sakata, Y.,** Grafting and attachment of antiplatelet agents to polyetherurethanes, *Polym. Prepr.,* 25, 27, 1984.

288. **Llanos, G. and Sefton, M. V.,** Immobilization of prostaglandin PGF2a on poly(vinyl alcohol), *Biomaterials,* 9, 429, 1988.

289. **Miura, Y., Hirota, K., and Yagi, K.,** Antithrombogenic biomaterials hybridized with bioactive molecules derived from blood vessel wall, Trans. 3rd World Biomater. Congr. April 21—25, 1988, Kyoto, Japan, 457.

290. **Benslimane, S., Guidoin, R., Marceau, D., King, M., Mehri, Y., Rao, T. J., Martin, L., Lafreniere-Gagnon, D., and Gosselin, C.,** Albumin-coated polyester arterial prosthesis: is exogenic albumin safe?, *Biomater. Artif. Cells Artif. Org.,* 15, 453, 1987.

291. **Packham, M. A., Evans, G., Glynn, M. E., and Mustard, J. E.,** The effect of plasma proteins on the interaction of platelets with glass surfaces, *J. Lab. Clin. Med.,* 73, 686, 1969.

292. **Ishikawa, Y., Sasakawa, S., Takase, M., and Osada, Y.,** Effect of albumin immobilization by plasma polymerization on platelet activity, *Thromb. Res.,* 35, 193, 1984.

293. **Lyman, D. J., Klein, K. G., Brash, J. J., Fritzinger, B. K., Andrade, J. D., and Banoma, F. S.,** Platelet interaction with protein coated surfaces, *Thromb. Diath. Haemor. Proc.* (Suppl. 42), 109, 1970.

294. **Chang, T. M. S.,** Platelet-surface interaction: effect of albumin coating or heparin complexing on thrombogenic surfaces, *Can. J. Physiol. Pharmacol.,* 52, 275, 1974.

295. **Addonizio, V. P., Jr., Macarak, E. J., Nicolaou, K. C., Edmunds, L. H., Jr., and Colman, R. W.,** Effects of prostacyclin and albumin on platelet loss during *in vitro* simulation of extracorporeal circulation, *Blood,* 53, 1033, 1979.

296. **Gyurko, G., Scucs, J., Banhegyi, J., Kenyeres, I., Szikszai, P., Noviczki, M., and Posze, J.,** Experimental experience with polyester velour in vascular plastic surgery, *Acta Chir. Acad. Sci. Hung.,* 15, 35, 1974.

297. **Guidoin, R., Snyder, R., Martin, L., Botsko, Marois, M., Awad, J., King, M., Domurado, D., Bedros, M., and Gosselin, C.,** Albumin coating of a knitted polyester arterial prosthesis: an alternative to preclotting, *Ann. Thorac. Surg.,* 37, 457, 1984.

298. **Merhi, Y., Roy, R., Guidoin, R., Hebert, J., Mourad, W., and Benslimane, S.,** Cellular reactions to polyester arterial prostheses impregnated with cross-linked albumin: *in vivo* studies in mice, *Biomaterials,* 10, 56, 1989.

299. **Bascom, J. V.,** Gelatin sealing to prevent blood loss from knitted arterial grafts, *Surgery,* 50, 504, 1961.

300. **Bordenave, L., Caix, J., Basse-Cathalinat, B., Baquey, C., Midy, D., Baste, J. C., and Constans, H.,** Experimental evaluation of a gelatin-coated polyester graft used as an arterial substitute, *Biomaterials,* 10, 235, 1989.

301. **Guidoin, R., Marceau, D., Rao, T. J., King, M., Merhi, Y., Roy, P.-E., Martin, L., and Duval, M.,** *In vitro* and *in vivo* characterization of an impervious polyester arterial prosthesis: the Gelseal Triaxial R graft, *Biomaterials,* 8, 433, 1987.

302. **Urry, D. W., Harris, R. D., and Long, M. M.,** Irradiation cross-linking of the polytetrapeptide of elastin and compounding to Dacron to produce a potential prosthetic material with elasticity and strength, *J. Biomed. Mater. Res.,* 16, 11, 1982.

303. **Nose, Y., Tajima, K., Imai, Y., Klain, M., Mrava, G., Schriber, K., Urbanek, K., and Ogawa, H.,** Artificial heart constructed with biological material, *Trans. Am. Soc. Artif. Intern. Organs,* 17, 482, 1971.

304. **Jacobs, H. and Kim, S. W.,** *In vitro* bioactivity of a synthesized PGE_1-heparin conjugate, *J. Pharm. Sci.,* 75, 172, 1986.

305. **Jacobs, H., Okano, T., Lin, J. Y., and Kim, S. W.,** PGE_1-heparin conjugate releasing polymers, *J. Contr. Rel.,* 2, 313, 1985.

306. **Jacobs, H. A., Okano, T., and Kim, S. W.,** Antithrombogenic surfaces: characterization and bioactivity of surface immobilized PGE_1-heparin conjugate, *J. Biomed. Mater. Res.,* 23, 611, 1989.

307. **Akashi, M., Takeda, S., Miyazaki, T., Yashima, E., Miyauchi, N., Maruyama, I., Okadome, T., and Murata, Y.,** Antithrombogenic poly(vinyl chloride) with heparin- and/or prostaglandin I_2-immobilized in hydrogels, *J. Bioact. Comp. Polym.,* 4, 4, 1989.

308. **Hennink, W. E., Feijen, J., Ebert, C. D., and Kim, S. W.,** Covalently bound conjugates of albumin and heparin: synthesis, fractionation and characterization, *Thromb. Res.,* 29, 1, 1983.

309. **Hennink, W. E., Feijen, J., Dost, L., and Kim, S. W.,** Interaction of albumin-heparin conjugate preadsorbed surfaces with blood, *Trans. Am. Soc. Artif. Intern. Organs,* 29, 200, 1983.

310. **Hennink, W. E., Feijen, J., and Kim, S. W.,** Inhibition of surface induced coagulation by preadsorption of albumin-heparin conjugate, *J. Biomed. Mater. Res.,* 18, 911, 1984.

311. **Miura, Y., Aoyagi, S., Ikeda, F., and Miyamoto, K.,** Anticoagulant activity of artificial biomedical materials with co-immobilized antithrombin III and heparin, *Biochimie,* 62, 595, 1980.

312. **Kusserow, B. K., Lallow, R., and Nichols, N.,** The urokinase-heparin bonded synthetic surface. An approach to the creation of a prosthetic surface possessing composite antithrombogenic and thrombolytic properties, *Trans. Am. Soc. Artif. Intern. Organs,* 17, 1, 1971.

313. **Aoshima, R., Kanda, Y., Takada, A., and Yamashita, A.,** Sulfonated poly(vinylidene fluoride) as a biomaterial: immobilization of urokinase and biocompatibility, *J. Biomed. Mater. Res.,* 16, 289, 1982.

314. **Shankar, H., Senatore, F., Wu, D. R., and Avantsa, S.,** Coimmobilization and interaction of heparin and plasmin on collagenoelastic tubes, *Biomater. Artif. Cells, Artif. Org.,* 18, 59, 1990.

315. **Raghunath, K., Biswas, G., Rao, K. P., Joseph, K. J., and Chvapil, M.,** Some characteristics of collagen-heparin complex, *J. Biomed. Mater. Res.,* 17, 613, 1983.

316. **Venkataramani, E. S., Senatore, F., Feola, M., and Tran, R. M.,** Nonthrombogenic small-caliber human umbilical vein vascular prosthesis, *Surgery,* 99, 735, 1986.

317. **Venkataramani, E. S., Senatore, F., Lokapur, A., Shankar, H., and Feola, M.,** Studies in heparin immobilization to collagen, *Res. Commun. Chem. Pathol. Pharmacol.,* 54, 421, 1986.

318. **Esquivel, C. D., Bjork, C. G., Bergqvist, D., Rothman, U., and Bergentz, S. E.,** Heparinized human umbilical vein grafts, *Eur. Surg. Res.,* 15, 289, 1983.

319. **Noishiki, Y. and Miyata, T.,** A simple method to heparinize biological materials, *J. Biomed. Mater. Res.,* 20, 337, 1986.

320. **Senatore, F., Shankar, H., Chen, J.-H., Avantsa, S., Feola, M., Posteraro, R., and Blackwell, E.,** *In vitro* and *in vivo* studies of heparinized-collagenous-elastic tubes, *J. Biomed. Mater. Res.,* 24, 939, 1990.

321. **Mohammad, S. F. and Olsen, D. B.,** Reduced platelet adhesion and activation of coagulation factors on polyurethane treated with albumin-IgG complex, *Trans. Am. Soc. Artif. Intern. Organs,* 32, 323, 1986.

322. **Mohammad, S. F., Kolff, W. J., and Olsen, D. B.,** Effect of polymeric albumin-IgG on blood-material interaction, *Trans. Am. Soc. Artif. Intern. Organs,* 33, 588, 1987.

323. **Chandy, T. and Sharma, C. P.,** The antithrombotic effect of prostaglandin E1 immobilized on albuminated polymer matrix, *J. Biomed. Mater. Res.,* 18, 1115, 1984.

324. **Kono, K., Ito, Y., Kimura, S., and Imanishi, Y.,** Platelet adhesion onto polyamide microcapsules coated with lipid bilayer membrane, *Biomaterials,* 10, 455, 1989.

325. **Hayward, J. A. and Chapman, D.,** Biomembranes as models for polymer design: the potential for haemocompatibility, *Biomaterials,* 4, 135, 1984.

326. **Durrani, A. A., Hayward, J. A., and Chapman, D.,** Biomembrane as models for polymer surfaces. II. The synthesis of reactive species for covalent coupling of phosphorylcholine to polymer surfaces, *Biomaterials,* 7, 121, 1986.

327. **Hayward, J. A., Durrani, A. A., Shelton, C. J., Lee, D. C., and Chapman, D.,** Biomembranes as models for polymer surfaces. III. Characterization of a phosphoryl-choline surface covalently bound to glass, *Biomaterials,* 7, 126, 1986.

328. **Hayward, J. A., Durrani, A. A., Lu, Y., Clayton, C., and Chapman, D.,** Biomem-branes as models for polymer surfaces. IV. ESCA analysis of the covalently-bound phosphorylcholine headgroups on polymers, *Biomaterials,* 7, 252, 1986.

329. **Hall, B., Bird, R. R., Kojima, M., and Chapman, D.,** Biomembranes as models for polymer surfaces. V. Thromboelastographic studies for polymeric lipids and polyesters, *Biomaterials,* 10, 219, 1989.

330. **Ishihara, K., Aragaki, R., Ueda, T., Watanabe, A., and Nakabayashi, N.,** Reduced thrombogenicity of polymers having phospholipid polar groups, *J. Biomed. Mater. Res.,* 24, 1069, 1990.

331. **Takagi, A., Tada, Y., and Idezuki, Y.,** Current trends of hybrid vascular prostheses, *Jpn. J. Artif. Organs,* 19, 1461, 1990.

332. **Herring, M., Gardner, A., and Glover, J.,** A single-staged technique for seeding vascular grafts with autogenous endothelium, *Surgery,* 84, 498, 1978.

333. **Graham, L. M., Vinter, D. W., and Ford, J. W.,** Cultured autogenous endothelial cell seeding of prosthetic vascular grafts, *Surg. Forum,* 30, 204, 1979.

334. **Rosenman, J. E., Kempczinski, R. F., Pearce, W. H., and Silverstein, E. B.,** Kinetics of endothelial cell seeding, *J. Vasc. Surg.,* 2, 778, 1985.

335. **Dilley, R. S. and Herring, M. B.,** Endothelial seeding of vascular prostheses, in *Biology of Endothelial Cells,* Jaffe, E. A., Ed., Martinus Nijhoff Pub., Boston, 1984, 401.

336. **Gerlach, J., Schauwecker, H. H., Henning, E., and Bucherl, E. S.,** Endothelial cell seeding on different polyurethanes, *Artif. Organs,* 13, 144, 1989.

337. **Shindo, S., Takagi, A., and Whittemore, A. D.,** Improved patency of collagen-im-pregnated grafts after *in vitro* autogenous endothelial cell seeding, *J. Vasc. Surg.,* 6, 325, 1987.

338. **Anderson, J. S., Price, T. M., Hanson, S. R., and Harker, L. A.,** *In vitro* endo-thelialization of small-caliber vascular grafts, *Surgery,* 101, 577, 1987.

339. **Sentissi, J. M., Ramberg, K., O'Donnell, T. F., Jr., Connolly, R. J., and Callow, D.,** The effect of flow on vascular endothelial cells grown in tissue culture on polyte-traethylene grafts, *Surgery,* 99, 337, 1986.

340. **Lindblad, B., Wright, S. W., Sell, R. L., Burkel, W. E., Graham, L. M., and Stanley, J. C.,** Alternative techniques of seeding cultured endothelial cells to ePTFE grafts on different diameters, porosities, and surfaces, *J. Biomed. Mater. Res.,* 21, 1013, 1987.

341. **Shirakawa, M., Miyata, T., Shindo, S., Takayama, Y., Sato, O., Takagi, A., Tada, Y., Idezuki, Y., Kurumatani, H., and Nagaoka, S.,** Simplified evaluation of prelined vascular endothelial cells onto synthetic graft by light microscopy, *Jpn. J. Artif. Organs,* 18, 294, 1989.

342. **Kesler, K. A., Herring, M. B., Arnold, M., Glover, J. L., Park, H.-M., Helmus, M. N., and Bendick, P. J.,** Enhanced strength of endothelial attachment on polyester elastomer and polytetrafluoroethylene graft surfaces with fibronectin substrate, *J. Vasc. Surg.,* 3, 58, 1986.

343. **Ramalanjaona, G., Kempczinski, R. F., Rosenman, J. E., Douville, E. C., and Silberstein, E. B.,** The effect of fibronectin coating on endothelial cell kinetics in polytetrafluoroethylene grafts, *J. Vasc. Surg.,* 3, 264, 1986.

344. **Seeger, J. M. and Klingman, N.,** Improved *in vivo* endothelialization of prosthetic grafts by surface modification with fibronectin, *J. Vasc. Surg.,* 8, 476, 1988.

345. **Schneider, P. A., Hanson, S. R., Price, T. M., and Harker, L. A.,** Preformed confluent endothelial cell monolayers prevent early platelet deposition on vascular prostheses in baboons, *J. Vasc. Surg.,* 8, 229, 1988.

346. **Scott, S. M., Gaddy, L. R., Sahmel, R., and Hoffman, H.,** A collagen coated vascular prosthesis, *J. Cardiovasc. Surg.,* 28, 498, 1987.

347. **Koveker, G. B., Petzke, K. H., Borg, M., and Nebendahl, K.,** Reduction of thrombogenicity in small-diameter vascular prostheses seeded with autologous endothelial cells, *Thorac. Cardiovasc. Surg.,* 34, 49, 1986.

348. **Kaehler, P., Zilla, P., Fasol, R., Deutsch, M., and Kadletz, M.,** Precoating substrate and surface configuration determine adherence and spreading of seeded endothelial cells on polytetrafluoroethylene grafts, *J. Vasc. Surg.,* 9, 535, 1989.

349. **Tannenbaum, G., Ahlborn, T., Benvenisty, A., Reemtsma, K., and Nowygrod, R.,** High-density seeding of cultured endothelial cells leads to rapid coverage of polytetrafluoroethylene grafts, *Curr. Surg.,* 44, 318, 1987.

350. **Greisler, H. P., Klosak, J., Dennis, J. W., Ellinger, J., Kim, D. U., Burgess, W., and Maciag, T.,** Endothelial cell growth factor attachment to biomaterials, *Trans. Am. Soc. Artif. Intern. Organs,* 32, 346, 1986.

351. **Greisler, H. P., Klosak, J. J., Dennis, J. W., Karesh, S. M., Ellinger, J., and Kim, D. U.,** Biomaterial pretreatment with ECGF to augment endothelial cell proliferation, *J. Vasc. Surg.,* 5, 393, 1987.

352. **Liu, S. Q., Ito, Y., and Imanishi, Y.,** Endothelialization of polyurethane tube immobilized with insulin, *J. Biomed. Mater. Res.,* in preparation.

353. **Kowligi, R. R., von Malzahn, W. W., and Eberhart, R. C.,** Fabrication and characterization of small-diameter vascular prostheses, *J. Biomed. Mater. Res., Appl. Biomater.,* 22, A3, 245, 1988.

354. **Yates, S. G., Barros d'Da, A. A. B., Berger, K., Fernandez, L. G., Wood, S. J., Rittenhouse, E. A., Davis, C. C., Mansfield, P. B., and Sauvage, L. R.,** The preclotting of porous arterial prostheses, *Ann. Surg.,* 188, 611, 1978.

355. **Sheehan, S. J., Rajah, S. M., and Kester, R. C.,** Effect of preclotting on the porosity and thrombogenicity of knitted Dacron grafts, *Biomaterials,* 10, 75, 1989.

356. **Sedlarik, K. M., van Wachem, P. B., Bartels, H., and Schakenraad, J. M.,** Rapid endothelialization of microporous vascular prostheses covered with meshed vascular tissue: a preliminary report, *Biomaterials,* 11, 4, 1990.

357. **Douville, E. C., Kempczinski, R. F., Birinyi, L. K., and Ramalanjaona, G. R.,** Impact of endothelial cell seeding on long-term patency and subendothelial proliferation in a small-caliber highly porous polytetraethylene graft, *J. Vasc. Surg.,* 5, 544, 1987.

358. **Boyd, K. L., Schmidt, S., Pippert, T. R., Hite, S. A., and Sharp, W. V.,** The effects of pore size and endothelial cell seeding upon the performance of small-diameter e-PTFE vascular grafts under controlled flow conditions, *J. Biomed. Mater. Res.,* 22, 163, 1988.

359. **Sharefkin, J. B., Latker, C., Smith, M., Cruess, D., Clagett, G. P., and Rich, N. M.,** Early normalization of platelet survival by endothelial seeding of Dacron arterial prostheses in dogs, *Surgery,* 92, 385, 1982.

360. **Allen, B. T., Long, J. A., Clark, R. E., Sicard, G. A., Hopkins, K. T., and Welch, M. J.,** Influence of endothelial cell seeding on platelet deposition and patency in small-diameter Dacron arterial grafts, *J. Vasc. Surg.,* 1, 224, 1984.

361. **Clagett, G. P., Burkel, W. E., Sharefkin, J. B., Ford, J. W., Hufnagel, H., Vinter, D. W., Kahn, R. H., Stanley, L. M., and Ramwell, P. W.,** Platelet reactivity *in vivo* in dogs with arterial prostheses seeded with endothelial cells, *Circulation,* 69, 632, 1984.

362. **Sharp, W. V., Schmidt, S. P., and Donovan, D. L.,** Prostaglandin biochemistry of seeded endothelial cells on Dacron prostheses, *J. Vasc. Surg.,* 3, 256, 1986.

363. **Schmidt, S. P., Hunter, T. J., Sharp, W. V., Malindzak, G. S., and Evancho, M. M.,** Endothelial cell-seeded four-millimeter Dacron vascular grafts: effects of blood flow manipulation through the grafts, *J. Vasc. Surg.,* 1, 434, 1984.

364. **Emerick, S., Herring, M., Arnold, M., Baughman, S., Reilly, K., and Glover, J.,** Leucocyte depletion enhances cultured endothelial retention on vascular prostheses, *J. Vasc. Surg.,* 5, 342, 1987.

365. **Hawthorn, L. A., Williams, A., Haywood, S., and Absolom, D. R.,** Erythrocyte and granulocyte interactions with endothelialized polymer surfaces, *Trans. Am. Soc. Artif. Intern. Organs,* 33, 501, 1987.

366. **Hussain, S., Glover, J. L., Augelli, N., Bendick, P. J., Maupin, D., and McKain, M.,** Host response to autologous endothelial seeding, *J. Vasc. Surg.,* 9, 656, 1989.

367. **Schmidt, S. P., Hunter, T. J., Falkow, L. J., Evancho, M. M., and Sharp, W. V.,** Effects of antiplatelet agents in combination with endothelial cell seeding on small diameter Dacron vascular graft performance in the canine carotid artery model, *J. Vasc. Surg.,* 2, 898, 1985.

368. **Hirko, M. K., Shmidt, S. P., Hunter, T. J., Evancho, M. M., Sharp, W. V., and Donovan, D. J.,** Endothelial cell seeding improves 4 mm PTFE vascular graft performance in antiplatelet medicated dogs, *Artery,* 14, 137, 1987.

369. **Campbell, J. B., Glover, J. L., and Herring, B.,** The influence of endothelial seeding and platelet inhibition on the patency of ePTFE grafts used to replace small arteries — an experimental study, *Eur. J. Vasc. Surg.,* 2, 365, 1988.

370. **Bomberger, R., DePalma, R., Ambrose, T., and Manalo, P.,** Aspirin and dipyridamole inhibit endothelial healing, *Arch. Surg.,* 117, 1450, 1982.

371. **Evans, D. K., Campbell, J., Herring, M., and Glover, J.,** Effect of antiplatelet drugs on human endothelial cell growth in tissue culture, *Trans. Am. Soc. Artif. Intern. Organs,* 32, 184, 1988.

372. **Hunter, T. J., Schmidt, S. P., Sharp, W. V., Debski, R. F., Evancho, M. M., Clark, R. E., and Falkow, L. J.,** Endothelial cell-seeded artificial prostheses for coronary bypass grafting, *ASAIO Trans.,* 32, 339, 1986.

373. **Herring, M., Gardner, A., Peigh, P., Madison, D., Baughman, S., Brown, J., and Glover, J.,** Patency in canine inferior vena cava grafting: effects of graft material, size, and endothelial seedings, *J. Vasc. Surg.,* 1, 877, 1984.

374. **Plate, G., Hollier, L. H., Fowl, R. J., Sande, J. R., and Kaye, M. P.,** Endothelial seeding of venous prostheses, *Surgery,* 96, 929, 1984.

375. **Koveker, G. B., Burkel, W. E., Graham, L. M., Wakefield, T. W., and Stanley, J. C.,** Endothelial cell seeding of expanded polytetrafluoroethylene vena cava conduits: effects on luminal production of prostacyclin, platelet adherence, and fibrinogen accumulation, *J. Vasc. Surg.,* 7, 600, 1988.

376. **Zamora, J. L., Navarro, L. T., Ives, C. L., Weilbaecher, D. G., Gao, Z. R., and Noon, G. P.,** Seeding of arteriovenous prostheses with homologous endothelium. A preliminary report, *J. Vasc. Surg.,* 3, 860, 1986.

377. **Shepard, A. D., Jorgensen, J. E., Keough, E. M., Foxall, T. F., Ramberg, K., Connolly, R. J., Mackey, W. C., Gavris, V., Auger, K. R., Libby, P., O'Donnell, T. F., and Callow, A. D.,** Endothelial cell seeding of small-caliber synthetic grafts in the baboon, *Surgery,* 99, 318, 1986.

378. **Hollier, L. H., Fowl, R. J., Pennell, R. C., Heck, C. F., Winter, K. A. H., Fass, D. N., and Kaye, M. P.,** Are seeded endothelial cells the origin of neointima on prosthetic vascular grafts?, *J. Vasc. Surg.,* 3, 65, 1986.

379. **Ortenwall, P., Bylock, A., Kjellstrom, B. T., and Risberg, B.,** Seeding of ePTFE carotid interposition grafts in sheep and dogs: species-dependent results, *Surgery,* 103, 199, 1988.

380. **Kent, K. C., Shindo, S., Ikemoto, T., and Whittemore, A. D.,** Species variation and the success of endothelial cell seeding, *J. Vasc. Surg.,* 9, 271, 1989.

381. **Vo, N. M., Arbogast, L., Arbogast, B., and Stanton, P. E.,** Effect of preclotting and precoating agents on seeded venous grafts, *J. Cardiovasc. Surg.,* 30, 604, 1989.

382. **Bill, H. A., Pittilo, R. M., Drury, J., Pollock, J. G., Clarke, J. M. F., Woolf, N., Marston, A., and Machin, S. J.,** Effects of autologous mesothelial cell seeding on prostacyclin production within Dacron arterial prostheses, *Br. J. Surg.,* 75, 671, 1988.

383. **Weinberg, C. B. and Bell, B.,** A blood vessel model constructed from collagen and cultured vascular cells, *Science,* 231, 397, 1986.

384. **Shindo, S., McCarthy, S. J., Ivarsson, B. L. W., and Whittemore, A. D.,** The autologous vascular bioprosthesis, *Surg. Forum,* 37, 451, 1986.

385. **Yue, X., van der Lei, B., Schakenraad, J. M., van Oene, G. H., Kuit, J. H., Feijen, J., and Wildevuur, Ch. R. H.,** Smooth muscle cell seeding in biodegradable grafts in rats: a new method to enhance the process of arterial wall regeneration, *Surgery,* 103, 206, 1988.

386. **Foxall, T. L., Auger, K. R., Callow, A. D., and Libby, P.,** Adult human endothelial cell coverage of small-caliber Dacron and polytetrafluoroethylene vascular prostheses *in vitro, J. Surg. Res.,* 41, 158, 1986.

387. **Radomski, J. S., Jarrell, B. E., Williams, S. K., Koolpe, E. A., Greener, D. A., and Carabasi, R. A.,** Initial adherence of human capillary endothelial cells to Dacron, *J. Surg. Res.,* 42, 133, 1987.

388. **Herring, M. B., Compton, R. S., LeGrand, D. R., Gardner, A. L., Madison, D. L., and Glover, J. L.,** Endothelial seeding of polytetraethylene popliteal bypasses. A preliminary report, *J. Vasc. Surg.,* 6, 114, 1987.

389. **Ortenwall, P., Wadenvik, H., Kutti, J., and Risberg, B.,** Reduction in deposition of indium 111-labeled platelets after autologous endothelial cell seeding of Dacron aortic bifurcation grafts in humans: a preliminary report, *J. Vasc. Surg.,* 6, 17, 1987.

390. **Zilla, P., Fasol, R., Deutsch, M., Fischlein, T., Minar, E., Hammerle, A., Krupicka, O., and Kadletz, M.,** Endothelial cell seeding of polytetrafluoroethylene vascular grafts in humans, a preliminary report, *J. Vasc. Surg.,* 6, 535, 1987.

391. **Ortenwall, P., Wadenvik, H., Kutti, J., and Risberg, B.,** Endothelial cell seeding reduces thrombogenicity of Dacron grafts in humans, *J. Vasc. Surg.,* 11, 403, 1990.

392. **Fasol, R., Zilla, P., Deutsch, M., Grimm, M., Fischlein, T., and Laufer, G.,** Human endothelial cell seeding: evaluation of its effectiveness by platelet parameters after one year, *J. Vasc. Surg.,* 9, 432, 1989.

393. **Wilson, J. M., Birinyi, L. K., Salomon, R. N., Libby, P., Callow, A. D., and Mulligan, R. C.,** Implantation of vascular grafts lined with genetically modified endothelial cells, *Science,* 244, 1344, 1989.

394. **Niu, S., Matsuda, T., and Oka, T.,** The substrate dependency of neoendothelialization: cell-behavior aspects, *Jpn. J. Artif. Organs,* 20, 438, 1991.

395. **Waxman, L., Smith, D. E., Arcuri, K. E., and Vlasuk, G. P.,** Tick anticoagulant peptide (TAP) is a novel inhibitor of blood coagulation factor Xa, *Science,* 248, 593, 1990.

396. **Rydel, T. J., Ravichandran, K. G., Tulinsky, A., Bode, W., Huber, R., Roitsch, C., and Fenton, J. W., II,** The structure of a complex of recombinant hirudin and human α-thrombin, *Science,* 249, 277, 1990.

2.2 ENZYME MODIFICATION WITH SYNTHETIC POLYMERS

Yuji Inada, Yoh Kodera, Ayako Matsushima, and Hiroyuki Nishimura

TABLE OF CONTENTS

I. INTRODUCTION

Proteins can be modified chemically by attaching synthetic and natural macromolecules to the surface of the protein molecule.[1,2] This manipulation of protein modification with polymers can eliminate some drawbacks of native proteins and improve their properties. These are important aspects for future application in the use of protein drugs and catalysts in bioreactors. This concept can be further expanded by conjugates with various combinations as shown below. Among the conjugates, this chapter focuses on recent work of proteins conjugated with synthetic macromolecule, polyethylene glycol (PEG), and their potential medical and biotechnological processes.[3-5] PEG, with a history as plasma expander, being nonimmunogenic, nontoxic, and amphipathic,[6,7] has been applied to various proteins as a superior agent for modification.[8] No disruption of protein conformation including secondary and tertiary structures by high PEG concentration has taken place.

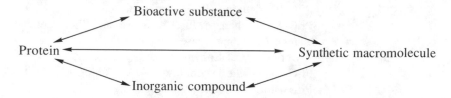

II. SYNTHESIS OF POLYETHYLENE GLYCOL-MODIFIED PROTEINS, PEG PROTEINS

Proteins can be readily modified by the substitution method with cyanuric chloride, 2,4,6-trichloro-*s*-triazine and PEG. Another method[10] is the formation of ester and acid-amide bonds between protein and PEG. These are shown in Figure 1.

Monomethoxypolyethylene glycol (with a molecular weight of ca. 5,000) and cyanuric chloride are allowed to react at a molar ratio 2:1; two of the three chlorine atoms on the cyanuric chloride molecule are replaced with PEG to yield 2,4-*bis*[*O*-methoxypoly(ethylene glycol)]-6-chloro-*s*-triazine, which is designated as activated PEG$_2$[9] (Figure 1b). We have recently established a revised method to synthesize activated PEG$_2$ using zinc oxide as a catalyst. The product was identified as activated PEG$_2$ without contamination of by-products.[11] Figure 2 shows the selective synthesis of activated PEG$_2$ during the course of the reaction. Activated PEG$_2$ can be combined with various proteins by replacing the remaining chlorine atom on each molecule of cyanuric chloride-*bis* PEG residue by an amino group on the protein molecule; this reaction proceeds effectively at a pH of 9.5 to 10.0 with the N terminal amino group or ε-amino group on a lysine residue of a protein. Activated PEG$_1$ (Figure 1a), 2-[*O*-methoxypoly(ethylene glycol)]-4,6-dichloro-*s*-triazine has been extensively used to make PEG-protein conjugates.[12,15] In this

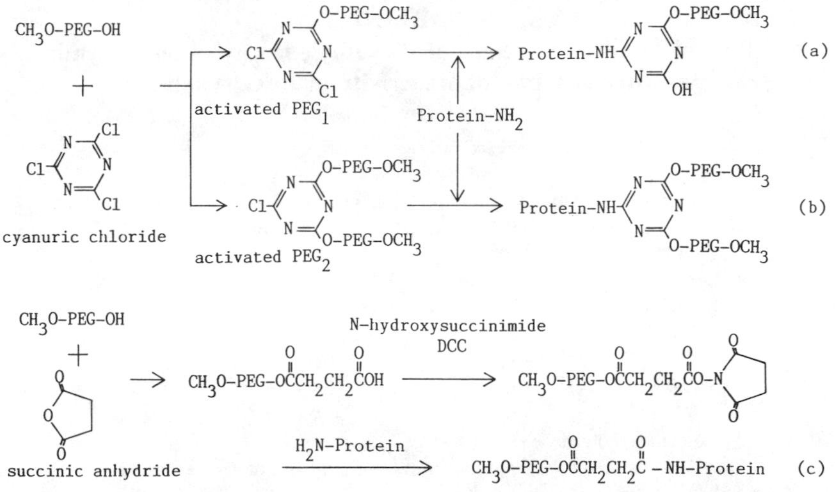

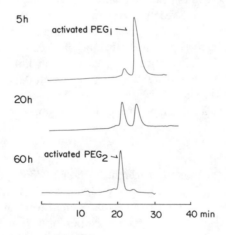

FIGURE 1. Synthesis of polyethylene glycol-modified proteins.

FIGURE 2. Time course of the activated PEG_2 synthesis using zinc oxide.[11] The column is TSK-gel G3000SW (7.5 mm × 60 cm) and the eluent is a mixture of ethanol and phosphate buffer (ethanol: 10 mM phosphate [pH 7.0] containing 0.2 M NaCl = 1:19). The elution profile was measured by the absorbance at 254 nm at a flow rate of 0.7 ml/min.

case, one PEG chain was introduced with one amino acid residue onto the triazine ring. Our experience indicates that activated PEG_2 is a far more effective modifier of proteins than activated PEG_1. Uricase modified with activated PEG_2 retained three times greater activity (45% of the original activity) than that with activated PEG_1 (15%) when the enzyme became nonimmunoreactive by the modification.[16,17] A similar trend was observed for asparaginase.[9,14] PEG protein synthesized with activated PEG (Figure 1c) may be split by hydrolytic enzymes at ester and/or acid-amide bonds in plasma.

TABLE 1
**Effect of Molecular Weight of Polyethylene Glycol on Enzymic
Activity and Immunoreactivity of PEG$_2$-Modified
L-Asparaginase[19]**

Average molecular weight of PEG[a]	Degree of modification[b] (%)	Enzymic activity (IU/mg of protein)	Immunoreactivity[c] (%)
Control	0	250	100
350	29	23	11.0
750	29	43	9.0
2000	34	63	1.9
5000	35	73	0.9

[a]   Monomethoxypolyethylene glycol.
[b]   The total number of amino groups in the asparaginase molecule was 92.
[c]   Relative values obtained from the precipitin reaction.

The degree of modification can be controlled by changing the molar ratio of modifier to protein in the reaction system. It is important to determine how many reactive groups of the protein need be modified for optimum performance in a given case. This must be done on a case-by-case basis. Aside from activated PEG$_2$, alternative reagents that can link through the amino groups are carbodiimide, glutaraldehyde, acid anhydride, and N-succinimidyl-3-(2-pyridyldithio)-propionate.[8,18]

In cases in which the amino group plays an important role in the active site of an enzyme, the carboxyl groups on the aspartic acid or glutamic acid residue can be utilized to attach to hydroxyl or amino groups using carbodiimide.

The effect of the molecular weight of monomethoxypolyethylene glycol on the PEG$_2$-modified asparaginase is summarized in Table 1.[19] Modification of asparaginase with the higher molecular weight form of monomethoxypolyethylene glycol resulted in a greater reduction of the immunoreactivity with antiasparaginase antibodies and retention of the enzymic activity. The lower molecular weight form may easily penetrate to the inside of the enzyme and react with inner amino groups of the enzyme. The active conformation of the enzyme would then be damaged, resulting in lower enzymic activity. These results revealed that the PEG$_2$-modified enzyme which was synthesized with monomethoxypolyethylene glycol having a molecular weight of 5000 had the highest enzymic activity (73 IU/mg of protein) and had almost lost its immunoreactivity.

III. MEDICAL APPLICATION OF PEG-PROTEIN CONJUGATES

When enzymes are administered to humans directly into the blood as drugs, there is the possibility that the protein will be recognized as a foreign substance, since most experimental enzymes originate from other animal species, plants, or microorganisms. Thus, the patient's immune system could produce antibodies specific for the administered protein, and a later administration of the same substance could cause immune reactions, such as anaphylaxis, which can have fatal results. Furthermore, these immune reactions could reduce the clearance time of the drug in the body or could eliminate its function. In addition to humoral immune reactions, cellular immune reactions also could be elicited by foreign immunogenic materials. These immunologic problems encountered with enzyme drugs or immunoglobulins can be reduced by using PEG-protein drugs. Therefore, expected properties of PEG-protein conjugates are (1) high enzymic activity, (2) reduction of immunoreactivity, (3) reduction of immunogenicity, (4) prolongation of clearance time, and (5) induction of immunotolerance.

A. CANCER THERAPY
1. PEG-Asparaginase Conjugate

L-Asparaginase is an enzyme used to treat lymphocytic leukemia and malignant lymphosarcoma.[20] Ordinarily, it is isolated from *E. coli* and is, therefore, a foreign protein. L-Asparaginase catalyzes the hydrolysis of L-asparagine to form L-aspartic acid and ammonia, and asparaginase therapy is responsible for eliminating L-asparagine from the blood of leukemia patients. PEG asparaginase, in which approximately 50% of the 92 amino groups in the asparaginase molecule were coupled with activated PEG_2, exhibits marked reduction of immunoreactivity toward antiasparaginase antibodies and yet retains 11% of the enzymic activity.[9]

PEG asparaginase showed extremely high antitumor activity in a dog with a spontaneous lymphosarcoma.[21] It did not react with antiasparaginase antibodies, and it did not induce the production of new antibodies. PEG-asparaginase administration prior to native asparaginase suppressed the primary and secondary immune responses to asparaginase, and it was clarified that the suppression was caused by suppressor T cells.[22] Moreover, PEG-asparaginase conjugate is extremely stable in the blood;[23] the half-life of PEG asparaginase is 20 times longer than that of the native enzyme in blood (Figure 3). PEG asparaginase was administered to three children with leukemia who had developed an allergic reaction to the native enzyme during the methotrexate-asparaginase sequential therapy.[19] They were placed on a regimen with PEG asparaginase for more than 5 months without any allergic reaction, and had no side effects such as liver dysfunction, coagulation disorder, or diabetes mellitus attributable to PEG asparaginase. This success opened up the possibility of developing a new type of enzyme drug. Other basic studies and clinical applications of PEG asparaginase have been reported by Davis et al. and others.[10,24-26]

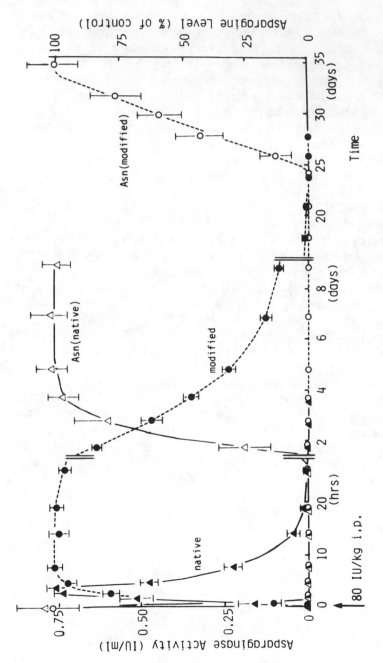

FIGURE 3.   L-Asparaginase activity and L-asparagine concentration in rat sera after intraperitoneal injection of native or PEG$_2$-modified L-asparaginase.[4] Asparaginase activity (closed symbols): ▲, native enzyme; ●, modified enzyme. Asparaginase concentration (open symbols): △, with native enzyme; ○, with modified enzyme.

2. PEG-Interleukin 2(IL-2) Conjugate

Interleukin 2 (IL-2), glycosylated lymphokine with an approximate molecular weight of 15,000 has therapeutic potential in treating cancers and infectious diseases. Human IL-2 was obtained from genetically engineered *E. coli* as an unglycosylated protein, rIL-2, with biological activities equivalent to those of native glycosylated IL-2. Katre et al.[27] reported PEG-rIL-2 conjugate which was modified by an active ester of polyethylene glycol. Covalently, attachment of the amphiphilic polymer enhanced the solubility of IL-2, prolonged its plasma clearance, and increased its antitumor potency in the Meth A murine sarcoma model (Table 2). The conjugated rIL-2 had high resistance toward various kinds of proteases and had a long clearance time, 20 times longer than that of nonmodified interleukin.[29] PEG-rIL-2 conjugate may facilitate the use of adoptive immunotherapy in the treatment of metastatic cancer.[28]

3. PEG-Tryptophanase,[30] PEG-Arginase,[31,32] and PEG-Phenylalanine Ammonia-Lyase[33] Conjugates

Tryptophanase from *E. coli* is a pyridoxal phosphate dependent enzyme with a molecular weight of 220,000 and four subunits. PEG tryptophanase, in which approximately 43% of the total amino groups and 38% of the total sulfhydryl groups in the molecule were coupled, completely lost the immunoreactivity and retained 10% of the enzymic activity. The physicochemical properties of PEG-tryptophanase [K_m] value, optimum pH, and optimum temperature) were the same as those of native tryptophanase. The conjugate was more resistant than the native tryptophanase against proteolytic digestion with trypsin.[30] Beef liver arginase has a molecular weight of 120,000 and contains 88 lysine residues. Upon modification of arginase with PEG in which 53% of amino groups were coupled, PEG-arginase was rendered to be both nonimmunogenic and nonantigenic and altered its kinetic properties. The conjugate was more effective than arginase in the *in vitro* destruction of L5178Y mouse leukemia.[31,32]

Phenylalanine ammonia-lyase from *Rhodotorula glutinis* has a molecular weight of 330,000 and 138 amino groups in the molecule. The PEG-phenylalanine ammonia-lyase conjugate demonstrates altered catalytic properties such as a shift in the pH and temperature optima, an increase in the K_m value, and a lowered V_{max} in comparison with the native enzyme. The conjugate had a much longer blood-circulating life in mice than did the native enzyme and was a poor antigen in mice and rabbits.[33]

B. INHERITED DEFICIENCY OF ENZYMES
1. PEG-Adenosine Deaminase (ADA) Conjugate

Severe combined immunodeficiency disease associated with an inherited deficiency of adenosine deaminase (ADA) is usually fatal unless affected children are kept in protective isolation or unless the immune system is

TABLE 2
Inhibition of Meth A Sarcoma in BALB/c Mice[27]

IL-2	Dose[a] (µg)	Tumor volume, mean ± SEM, (mm³)					
		Day 0[b]	Day 3	Day 4	Day 5	Day 7	Day 10
None	—	45 ± 5	160 ± 22	330 ± 70	630 ± 172	1100 ± 360	2400 ± 360
rIL-2	100	48 ± 14	66 ± 18	61 ± 20	59 ± 24	82 ± 28	110 ± 24
	25	56 ± 6	92 ± 16	130 ± 22	130 ± 23	300 ± 68	200 ± 47
	6.35	56 ± 10	150 ± 40	170 ± 38	170 ± 23	330 ± 130	440 ± 400
	1.6	62 ± 13	220 ± 34	290 ± 54	300 ± 77	550 ± 140	880 ± 200
PEG IL-2	1.6	52 ± 12	85 ± 19	80 ± 17	77 ± 25	57 ± 9	29 ± 21
	0.4	52 ± 10	120 ± 16	130 ± 18	120 ± 23	200 ± 52	170 ± 15

[a] µg Injected intraperitoneally daily for 5 d.
[b] Represents the start of the injection protocol.

reconstituted by transplanting bone marrow. Hershfield et al.[34] treated two children who had ADA deficiency and severe combined immunodeficiency disease by injection of bovine ADA modified with PEG. PEG-ADA was rapidly absorbed after intramuscular injection and had a half-life in plasma of 48 to 72 h. Weekly doses of 15 U/kg of body weight maintained plasma ADA activity at two to three times the level of erythrocyte ADA in normal subjects. The principal biochemical consequences of ADA deficiency were almost completely revealed. Neither toxic effects nor hypersensitivity reactions were observed. *In vitro* tests of the cellular immune function of each patient showed marked improvement, along with an increase in circulating T lymphocytes.

2. PEG-Bilirubin Oxidase Conjugate[35,36]

Hyperbilirubinemia is associated with various liver disorders and hemolytic jaundice. Bilirubin, the end product of heme catabolism, is generally regarded as a potentially cytotoxic agent that must be excreted. Its removal from the bloodstream is vitally important for patients with such diseases. When PEG-bilirubin oxidase conjugate was given intravenously to rats, its plasma half-life was 20 times longer than that of native enzyme. The plasma bilirubin level in experimentally jaundiced rats dropped to normal after only one injection, and the low bilirubin level could be maintained for 12 to 18 h. The antigenicity of the conjugate was greatly reduced. PEG-bilibrubin oxidase appears to have potential value for the treatment of hyperbilirubinemia observed in such diseases.

3. PEG-β-Glucuronidase,[37] PEG-β-Glucosidase, PEG-α-Galactosidase,[38] and PEG-β-Galactosidase Conjugates[39]

A deficiency in the lysosomal hydrolase β-glucuronidase produced a disease characterized biochemically by the excessive intracellular accumulation of acid mucopolysaccharides. PEG-β-glucuronidase exhibited reasonable retention of catalytic activity with an unaltered K_m and a decreased V_{max}, in addition to a shift in the pH optimum as compared to the native enzyme. The conjugate demonstrated an increased resistance of proteolytic digestion, an extended single injection plasma clearance rate, and less immunogenic and antigenic than the native enzyme.[37] Similar phenomena were observed for β-glucosidase and α-galactosidase conjugates coupled with PEG.[38] PEG-β-galactosidase caused the alteration of the substrate specificity in comparison with that of the native enzyme.[39]

4. PEG-Uridase Conjugate[17,40,41]

The disease, gout and hyperuricemia, is due to overproduction of uric acid in human subjects. Attachment of PEG (molecular weight of 5000) to uricase from hog liver and *Candida utilis* rendered the enzyme incapable of eliciting antibody production in mice and of reacting with antibodies to the

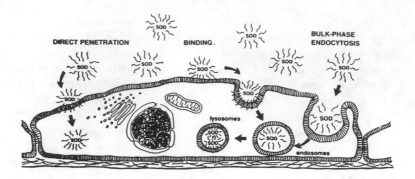

FIGURE 4.   Three possible mechanisms for PEG-conjugated antioxidant enzyme uptake by endothelial cells.[48] Direct membrane penetration of PEG enzymes due to the amphiphilic nature of PEG is considered to be unlikely because PEG would be insoluble in the aliphatic core of the membrane. However, evidence suggests that PEG-conjugated proteins may bind to the polar head groups of phospholipid membranes as well as being taken up by endocytosis. The same mechanisms apply equally to PEG catalase. SOD, superoxide dismutase.

native enzyme. The blood-circulating lives of PEG-urinase following intravenous injection are much longer than those of the native uricase; repetitive injections over a period of 90 d did not alter the blood-circulating lives of the PEG-uricase conjugate.

C. ANTI-INFLAMMATION

1. PEG-Superoxide Dismutase (SOD)[42-49] and PEG-Catalase Conjugates[13,46,48-50]

Superoxide dismutase catalyzes the dismutation of the highly reactive superoxide free-radical anion: $2O_2^- + 2H^+ \rightarrow O_2 + H_2O_2$. It has been used as a novel anti-inflammatory metallo-protein drug capable of alleviating many of the inflammatory symptoms typifying free-radical superoxide pathologies such as rheumatoid arthritis, osteoarthritis, and radiation cystitis.

Davis et al. synthesized first PEG-superoxide dismutase (PEG-SOD)[42] and PEG-catalase conjugates.[13] These conjugates exhibited a sharply enhanced serum circulation life, and a reduced immunogenicity in comparison with the native enzymes. Consequently, PEG-SOD can prevent myocardial necrosis due to oxygen radicals produced during the ischemic intervals[44] and acute pancreatitis.[47] The protective effect of PEG-SOD and/or PEG-catalase, free-radical scavengers, on ischemic brain injury,[46,49] carrageenan pleurisy,[45] and asbestos-induced lung disease[50] were also reported. Beckman et al.[48] hypothesized that PEG conjugate could enhance cell binding and association of normally membrane-impermeable enzymes (Figure 4). Incubation of cultured porcine aortic endothelial cells [^{125}I]PEG-catalase or [^{125}I]PEG-SOD produced a linear, concentration dependence in cellular enzymic activity and radioactivity. Mechanical injury to cell monolayers, which is known to stimulate endocytosis, increases the uptake of PEG-SOD. Endothelial cell cultures incu-

TABLE 3
Radioprotection by SOD and PEG
SOD in Mice Irradiated by γ-Rays[43]

	N	Mortality
Saline	18	15/18
SOD	18	10/18
PEG SOD	18	1/18

Note: Mice were given SOD or PEG SOD i.v. at 40 mg as SOD/kg/d for 3 d and subsequently irradiated with [⁶⁰Co] at 729 rem/mouse. The mortality of the mice was measured for 2 weeks after the irradiation.

bated with PEG-SOD and PEG-catalase were protected from the damaging effect of reactive oxygen species. Miyata et al.[43] reported that PEG-SOD from *Serratia* induced immunotolerance to SOD and enhanced its pharmacological activities as an anti-inflammatory and radioprotective agent (Table 3).

D. ANTITHROMBOSIS
1. PEG-Streptokinase (PEG-SK) Conjugate

Streptokinase (SK), a streptococcal clot-lysing protein, has been used as an intravenous agent for the therapy of deep vein thrombosis and pulmonary embolism. It acts by forming an equimolar complex with plasminogen. PEG-SK exhibited a reduced antigenicity and a retained amidolytic activity against chromogenic substrates while its fibrinolytic activity was 5 to 20% of the native enzyme.[51,52] PEG-SK was protected from proteolytic degradation and was marked increased in clearance time.[53]

Urokinase[54,55] and plasminogen activator[55] were also coupled with PEG, and these conjugates retained the enzymic activity with longer clearance time *in vivo*.

2. PEG-Batroxobin Conjugate[56]

Batroxobin from venom of a snake *Bathrops atrox* is a thrombin-like protease with molecular weight of 36,000 and catalyzes the release of fibrinopeptides A from fibrinogen. Introduction of the enzyme to animals caused selective depletion of fibrinogen in plasma. Its therapeutic study for patients with thrombotic disorders, batroxobin has been tested as a very promising medicine. PEG batroxobin had the reduced binding ability toward the anti-batroxobin antibody, but retained its enzymic activity *in vitro* and *in vivo*. Administration of modified batroxobin (in which 29% of the total amino groups in the molecule had been modified) to beagle dogs preimmunized with native batroxobin gave rise to a marked reduction of the fibrinogen level in plasma, accompanied with an increased level of fibrinogen (fibrin) degradation products, FDP. On the other hand, no reduction of fibrinogen level was observed when native batroxobin instead of modified batroxobin was injected to immunized dogs.

3. PEG-Elastase Conjugate[57]

Elastase is a proteolytic enzyme of animal tissue which catalyzes elastin, a fibrous insoluble protein of connective tissue. Because of its unique substrate specificity, the enzyme has been used for connective tissue diseases of atherosclerosis and aging. PEG elastase gave rise to the complete loss of immunoreactivity and retained 35 and 17% of the original activities for a synthetic substrate and casein, respectively. However, the enzymic activity toward elastin was lost, and the inhibition of PEG-elastase with α_2-macroglobulin was less than that of the native elastase.

4. Magnetite-PEG-Urokinase Conjugate (Magnetic Urokinase)

For the purpose of selective delivery or targeting urokinase to the site of lesion, magnetic urokinase was prepared by conjugating magnetite (Fe_3O_4), PEG, and urokinase.[58] *In vitro* selective fibrinolysis took place by adding magnetic urokinase into circulating plasma with fibrin clots and by setting a magnet (250 Oe) on the fibrin clot (Figure 5).[59] No fibrinolysis occurred without the magnet. In this system, the fibrinogen, precursor of fibrin, was not degraded by the addition of magnetic urokinase. Moreover, the magnetic urokinase was rather stable in plasma containing various kinds of inhibitors such as plasminogen activator inhibitor, α_2-macroglobulin, and α_2-plasmin inhibitor, in comparison with native urokinase. A similar phenomenon was also observed for magnetic plasmin.[60]

Targeting of magnetic urokinase labeled with [^{125}I] by a magnetic field (0 to 12,000 G) was tested in mice.[61] The magnetic urokinase was injected into the tip of a tail vein of a mouse, while a magnetic field was generated around the root part of the tail or heart as target sites (Table 4). In a magnetic field of 12,000 G, approximately as much as 60% of magnetic urokinase injected was localized in the target, the tail root. By applying a magnetic field of 12,000 G to the heart, approximately 20% of magnetic urokinase injected was attracted to the heart. There was almost no retention of magnetic urokinase in the tail root when magnetic intensity of 12,000 G was applied even 10 s after the injection of the magnetic urokinase. It seems that if the magnetic urokinase misses the chance to be attracted by a magnetic field and passes through the lung and the liver, it is readily trapped there. It is possible that magnetic urokinase might be used for targeting in the future.

E. BLOOD SUBSTITUTION
1. PEG-Hemoglobin Conjugate

Recently, the transfusion system faced serious problems such as viral

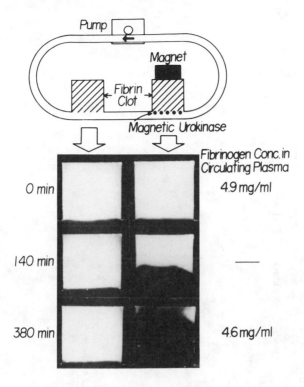

FIGURE 5. Selective fibrinolysis in circulating plasma by magnetic urokinase.[60] An apparatus was prepared by connecting a silicone tube with two separate square chambers containing fibrin clots (upper part). Plasma (6.6 ml) containing plasminogen was circulated with flow rate of 11 ml/min by a pump at 25°C. Then 100 μl of magnetic urokinase (1100 IU/ml) was injected to the circulating system and a magnet of 250 Oe was placed at the top of the square chamber on the right side. No magnet was placed on another chamber on the left side. Photographs of these chambers were taken at given times (lower, left part). The light part in the photographs represents fibrin clot and the dark part represents the dissolved part by plasmin formed by the activation of plasminogen with magnetic urokinase. At 0 and 380 min, the concentration of clottable fibrinogen in the circulating plasma was determined (lower, right part).

infection through blood transfusion and shortage of blood. Thus, more efficient utilization of the collected blood, especially hemoglobin isolated from red cells, should be studied. In order to use an oxygen carrier as red cell substitution, the following properties are desirable: (1) it should be capable of loading oxygen in lungs and also unload in tissues; (2) it should survive in circulation long enough to support anemic animals; and (3) it should be applicable as a drug in terms of its characteristics, stability, safety, and efficacy. Based on these concepts, Ajisaka and Iwashita prepared human hemoglobin modified with PEG with a molecular weight of 750, 1900, 4000, or 5000. The conjugates exhibited P_{50} value of 10 to 15 mmHg, which is enough to deliver oxygen from the lungs to tissues. The longest half-life of these derivatives was about 180 min in contrast to free hemoglobin.[62] He-

TABLE 4
Targeting of Urokinase by Magnetic Field[61]

Magnetic intensity (G)	Tail root	Tail tip	Liver	Spleen	Kidney	Lung	Heart
			Percent distribution (%), mean ± S.E.				
			Magnetic Urokinase				
0	0.7 ± 0.2	0.6 ± 0.2	32.1 ± 6.9	2.5 ± 0.5	4.5 ± 1.0	50.8 ± 6.5	0.6 ± 0.1
			(1) Targeting to tail root				
4,000	14.2 ± 3.6	2.9 ± 1.3	20.0 ± 3.1	1.5 ± 0.6	3.7 ± 0.7	43.3 ± 8.3	0.4 ± 0.1
8,000	32.9 ± 8.3	2.2 ± 0.8	17.9 ± 3.0	1.8 ± 0.8	5.9 ± 2.9	28.7 ± 8.9	0.5 ± 0.1
12,000	60.8 ± 7.2	1.5 ± 0.2	8.2 ± 1.7	0.7 ± 0.1	1.6 ± 0.3	20.4 ± 8.3	0.7 ± 0.4
			(2) Targeting to heart				
12,000	2.0 ± 0.6	2.0 ± 0.5	7.9 ± 1.9	0.3 ± 0.2	1.9 ± 0.6	39.0 ± 5.7	18.2 ± 2.2
			(3) Post targeting to tail root[a]				
12,000	0.8 ± 0.1	0.5 ± 0.2	30.6 ± 3.9	1.6 ± 0.3	3.8 ± 0.9	50.8 ± 4.4	0.4 ± 0.1
			Native Urokinase				
0	1.5 ± 0.4	2.5 ± 1.0	42.1 ± 1.9	1.3 ± 0.2	21.7 ± 0.7	1.5 ± 0.4	0.5 ± 0.2

[a] The magnetic field was applied at 10 s after the injection of magnetic urokinase.

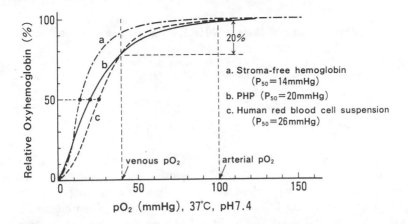

FIGURE 6.   Oxygen equilibrium curves of oxygen carriers.

moglobin modified in the deoxy state exhibited a lower oxygen affinity of polymer-linked hemoglobin. When deoxy hemoglobin was coupled with organic phosphate, pyridoxal phosphate during the condensation reaction, the resulting conjugates exhibited oxygen-binding characteristics quite similar to those of native hemoglobin.[63] Iwashita and co-workers,[62,64-66] have been extensively studied on the pyridoxylated hemoglobin-PEG conjugate (PHP) as an oxygen carrier. Figure 6 shows oxygen equilibrium curves of PHP together with stroma-free hemoglobin and human red cell suspension. The P_{50} value of PHP, 20 mmHg at 37°C and pH 7.4, is between the value of the red cell (26 mmHg) and that of the stroma-free hemoglobin (14 mmHg). The amount of released oxygen from PHP is 20% of the loaded oxygen, assuming the partial pressure in the vein to be 40 mmHg and that in the artery, 100 mmHg. The amount of oxygen unloaded from PHP is about the same as the amount of oxygen transported by the red cell per weight hemoglobin, and it is five times greater than in the case of stroma-free hemoglobin. In order to show how PHP solution supports animals with anemia, the exchange-transfusion experiments were done using rats. When stroma-free hemoglobin solution (SFH) (6%) was used at the exchange ratio 90%, one of the six animals was alive over 1 d. On the other hand, all six rats were alive over 24 h in the case where the rats were given an exchange transfusion with the PHP solution at 90% (Table 5).

2. PEG-Immunoglobulin G(IgG) Conjugate

Human immunoglobulin G (IgG) is valuable in passive immunotherapy for immunodeficient patients. However, intravenous administration of IgG preparations has posed serious side effects (caused by aggregation of IgG) which are formed during production and/or storage processes. In order to obtain a stable human IgG preparation for clinical use, Tomono and co-

TABLE 5
Survival Time of Rats in 90% Exchange Transfusion[66]

Oxygen carriers	Number of animals	Exchange ratio (%)	Survival time (h)
Lactate ringer solution	6	91.8 ± 0.5	0.6 ± 0.1
SFH	6	92.7 ± 0.7	2.1 ± 0.1
PHP66E	10	91.6 ± 0.5	>24

workers[67,68] prepared IgG modified with PEGs, each with a different molecular weight. Their physicochemical and biological properties were revealed in terms of molecular structure, surface activity, interfacial aggregability, heat-aggregability, and antigen activity. The higher the molecular weight of PEG coupled, the more stable was the PEG-IgG conjugate obtained concerning interfacial exposure and heating. The PEG-IgG conjugates were assessed as effective stabilizers for IgG preparations.

3. PEG-Serum Albumin Conjugate[12]

Bovine serum albumin, to which PEG (molecular weight of 1900 or 5000) has been attached, exhibited reduction of immunogenicity and immunoreactivity. PEG-serum albumin iodinated with ^{125}I or prepared with $[^{14}C]$-activated PEG were injected intravenously in rabbits. Both labels appeared almost quantitatively in the urine after 30 d.

4. PEG-Insulin Conjugate

Ehrat and Luisi[69] prepared insulin coupled with PEG at specific positions. Porcine insulin consists of A-chain with 21 amino acid residues and B-chain with 30 amino acid residues and contains three amino groups in the molecule; two N terminal amino groups, $GlyA_1$ and $PheB_1$; and one ϵ-amino group of $LysB_{29}$. Two kinds of PEG-insulin conjugates, in which $PheB_1$ and $LysB_{29}$ were coupled with PEG and $PheB_1$ was coupled with PEG, were separately synthesized after protecting amino group(s) with methylsulfonylethyloxycar-bonyl-*para*-nitrophenylester(Msc-pNp). These conjugates were subjected to spectrophotometric analyses (UV and CD).

5. PEG-Lactoferrin and PEG-α_2-Macroglobulin Conjugates[70]

Lactoferrin is an iron-binding protein found in the granules of neutrophils which may play an important role in tissue injury mediated by phagocytes. The plasma clearance half-life of lactoferrin was increased 5- to 20-fold by PEG modification. α_2-Macroglobulin is a high molecular weight plasma protein inhibitor with a molecular weight of 800,000. PEG-a_2-macroglobulin showed decreased protease binding activity but PEG-conjugate of preformed a_2-macroglobulin-trypsin demonstrated no loss of amidolytic activity.

IV. BIOTECHNOLOGICAL APPLICATION OF PEG-ENZYME CONJUGATES[5,71-78]

Recently, we have been able to make various enzymes soluble and active in highly hydrophobic organic solvents. The key to this success is the chemical modification of enzymes with an amphipathic polymer, polyethylene glycol. The activated PEG_2 can be attached to enzymes in aqueous buffer solution and, once enzymes are modified, they become soluble and active in various organic solvents such as benzene, toluene, and chlorinated hydrocarbons (Table 6). Catalase, peroxidase, cholesterol oxidase, chymotrypsin, papain, and lipase modified with PEG catalyzed the respective reactions efficiently in organic solvents. Polyethylene glycol (PEG)-catalase, in which 42% of the total 112 amino groups in the molecule were coupled with PEG, exhibited high activity in benzene; the activity in benzene was 1.6 times higher than that of the native enzyme in an aqueous system.

Modified peroxidase and cholesterol oxidase catalyzed a coupled reaction in organic solvents. Modified enzymes catalyzed the reverse reaction of hydrolysis in organic solvents, formation of acid-amide bonds by modified proteases, and ester synthesis and ester exchange reactions by modified lipase. Quite recently, geranyl farnesylacetate (gefarnate), which is used as an antipeptic ulcer drug, was synthesized from farnesylacetic acid and geraniol in 1,1,1-trichloroethane at 25°C using lipase from *Pseudomonas fragi* 22.39B modified with activated PEG_2. The PEG-lipase conjugate, in which 50% of the total amino groups in the molecule were coupled with PEG, retained 40% of the hydrolytic activity of native lipase in aqueous solution. The conjugate exhibited the ester synthesis activity of 0.13 μmol min^{-1} mg^{-1} protein and 87% yield for gefarnate synthesis in 1,1,1-trichloroethane. PEG-lipase was easily recovered by adding *n*-hexane to the reaction system without loss of the enzymic activity.[74]

The PEG-lipase conjugate catalyzed esterification of chiral secondary alcohols with fatty acids in benzene and exhibited preference for *R* isomers over *S* isomers. The values of K_m and V_{max} for each isomer of various alcohols were obtained by kinetic study. PEG-lipase-catalyzed esterification leads to optical resolution of a racemic alcohol.[75]

In the case of peptide synthesis using proteases — trypsin, chymotrypsin, papain, thermolysin, and pepsin were modified with activated PEG_2. Each PEG protease became soluble and active in organic solvents and catalyzed peptide bond formation with its substrate specificity. Using five kinds of PEG-protease conjugates, various kinds of tripeptides were synthesized by a solid-phase system in methanol or in dimethylformamide.[76]

Tada et al. reported that ornithyl-β-alanine, obtained by conventional organic synthesis, exhibits a salty taste equal to sodium chloride.[77] N^α,N^δ-Dicarbobenzoxy-L-ornithyl-β-alanine benzyl ester, a derivative of salty peptide, was synthesized in good yield from N^a,N^δ-dicarbobenzoxy-L-ornithine

TABLE 6-1
Properties of PEG-Oxidoreductase Conjugates

	Catalase	Peroxidase	Cholesterol oxidase
EC	1.11.1.6	1.11.1.7	1.1.3.6
Source	Bovine liver	Horseradish	*Nocardia* sp.
MW	248,000	40,000	68,000
Subunit	4	1	1
Active site	Heme	Heme	FAD
Number of amino groups	112	6	Not determined
Degree of modification (%)	42	60	64
Solvent	Benzene, toluene, chloroform, perchloroethylene, 1,1,1-trichloroethane, trichloroethylene		
Reaction	Decomposition of H_2O_2	Oxidation with H_2O_2	Oxidation
Activity (μmol/min/mg)	70,000	30	0.6

ethyl ester and β-alanine benzyl-ester in 1,1,1-trichloroethane using papain modified with polyethylene glycol.[78]

V. FUTURE PROSPECTS

PEG protein drugs have great potential as therapeutic agents for diseases such as cancer and for genetic disorders, with the proteins derived from genetic engineering, fermentation, or animal sources. Modification of proteins with PEG, being nontoxic and nonimmunogenic, causes nonimmunogenicity and nonimmunoreactivity of protein, and prolongation of clearance time in clinical uses. Quite recently, PEG-adenosine deaminase has been approved by the Food and Drug Administration (FDA) in the U.S. as an orphan drug.

Another application of PEG-enzyme conjugates is biotechnological processes. The conjugates become soluble and active in organic solvents so that various kinds of organic substances can be synthesized by PEG enzymes.

TABLE 6-2
Properties of PEG-Hydrolase Conjugates

	Chymotrypsinogen	Papain	Lipase		
			Pseudomonas fluorescens	*Pseudomonas fragi* 22.39B	*Candida cylindracea*
EC	3.4.21.1	3.4.22.2	3.1.1.3		
Source	Bovine pancreas	Papaya	*Pseudomonas fluorescens*	*Pseudomonas fragi* 22.39B	*Candida cylindracea*
MW	26,000	23,000	33,000	30,000	68,000
Subunit	1	1	1	1	1
Active site	Ser	Cys	Ser	Ser	Ser
Number of amino groups	112	6	8	7	34 Lys/1000 residues
Degree of modification (%)	80	37	50	55	47
Solvent	Benzene, toluene, chloroform, 1,1,1-trichloroethane, trichloroethylene, perchloroethylene				
Reaction	Acid-amide bond formation		Ester synthesis and ester exchange		
Activity (μmol/min/mg)	0.8	0.1	28.4[a] 2.6[b]	30.5[a]	0.05[a] +[c]

[a] Ester synthesis.
[b] Ester exchange in substrates.
[c] Ester exchange between phosphatidylcholine and fatty acid.

REFERENCES

1. **Regelson, W.,** Heparin, heparinoids, synthetic polyanions and anionic dyes, *J. Polym. Sci.,* in press.
2. **Maeda, H., Oda, T., Matsumura, Y., and Kimura, M.,** Improvement of pharmacological properties of protein-drugs by tailoring with synthetic polymers, *J. Bioactive Biocompatible Polym.,* 3, 27, 1988.
3. **Inada, Y., Matsushima, A., Kodera, Y., and Nishimura, H.,** PEG-protein conjugates; application to biomedical and biotechnological processes, *J. Bioactive Compatible Polym.,* 5, 343, 1990.
4. **Inada, Y., Matsushima, A., Saito, Y., Yoshimoto, T., Nishimura, H., and Wada, H.,** Synthetic polymer-protein hybrid; modification of L-asparaginase with PEG, in *Multiphase Biomedical Materials,* Tsuruta, T. and Nakajima, A., Eds., VSP, Utrecht, 1989, chap. 10.
5. **Inada, Y., Matsushima, A., Takahashi, K., and Saito, Y.,** PEG-modified lipase soluble and active in organic solvents, *Biocatalysis,* 3, 317, 1990.
6. **Curme, G. O.,** *Glycols,* Reinhold Pub., New York, 1952.
7. **Carpenter, C. P., Woodside, M. D., Kinkead, E. R., King, J. M., and Sullivan, L. J.,** Response of dogs to repeated intravenous injection of polyethylene glycol 4000, *Toxicol. Appl. Pharmacol.,* 18, 35, 1971.
8. **Harris, J. M.,** Laboratory synthesis of polyethylene glycol derivertives, *Macromol. Chem. Phys.,* C25, 325, 1985.
9. **Matsushima, A., Nishimura, H., Ashihara, Y., Yokota, Y., and Inada, Y.,** Modification of *E. coli* asparaginase with 2,4-bis(*O*-methoxypolyethylene glycol)-6-chloro-*s*-triazine (activated PEG$_2$): disappearance of binding ability towards anti-serum and retention of enzymic activity, *Chem. Lett.,* 773, 1980.
10. **Abuchowski, A., Kazo, G. M., Verhoest, C. R., van Es, T., Kafkewitz, D., Nucci, M. L., Viau, A. T., and Davis, F. F.,** Cancer therapy with chemically modified enzymes. I. Antitumor properties of PEG-asparaginase conjugates, *Cancer Biochem. Biophys.,* 7, 175, 1984.
11. **Ono, K., Kai, Y., Maeda, H., Samizo, F., Sakurai, K., Nishimura, H., and Inada, Y.,** Selective synthesis of 2,4-bis(*O*-methoxypolyethylene glycol)-6-chloro-*s*-triazine as a protein modifier, *J. Biomater. Sci.,* 2, 61, 1991.
12. **Abuchowski, A., van Es, T., Palczuk, N. C., and Davis, F. F.,** Alteration of immunological properties of bovine serum albumin by covalent attachment of PEG, *J. Biol. Chem.,* 252, 3578, 1977.
13. **Abuchowski, A., McCoy, J. R., Palczuk, N. C., van Es, T., and Davis, F. F.,** Effect of covalent attachment of polyethylene glycol on immunogenicity and circulating life of bovine liver catalase, *J. Biol. Chem.,* 252, 3582, 1977.
14. **Ashihara, Y., Kono, T., Yamazaki, S., and Inada, Y.,** Modification of *E. coli* L-asparaginase with polyethylene glycol, disappearance of binding ability to anti-asparaginase serum, *Biochem. Biophys. Res. Commun.,* 83, 385, 1978.
15. **Jackson, C. C., Charlton, J. L., Kuzminski, K., Lang, G. M., and Sehon, A. H.,** Synthesis, isolation, and characterization of conjugates of ovalbumin with monomethoxypolyethylene glycol using cyanuric chloride as the coupling agent, *Anal. Biochem.,* 165, 114, 1987.
16. **Nishimura, H., Ashihara, Y., Matsushima, A., and Inada, Y.,** Modification of yeast uricase with PEG; disappearance of binding ability towards anti-uricase serum, *Enzyme,* 24, 261, 1979.
17. **Nishimura, H., Matsushima, A., and Inada, Y.,** Improved modification of uricase with PEG: accompanied with nonimmunoreactivity towards anti-uricase serum and high enzymic activity, *Enzyme,* 26, 49, 1981.

18. **Feeney, R. E.,** Tayloring proteins for food and medical uses: state of the art and inter-relationships, in *Protein Tailoring for Food and Medical Uses,* Feeney, R. E. and Whi-taker, J. R., Eds., Marcel-Dekker, New York, 1986, 1.

19. **Yoshimoto, T., Nishimura, H., Saito, Y., Sakurai, K., Kamisaki, Y., Wada, H., Sako, M., Tsujino, G., and Inada, Y.,** Characterization of PEG-modified L-asparaginase from *Escherichia coli* and its application to therapy of leukemia, *Jpn. J. Cancer Res. (Gann),* 77, 1264, 1986.

20. **Gallager, M. P., Marshall, R. D., and Wilson, R.,** Asparaginase as a drug for treatment of acute lymphoblastic leukemia, *Essays Biochem.,* 24, 1, 1989.

21. **Kamisaki, Y., Wada, H., Yagura, T., Nishimura, H., Matsushima, A., and Inada, Y.,** Increased antitumor activity of *Escherichia coli* L-asparaginase by modification with monomethoxypolyethylene glycol, *Gann,* 73, 470, 1982.

22. **Kawamura, K., Igarashi, T., Fujii, T., Kamisaki, Y., Wada, H., and Kishimoto, S.,** Immune responses of PEG-modified L-asparaginase in mice, *Int. Arch. Allergy Appl. Immunol.,* 76, 324, 1985.

23. **Kamisaki, Y., Wada, H., Yagura, T., Matsushima, A., and Inada, Y.,** Reduction of immunogenicity and clearance rate of *Escherichia coli* L-asparaginase by modification with monomethoxypolyethylene glycol, *J. Pharmacol. Exp. Ther.,* 216, 410, 1981.

24. **Abuchowski, A., Davis, F. F., and Davis, S.,** Immunosuppressive properties and circulating life in human of *Achromobactor* glutaminase-asparaginase covalently attached to PEG, *Cancer Treat. Rep.,* 65, 1077, 1981.

25. **Park, Y. K., Abuchowski, A., Davis, S., and Davis, F. F.,** Pharmacology of *E. coli* L-asparaginase-PEG adducts, *Anticancer Res.,* 1, 373, 1981.

26. **Ho, D. H., Brown, N. S., Yen, A., Holmes, R., Keating, M., Abuchowski, A., Newman, R. A., and Krakoff, I. H.,** Clinical pharmacology of PEG-L-asparaginase, *Drug Metab. Dispos.,* 14, 349, 1986.

27. **Katre, N. V., Knauf, M. J., and Laird, W. J.,** Chemical modification of recombinant interleukin 2 by PEG potency in the murine Meth A sarcoma model, *Proc. Natl. Acad. Sci. U.S.A.,* 84, 1487, 1987.

28. **Rosenberg, S. A., Spiess, P., and Lafreniere, R.,** A new approach to the adoptive immunotherapy of cancer with tumor-infiltrating lymphocytes, *Science,* 233, 1318, 1986.

29. **Knauf, M. J., Bell, D. P., Hirtzer, P., Luo, Z. P., Young, J. D., and Katre, N. V.,** Relationship of effective molecular size to systemic clearance in rats of recombinant interleukin 2 chemically modified with water soluble polymers, *J. Biol. Chem.,* 263, 15064, 1988.

30. **Yoshimoto, T., Chao, S. G., Saito, Y., Imamura, I., Wada, H., and Inada, Y.,** Chemical modification of tryptophanase from *E. coli* with PEG to reduce its immuno-reactivity towards anti-tryptophanase antibodies, *Enzyme,* 36, 261, 1986.

31. **Savoca, K. V., Davis, F. F., van Es, T., McCoy, J. R., and Palczuk, N. C.,** Cancer therapy with chemically modified enzymes. II. The therapeutic effectiveness of arginase, and arginase modified by covalent attachment of PEG, on the taper liver tumor and the L5178Y murine leukemia, *Cancer Biochem. Biophys.,* 7, 261, 1984.

32. **Savoca, K. V., Abuchowski, A., van Es, T., Davis, F. F., and Palczuk, N. C.,** Preparation of a non-immunogenic arginase by the covalent attachment of PEG, *Biochim. Biophys. Acta,* 578, 47, 1979.

33. **Wieder, K. J., Palczuk, N. C., van Es, T., and Davis, F. F.,** Some properties of PEG: phenylalanine ammonia-lyase adducts, *J. Biol. Chem.,* 254, 12579, 1979.

34. **Hershfield, M. S., Buckley, R. H., Greenberg, M. L., Melton, A. L., Schiff, R., Hatem, C., Kurtzberg, J., Markert, M. L., Kobayashi, R. H., Kobayashi, A. L., and Abuchowski, A.,** Treatment of adenosine deaminase deficiency with PEG-modified adenosine deaminase, *N. Engl. J. Med.,* 316, 589, 1987.

35. **Kimura, M., Matsumura, Y., Miyauchi, Y., and Maeda, H.,** A new tactic for the treatment of jaundice: an injectable polymer conjugated bilirubin-oxidase, *Proc. Soc. Exp. Biol. Med.,* 188, 364, 1988.

36. **Sugi, K., Inoue, M., and Morino, Y.,** Degradation of plasma bilirubin by a bilirubin oxidase derivative which has a relatively long half-life in the circulation, *Biochim. Biophys. Acta,* 991, 405, 1989.

37. **Lisi, P. J., van Es, T., Abuchowski, A., Palczuk, N. C., and Davis, F. F.,** PEG-β-glucuronidase conjugates as potential therapeutic agents in acid mucopolysaccharidosis, *J. Appl. Biochem.,* 4, 19, 1982.

38. **Wieder, K. J. and Davis, F. F.,** Enzyme therapy. II. Effect of covalent attachment of PEG on biochemical parameters and immunological determinants of β-glucosidase and α-galactosidase, *J. Appl. Biochem.,* 5, 337, 1983.

39. **Naoi, M., Kiuchi, K., Sato, T., Morita, M., Tosa, T., Chibata, I., and Yagi, K.,** Alteration of the substrate specificity of *Aspergillus orizae* β-galactosidase by modification with PEG, *J. Appl. Biochem.,* 6, 91, 1984.

40. **Chen, R. H.-L., Abuchowski, A., van Es, T., Palczuk, N. C., and Davis, F. F.,** Properties of two urate oxidases modified by covalent attachment of PEG, *Biochim. Biophys. Acta,* 660, 293, 1981.

41. **Abuchowski, A., Karp, D., and Davis, F. F.,** Reduction of plasma urate levels in the cockerel with PEG-uricase, *J. Pharmacol. Exp. Ther.,* 219, 352, 1981.

42. **Pyatak, P. S., Abuchowski, A., and Davis, F. F.,** Preparation of PEG-superoxide dismutase adducts, and an examination of its blood circulating life and anti-inflammatory activity, *Res. Commun. Chem. Pathol. Pharmacol.,* 29, 113, 1980.

43. **Miyata, K., Nakagawa, Y., Nakamura, M., Ito, T., Sugo, K., Fujita, T., and Tomoda, K.,** Altered properties of *Serratia* superoxide dismutase by chemical modification, *Agric. Biol. Chem.,* 52, 1575, 1988.

44. **Chi, L., Tamura, Y., Hoff, P. T., Macha, M., Gallagher, K. P., Schork, M. A., and Lucchesi, B. R.,** Effect of superoxide dismutase on myocardial infarct size in the canine heart after 6 hours of regional ischemia and reperfusion: a demonstration of myocardial salvage, *Circulation Res.,* 64, 665, 1989.

45. **Conforti, A., Franco, L., Milanino, R., and Velo, G. P.,** PEG-superoxide dismutase derivatives; anti-inflammatory activity in carrageenan pleurisy in rats, *Pharmacol. Res. Commun.,* 19, 287, 1987.

46. **Liu, T. H., Beckman, J. S., Freeman, B. A., Hogan, E. L., and Hsu, C. Y.,** PEG-conjugated superoxide dismutase and catalase reduced ischemic brain injury, *Am. J. Physiol.,* 256, 589, 1989.

47. **Wisner, J., Green, D., Ferrell, L., and Renner, I.,** Evidence of a role of oxygen derived free radical in the pathogenesis of caerulein-induced acute pancreatitis in rats, *Gut,* 29, 1516, 1988.

48. **Beckman, J. S., Minor, R. L., Jr., White, C. W., Repine, J. E., Rosen, G. M., and Freeman, B. A.,** Superoxide dismutase and catalase conjugated to PEG increased endothelial enzymic activity and oxidant resistance, *J. Biol. Chem.,* 263, 6884, 1988.

49. **Nucci, M. L., Olejarczyk, J., and Abuchowski, A.,** Immunogenicity of PEG-modified superoxide dismutase and catalase, *J. Free Radicals Biol. Med.,* 2, 321, 1986.

50. **Mossman, B. T., Marsh, J. P., Hardwick, D., Gilbert, R., Hill, S., Sesko, A., Shatos, M., Doherty, J., Weller, A., and Bergeron, M.,** Approaches to prevention of asbestos-induced lung disease using PEG-conjugated catalase, *J. Free Radicals Biol. Med.,* 2, 335, 1986.

51. **Newmark, J., Abuchowski, A., and Murano, G.,** Preparation and properties of adducts of streptokinase and streptokinase-plasmin complex with PEG, *J. Appl. Biochem.,* 4, 185, 1982.

52. **Koide, A., Suzuki, S., and Kobayashi, S.,** Preparation of PEG-modified streptokinase with disappearance of binding ability towards anti-serum and retention of activity, *FEBS Lett.,* 143, 73, 1982.

53. **Rajagopalan, S., Gonias, S. L., and Pizzo, S. V.,** A nonantigenic covalent streptokinase-PEG complex with plasminogen activator function, *J. Clin. Invest.,* 75, 413, 1985.

54. **Sakuragawa, N., Shimizu, K., Kondo, K., Kondo, S., and Niwa, M.,** Studies of the PEG-modified urokinase on coagulation-fibrinolysis using beagles, *Thromb. Res.,* 41, 627, 1986.
55. **German, A. J. and Kalindjian, S. B.,** The preparation and properties of novel reversible polymer-protein conjugates, *FEBS Lett.,* 223, 361, 1987.
56. **Nishimura, H., Takahashi, K., Sakurai, K., Fujinuma, K., Imamura, Y., Ooba, M., and Inada, Y.,** Modification of batroxobin with activated PEG: reduction of binding ability towards anti-batroxobin antibody and retention of defibrinogenation activity in circulation of preimmunized dogs, *Life Sci.,* 33, 1467, 1983.
57. **Koide, A. and Kobayashi, S.,** Modification of amino groups in porcine pancreatic elastase with PEG, in relation to binding ability towards anti-serum and to enzymic activity, *Biochem. Biophys. Res. Commun.,* 111, 659, 1983.
58. **Inada, Y., Ohwada, K., Yoshimoto, T., Kojima, S., Takahashi, K., Kodera, Y., Matsushima, A., and Saito, Y.,** Fibrinolysis by urokinase with magnetic property, *Biochem. Biophys. Res. Commun.,* 148, 392, 1987.
59. **Yoshimoto, T., Ohwada, K., Takahashi, K., Matsushima, A., Saito, Y., and Inada, Y.,** Magnetic urokinase: targeting of urokinase to fibrin clot, *Biochem. Biophys. Res. Commun.,* 152, 739, 1988.
60. **Takahashi, K., Ohwada, K., Yoshimoto, T., Saito, Y., Kodera, Y., Matsushima, A., and Inada, Y.,** Plasminogen endowed with magnetic properties, *J. Biotechnol.,* 8, 135, 1988.
61. **Yoshimoto, T., Saito, Y., Sugibayashi, K., Morimoto, Y., Tsukada, T., Kanmatsuse, K., Kodera, Y., Matsushima, A., and Inada, Y.,** Targeting of urokinase by magnetic force, *Drug Delivery Syst.,* 4, 121, 1989.
62. **Ajisaka, K. and Iwashita, Y.,** Modification of human hemoglobin with PEG: a new-candidate for bold substitution, *Biochem. Biophys. Res. Commun.,* 97, 1076, 1980.
63. **Leonard, M. and Dellacherie, E.,** Acylation of human hemoglobin with polyethylene derivatives, *Biochim. Biophys. Acta,* 791, 219, 1984.
64. **Iwasaki, K. and Iwashita, Y.,** Preparation and evaluation of hemoglobin-PEG conjugate (pyridoxylated PEG-hemoglobin) as an oxygen carrying resuscitation fluid, *Artif. Organs,* 10, 411, 1986.
65. **Iwashita, Y., Iwasaki, K., Ajisaka, K., Ikeda, K., and Uematsu, T.,** Use of new oxygen-carrying blood substitute, *Progr. Artif. Organs,* 867, 1983.
66. **Iwashita, Y.,** Pyridoxalated hemoglobin-polyoxyethylene conjugate (PHP) as an oxygen carrier, *Artif. Organ Today,* 1, 89, 1991.
67. **Suzuki, T., Kanbara, N., Tomono, T., Hayashi, N., and Shinohara, I.,** Physicochemical and biological properties of poly(ethylene glycol)-coupled immunoglobulin G., *Biochim. Biophys. Acta,* 788, 248, 1984.
68. **Suzuki, T., Ikeda, K., and Tomono, T.,** Physicochemical and biological properties of poly(ethylene glycol)-coupled immunoglobulin G. II. Effect of molecular weight of poly(ethylene glycol), *J. Biomater. Sci. Polym. Ed.,* 1, 71, 1989.
69. **Ehrat, M. and Luisi, P. L.,** Synthesis and spectroscopic characterization of insulin derivatives containing one or two poly(ethylene oxide) chains at specific positions, *Biopolymers,* 22, 569, 1983.
70. **Beauchamp, C. O., Gonias, S. L., Menapace, D. P., and Pizzo, S. V.,** A new procedure for the synthesis of polyethylene glycol-protein adducts; effects on function, receptor recognition, and clearance of superoxide dismutase, lactoferrin and α_2-macroglobulin, *Anal. Biochem.,* 131, 25, 1983.
71. **Inada, Y., Yoshimoto, T., Matsushima, A., and Saito, Y.,** Engineering physicochemical properties of proteins by chemical modification, *Trends Biotechnol.,* 4, 68, 1986.
72. **Inada, Y., Takahashi, K., Yoshimoto, T., Ajima, A., Matsushima, A., and Saito, Y.,** Application of polyethylene glycol-modified enzymes in biotechnological processes: organic solvent-soluble enzymes, *Trends Biotechnol.,* 4, 190, 1986.

73. **Inada, Y., Takahashi, K., Yoshimoto, T., Kodera, Y., Matsushima, A., and Saito, Y.,** Application of PEG-enzyme and magnetite-PEG-enzyme conjugates for biotechnological processes, *Trends Biotechnol.,* 6, 131, 1988.

74. **Mizutani, A., Takahashi, K., Aoki, T., Ohwada, K., Kondo, K., and Inada, Y.,** Synthesis of gefarnate with reusable lipase modified with polyethylene glycol, *J. Biotechnol.,* 11, 499, 1989.

75. **Kikkawa, S., Takahashi, K., Katada, T., and Inada, Y.,** Esterification of chiral secondary alcohols with fatty acid in organic solvents by polyethylene glycol-modified lipase, *Biochem. Int.,* 19, 1125, 1989.

76. **Sakurai, K., Kishimoto, K., Kodera, Y., and Inada, Y.,** Solid phase synthesis of peptides with polyethylene glycol-modified protease in organic solvents, *Biotechnol. Lett.,* 12, 685, 1990.

77. **Tada, M., Shinoda, I., and Okai, H.,** L-ornithyl taurin, a new salty peptide, *J. Agric. Food Chem.,* 32, 992, 1984.

78. **Ohwada, K., Aoki, T., Toyoda, A., Takahashi, K., and Inada, Y.,** Synthesis of N^{α},N^{δ}-dicarbobenzoxy-L-ornithyl-β-alanine benzyl ester, a derivative of salty peptide, in 1,1,1-trichloroethane, *Biotechnol. Lett.,* 11, 499, 1989.

Biocomposite Materials
for Drug Delivery Systems

3.1 MODIFICATION OF POLYPEPTIDE HORMONES FOR CONTROLLING THE RECEPTOR SELECTIVITY

S. Kimura

TABLE OF CONTENTS

I. RECEPTOR SUBTYPE DIVERSITY IN THE CENTRAL NERVOUS SYSTEM

It is now generally admitted that receptor subtype heterogeneity is a common feature of signal reception proteins involved in the central nervous system. For example, neuroreceptors and ion channels are categorized into three superfamilies: G protein-coupled receptors, ligand-gated ion channel receptors, and voltage-gated ion channels. Receptors in the category of each superfamily are characterized by the presence of seven, four, and six membrane-spanning domains in the protein structure, respectively. The presence of distinct subtypes has been shown for many receptors belonging to each superfamily by the molecular analysis of the genes encoding neuroreceptors as well as pharmacological analyses. Schofield et al.[1] have pointed out that the receptor subtype diversity should be important for the evolutionary survival and adaptation of species. Moreover, receptor subtype diversity contributes to the central nervous system by increasing the information-handling capacities, resulting in neural plasticity.

Opioid receptors also have been shown to be composed of δ-, μ-, and κ-types.[2,3] Despite intensive studies on the receptors, however, relatively little is yet known about them because of difficulties in purification of the receptor in an active form. So far, Ueda et al.[4] have extracted μ-receptors and succeeded in their reconstitution with G-protein in the presence of phospholipid membranes. The molecular weight of a μ-receptor was estimated to be 58,000. However, the primary sequence is not yet known.

Opioid peptides have attracted much attention because of their analgesic effect. For example, morphine possesses a therapeutic potential as an analgesic and is known to bind to μ-receptors. However, it has various problems such as abuse potential, dependence liability, tolerance, and respiratory depression. On the other hand, there is substantial evidence that δ- and κ-receptors can also mediate analgesia, however, with different side effects.[5,6] Therefore, it is of importance to develop highly specific opioid ligands not only for clarification of the distinct pharmacological action mediated by each receptor type but also for obtaining clinically available analgesics. In this chapter several strategies for attaining a specific peptide ligand are introduced as shown in Figure 1. They can be classified into two groups, the intrinsic and extrinsic modifications of opioid peptides. In the former, the amino acid residues, which are thought to be involved directly in binding to a receptor, are replaced so as to enhance the affinity for the specific receptor. In the latter, the sequence of the essential part of peptide hormones is retained but connected to other specific peptides or carrier molecules, which are designed to affect the activity by changing the peptide conformation, the affinity for phospholipid bilayer membrane, or the interaction mode with receptors from a monovalent to multivalent way.

Intrinsic Modification

- Replacement of Amino Acid in the Message Segment with Naturally Occurring Amino Acids or with Unnatural Amino Acids
- Conformational Constraints Induced by Using Bulky Amino Acids or by Cyclization of the Backbone

Extrinsic Modification

- Attachment of Membrane-Oriented Peptides to the Message Segment
- Bivalent Ligand
- Multivalent Ligand

FIGURE 1. Strategy for developing an analogue highly selective for a specific receptor.

TABLE 1
The Primary Sequence of Opioid Peptides

Peptide	Sequence	Receptor selectivity
[Leu]enkephalin	YGGFL	δ
[Met]enkephalin	YGGFM	δ
β-Endorphin	YGGFMTSEKSQTPLVTLFKNAIIKNAYKKGQ	δ – μ
Adrenorphin	YGGFMRRV-NH$_2$	μ
Dynorphin	YGGFLRRIRRIRRKLKWDNQ	κ

II. NATURALLY OCCURRING OPIOID PEPTIDES

In 1975, Hughes et al.[7] found endogenous opioid peptides, [Met]enkephalin and [Leu]enkephalin, by using guinea pig ileum (GPI) and mouse vas deferens (MVD) bioassays. Since then, various endogenous opioid peptides have been reported. Moreover, these peptides have been shown to originate from three precursors, preproenkephalin A, preprodynorphin, and proopiomelanochortin, by processing enzymes.[8-10] All of these peptides have a common sequence, Tyr-Gly-Gly-Phe-, at the N terminal (Table 1). Therefore, tyramine moiety of Tyr 1 and phenyl moiety of Phe 4 were regarded to be the major contributors to recognition between opioid peptides and receptors.[11]

The tyramine moiety was supposed to a key pharmacophore not only for opioid peptides but also for alkaloid opiates because of the resemblance in stereochemistry between them (Figure 2). In order to examine the hypothesis, several hybrid opioid peptides were synthesized (Figure 3).[12,13] The analogues, however, did not show any significant *in vitro* opioid agonist or antagonist activity. The reason for the absence of activity was ascribed to the tertiary or secondary amino terminus and conformational restriction of the phenolic moiety. Furthermore, it was suggested that the recognition site of opioid receptors should differ between peptidergic and alkaloid opioids. In other words, the stereochemical requirements for δ- and μ-receptors are different.

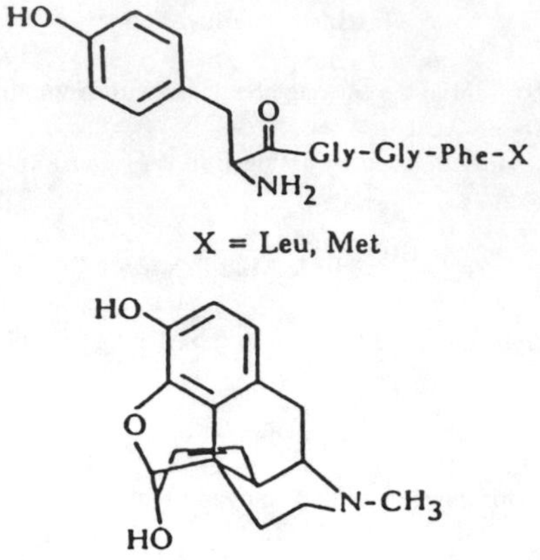

FIGURE 2.   Structural resemblance between Tyr moiety of opioid peptides and alkaloid opiates.

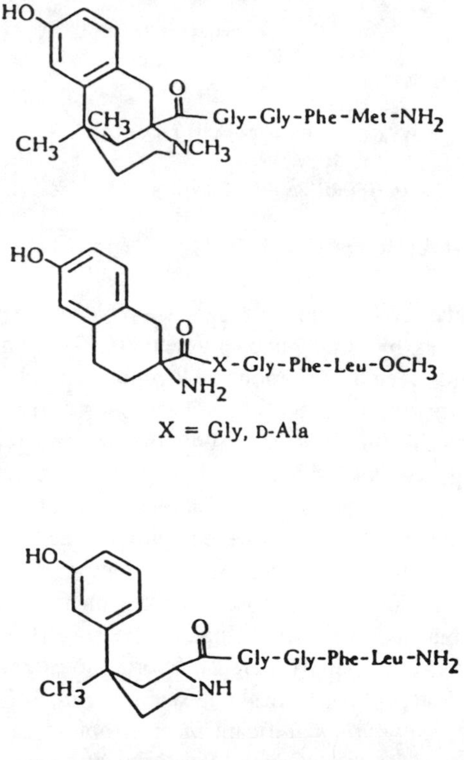

FIGURE 3.   Hybrid opioid peptide analogues.

In this chapter various approaches will be introduced for obtaining selective opioid ligands mainly by using peptide derivatives, because peptides and alkaloids are too different in chemical properties to discuss and compare the interactions with the receptors.

It is noteworthy that there are other peptides which show potent analgesic activity. One is dermorphin (Tyr-D-Ala-Phe-Gly-Tyr-Pro-Ser-NH_2) extracted from frog skin,[14] and other is casomorphin (Tyr-Pro-Phe-Pro-Gly-Pro-Ile) found in hydrolysates of casein.[15] These peptides indicate that Phe 4 was not a firm requirement for opioid activity.[15]

III. MODIFICATION OF THE SEQUENCE OF OPIOID PEPTIDES

One way to obtain a highly specific opioid peptide is to modify the component amino acids. The modification at Gly 2 or Gly 3 is especially interesting, because two successive glycine residues in Tyr-Gly-Gly-Phe-sequence of endogenous opioid peptides allow many low-energy conformations, which should result in binding to different receptor sites such as μ- and δ-receptors. Roques et al. have succeeded in obtaining a highly μ-selective ligand, TRIMU 4,[16] and two δ-probes, DSLET[17] and DTLET,[18] by modification at Gly 2 and Phe 4 in the sequence. The sequence of TRIMU 4 is Tyr-D-Ala-Gly-NHCH(CH_3)-CH_2-CH(CH_3)$_2$, in which Phe 4 of enkephalin was replaced by the hydrophobic aliphatic chain. The receptor affinities of the analogues are summarized in Table 2. The aromatic side chain of Phe 4 is important for μ-receptor selectivity against the δ-receptor, as shown by the shift toward μ-selectivity resulting in the replacement by the aliphatic hydrophobic residue. Replacement of D-Ala 2 by Aib or L-Ala decreased the receptor affinity remarkably. The energetically accessible regions of the ϕ, ψ-space should not be compatible with the biologically active conformations due to introduction of Aib or L-Ala to the 2nd position.[19] μ-Selectivity increased with using Ser or Thr at the 2nd position, probably because the moderately hydrophilic side chain is not favorable for the corresponding hydrophobic subsite of the δ-receptor.

The analgesic potency (AD_{50} in the hot plate test) of the analogues was correlated with the inhibitory effect on the GPI (IC_{50}) and the ability to inhibit [^{3}H]DAGO binding (K_i) in rat brain tissue. The results indicate that the antinociceptive activity of the analogues is related to the affinity for the μ-receptor.

IV. QUANTITATIVE STRUCTURE-ACTIVITY RELATIONSHIP

A series of enkephalin derivatives with the general structure, Tyr-D-Ala-Gly-X-Y-NH_2 (X and Y are variable amino acid residues) were synthesized

TABLE 2
Inhibitory Potencies (IC$_{50}$) of Morphine and Opioid Peptides on the Guinea Pig Ileum (GPI) and Mouse Vas Deferences (MVD)[a]

Compounds	IC$_{50}$, nM		IC$_{50}$ (MVD)/IC$_{50}$ (GPI)
	GPI	MVD	
1, Morphine	70 ± 8	309 ± 30	5.55
2, [Met]enkephalin	200 ± 19	13.0 ± 1.5	0.065
3, Tyr-D-Ala-Gly-NHCH(CH$_3$)CH$_2$CH$_2$CH(CH$_3$)$_2$(TRIMU 4)	237 ± 56	1050 ± 90	4.35
4, Tyr-Aib-Gly-NH-CH-(CH$_3$)-CH$_2$-CH(CH$_3$)$_2$	1850 ± 120	4720 ± 530	2.56
5, Tyr-L-Ala-Gly-NH-CH(CH$_3$)-CH$_2$-CH(CH$_3$)$_2$	≥50,000	≥50000	
6, Tyr-D-Ser-Gly-NH-CH(CH$_3$)-CH$_2$-CH(CH$_3$)$_2$	718 ± 48	990 ± 143	1.37
7, Tyr-D-Ser(O-t-Bu)-Gly-NH-CH(CH$_3$)-CH$_2$-CH(CH$_3$)$_2$	111 ± 29	105 ± 25	0.94
8, Tyr-D-Ala-Gly-N(Et)-CH(CH$_3$)-CH$_2$-CH(CH$_3$)$_2$	216 ± 15	1800 ± 40	8.33
9, Tyr-D-Ala-Gly-NH-CH$_2$-CH$_2$-CH(CH$_3$)$_2$ (TRIMU 5)	323 ± 62	3410 ± 310	10.53
10, Tyr-D-Met-Gly-NH-CH$_2$-CH$_2$-CH(CH$_3$)$_2$	154 ± 9	733 ± 61	4.76
11, Tyr-D-Ala-Gly-N(Me)-Phe-Gly-01 (DAGO)	11.5 ± 0.4	76.1 ± 6.7	6.66
12, Tyr-D-Ala-Gly-N(Me)-Leu-Gly-01	379 ± 62	3180 ± 150	8.33
13, Tyr-D-Ser-Gly-Phe-Leu-Thr (DSLET)	406 ± 46	0.40 ± 0.04	0.001
14, Tyr-Azgly-Gly-Phe-Leu-Thr	14000 ± 500	38.0 ± 3.3	0.003
15, Tyr-D-Thr-Gly-Phe-Leu-Thr (DTLET)	460 ± 60	0.150 ± 0.012	0.0003
16, Tyr-β-OH-Nva-Gly-Phe-Leu-Thr	21.3 ± 5.7	0.05 ± 0.01	0.002
17, Tyr-D-Ser-Gly-Phe-Leu-Leu (DSLEL)	257 ± 28	2.45 ± 0.80	0.009
18, Tyr-D-Thr-Gly-Phe-Leu-Leu	260 ± 38	0.410 ± 0.037	0.0015
19, Tyr-D-Ser-Gly-Phe-Leu-des-COOH-Thr	64 ± 12	3.45 ± 0.13	0.054
20, Tyr-D-Thr-Gly-Phe-Leu-Ser	99 ± 12	0.25 ± 0.04	0.0025
21, Tyr-D-Thr-Gly-Phe-Leu-Hyp (DSLEH)	88 ± 10	1.10 ± 0.07	0.012
22, Tyr-D-Thr-N(Me)-Phe-Leu-Thr	93.2 ± 15	7.30 ± 0.30	0.78
23, Tyr-D-Thr-Gly-Phe-N(Me)-Leu-Thr	224 ± 21	1.33 ± 0.09	0.006
24, Tyr-D-Pen-Gly-Phe-Pen	10000 ± 400	2.50 ± 0.50	0.00025

[a] The values are the means ± SEM of four to eight independent determinations.

From Gacel, G., Zajac, J. M., Delay-Goyet, P., Dauge, V., and Roques, B. P., J. Med. Chem., 31, 374, 1988. With permission.

and studied on GPI and MVD assays by Do, Fauchere, and Schwyzer.[20-22] Artificial "fat" amino acids such as *o*-carboranylalanine (Car), adamantylalanine (Ada), β-methylvaline (Bug), neopentylglycine (Neo), and *p*-nitrophenylalanine (Nip) were introduced into the 4th or 5th position of enkephalin analogues. The concept of using such amino acids is that they can become probes for elucidating structure-activity relationships in terms of electronic, steric, and hydrophobic factors. It was found that the electrophilic aromatic character of the 4th position and the hydrophobicity of the 5th position effectively enhanced the potency in the GPI assay. On the basis of these results, a quantitative structure-activity relationship (QSAR) has been proposed between the dose for half-maximal inhibition of the GPI assay (IC_{50}) and structural parameters which characterize the side chains of the 4th and 5th residues as shown in Equation 1[21]

$$\log(1/IC_{50}) = k_0 + \sum_i^n k_i P_i \tag{1}$$

in which P_i is the physicochemical parameter of side chains of the 4th and 5th residues, k_0 and k_i are the regression constants, and n is the number of parameters or parameter combinations involved. The model assumes additivity of the contributions from the 4th and 5th residues. Three kinds of side-chain parameters — hydrophobicity, electronic feature, and steric effect — were considered. An excellent statistic was obtained for the equation of regression according to an F test. However, despite the success of the QSAR method to account for the opiate activity in this series, QSAR does not seem to be helpful to explain the receptor selectivity of enkephalin analogues.[23]

V. DERMORPHIN ANALOGUES

Salvadori et al.[24] have synthesized many small dermorphin analogues and found that tetrapeptides Tyr-D-Ala-Phe-X are potent analgesics. Furthermore, Tyr-D-MetO-Phe-Gly-NH$_2$ was about 1500 and 17 times, respectively, as potent an analgesic as morphine by intracerebroventricular or subcutaneous administrations in mice. D-Methionine S-oxide at the 2nd position showed the remarkably enhanced activity more than D-Met and D-Ala at the 2nd position. For explanation of the result, D-MetO at the 2nd position is considered not only to induce the active conformer but also to provide an extra binding site for the receptor. The differences in the pharmacokinetic properties between D-MetO and D-Ala at the 2nd position are ascribed to the hydrophilic nature of the former residue.

Moreover, dermorphin tetrapeptides, W-Tyr-D-MetO-Phe-X-Y (W = H, H$_2$NC = (NH); X = Gly, Sar, D-Ala; Y = OH, OCH$_3$, NH$_2$), were prepared and tested for opioid activity. Tyr-D-MetO-Phe-Gly-OCH$_3$ was shown to be one of the most selective μ-selective ligand. However, Tyr-D-MetO-Phe-D-

Ala-OH exhibited the best antinociceptive effect following intracerebroventricular administration in mice (30 times more potent than morphine on a molar basis). Since the receptor affinities of the analogue were relatively low, no correlation was found between the affinity for each type of receptor and the intracerebroventricular antinociceptive activity of dermorphin analogues. It has been pointed out that the discrepancy may be explained by pharmacokinetic or transport parameters and/or low stability in plasma of the peptides. For example, Tyr-D-Ala-Phe-Gly-NH$_2$ is a very potent analgesic after intracerebroventricular injection, but not following subcutaneous administration. In order to enhance stability of the peptide against enzymatic degradation, one or more amide bonds are reversed (retro-inverso isomer).[25] However, the activity of the partially modified retro-inverso analogues by subcutaneous administration was not improved, indicating that the difference in the metabolic stability cannot fully explain the different potency between intracerebroventricular and subcutaneous administrations. Thus, low analgesic potency after subcutaneous injection is considered due to inadequate central nervous system permeation, diffusion from subcutaneous sites, entry into blood, and binding to plasma proteins.

VI. REGULATION OF DISTANCE BETWEEN Tyr AND Phe

As described before, the relative spatial disposition between an amino-terminal Tyr and a Phe aromatic ring is a key factor for determining μ-activity of opioid peptides. Jacobson et al.[26] synthesized a novel series of hybrid retro peptides in which Tyr was connected to *N*-acyl Phe through a variety of diamine spacers. The molecular design is based on the idea that the role of amino acid residues in between Tyr and Phe is to control the conformation rather than to present additional primary recognition contacts with the receptor. Indeed, Tyr-NHC(CH$_3$)$_2$CH$_2$NH-retro-Phe-Ac was found to be 21 times more potent than morphine in the GPI assay. The finding suggests that a very simple molecule containing only two amino acids, Tyr and Phe, connected by an appropriate spacer, is sufficient for eliciting potent opioid activity.

VII. CONFORMATIONALLY CONSTRAINED LINEAR ANALOGUES INDUCED BY BULKY SIDE CHAINS

The weak receptor selectivity of the endogenous short linear peptides can be ascribed to the molecular flexibility, which allows their easy adaptation to the various receptor types. Gacel et al.[27] have synthesized a series of linear conformationally constrained opioid peptides by incorporating bulky residues into the sequence of δ-specific agonist, DSLET or DTLET. The synthesized analogues, Tyr-D-X(OY)-Gly-Phe-Leu-Thr(OZ) (X = Ser or Thr, Y = *t*-butyl or benzyl, X = *t*-butyl), were tested by binding studies based on displacement of μ- and δ-opioid receptors selective radiolabeled ligands from rat brain membranes. The affinity of DSLET for the μ-receptor was hampered remarkably by incorporation of bulky residues such as Tyr-D-Ser(O-t-Bu)-Gly-Phe-Leu-Thr(O-t-Bu) (BUBU), though affinity for the δ-receptor was retained. As a result, BUBU was 10 times more δ-selective than DSLET.

VIII. CONFORMATIONALLY CONSTRAINED CYCLIC ANALOGUES

The number of available conformations of peptides can be enormously reduced by cyclization. Indeed, cyclic peptides have been intensively utilized for conformational analyses and construction of functional peptides such as enzyme and ionophore models.[28] Because the conformation of cyclic peptides in solution can be determined directly by various spectroscopic methods, cyclic peptides are highly advantageous molecules for discussions on structure-function relationships. Thus, cyclic analogues of peptide hormones are also expected to provide useful information on ligand-receptor interactions and receptor structures.[29]

Schiller et al.[30] have synthesized a cyclic analogue of enkephalin, Tyr-cyclo[-$N\gamma$-D-A$_2$bu-Gly-Phe-Leu-] (A$_2$bu represents α,γ-diaminobutyric acid), and compared the cyclic compound with its corresponding open-chain analogue, [D-A$_2$bu^2,Leu^5]enkephalinamide, in binding assays using rat brain homogenates (Figure 4). In the δ-receptor selective binding assay, the cyclic analogue showed a 28-fold decrease in affinity compared with the linear peptide, whereas in the μ-receptor selective binding assay, only a 2-fold reduction in affinity was observed (Table 3). Therefore, it is concluded that the semirigid 14-membered ring structure of the cyclic analogue is highly compatible with the topography of the μ-receptor, but causes a decrease in affinity for δ-receptors compared with the open-chain analogue. The results also permit the unambiguous conclusion that different opiate receptor classes have different conformational requirements. There are two theories to account for the difference in receptor topography: a single opiate receptor protein with different conformational forms regulated by allosteric effectors, and heterogeneous opiate receptors with minor changes in the primary sequence of the proteins.

A number of other cyclic enkephalin analogues have been prepared (Figure 5).[31-33] Cyclization has been found to change the receptor selectivity remarkably: analogues I-IV, μ-selective; analogue V, nonselective; analogue VI, δ-selective. Cyclic dermorphin analogues also have been synthesized, in which Phe residue is located in position 3.[34] The results of the binding assay are shown in Table 4. Tyr-D-Orn-Phe-Asp-NH$_2$ (containing a 13-membered ring) was shown to be one of the most selective μ-receptor ligands (K_i^δ/K_i^μ = 213). Cyclic lactam analogue Tyr-D-Asp-Phe-A$_2$bu-NH$_2$, containing an even smaller 12-membered ring structure, also showed high μ-receptor selectivity

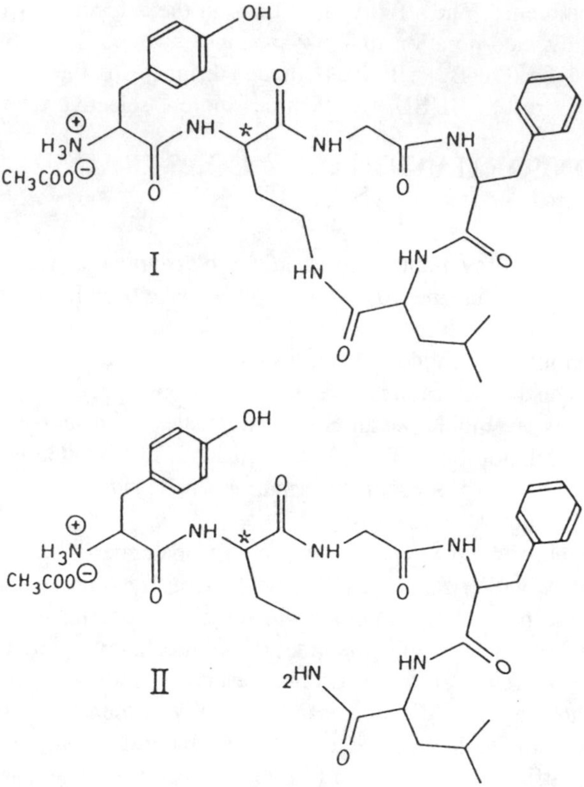

FIGURE 4.   Molecular structure of cyclic and linear enkephalin analogues.

$(K_i^\delta/K_i^\mu = 168)$. In contrast, the conformationally less restricted cyclic analogue, Tyr-D-Lys-Phe-Glu-NH$_2$ (15-membered ring) showed no receptor preference $(K_i^\delta/K_i^\mu = 3.05)$. On the other hand, the cystine-containing analogue Tyr-D-Cys-Phe-Cys-NH$_2$ (11-membered ring) was less selective $(K_i^\delta/K_i^\mu = 33.9)$, probably due to its more flexible ring structure. These results indicate that both ring size and ring flexibility affect receptor affinity and selectivity.

Interestingly, Tyr-D-Pen-Gly-Phe-D-Pen was shown to be more δ-selective $(K_i^\delta/K_i^\mu = 1/116)$ than DSLET and DTLET, although it exhibited a weaker affinity for the δ-receptor.[35] It is suggested that the interaction of the conformationally constrained and cyclic penicillamine-containing peptides with the μ-receptor is strongly hampered whereas δ-recognition is not as highly affected.

It is notable that cyclic opioid peptides sometimes show higher opioid activity in the bioassays than that expected on the basis of the results obtained in the binding assays, indicating enhanced intrinsic activity of cyclic analogues at the receptors.

TABLE 3

Inhibitory Effects of Peptide Analogues and [Leu^5]Enkephalin on the Binding of ^{3}H-Naloxone and ^{3}H[D-Ala^2, D-Leu^5]Enkephalin in Rat Brain Homogenates

No.	Compound	^{3}H-Naloxone binding, K_i^μ (nM)	^{3}H-[D-Ala^2, D-Leu^5]enkephalin binding, K_i^δ (nM)	$K_i^\delta K_i^\mu$
I	Tyr-cyclo[-N^γ-D-A$_2$bu-Gly-Phe-Leu-]	13.8 ± 3.4	115 ± 16	8.33
II	[D-Abu^2,Leu^5]enkephalinamide	5.60 ± 0.11	4.06 ± 1.46	0.725
III	[Leu^5]enkephalin	27.7 ± 5.7	8.05 ± 0.96	0.291

Note: Binding studies using rat brain membrane preparations were carried out as described elsewhere.[20] ^{3}H-naloxone and ^{3}H-[D-Ala^2,D-Leu^5]-enkephalin at concentrations of 0.40 and 0.96 nM, respectively, were used as radioligands and incubations were performed at 0°C for 1 h. K_i values were calculated based on Cheng and Prusoff's equation[24] using values of 0.5 and 0.8 nM for the dissociation of ^{3}H-naloxone and ^{3}H-[D-Ala^2,D-Leu^5]enkephalin, respectively. Values shown are the mean of three determinations (±SEM).

From Schiller, P. W. and DiMaio, J., *Nature (London)*, 297, 74, 1982. With permission.

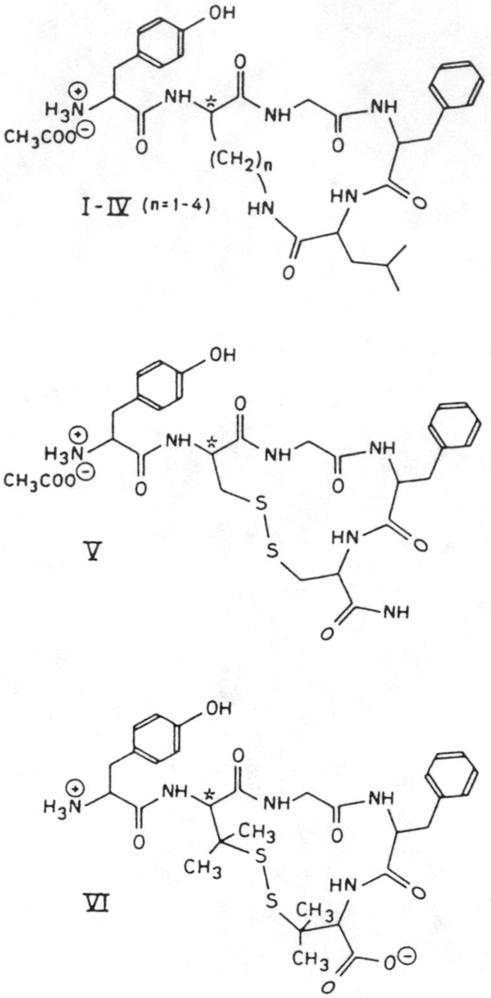

FIGURE 5.   Various cyclic enkephalin analogues.

IX. MESSAGE-ADDRESS CONCEPT

The molecular organization of peptide hormones has been classified by Schwyzer[36] into two types: sychnologic and rhegnylogic. In the former, peptides are composed of a "message" sequence (triggering the receptor) and an "address" sequence (delivering to the specific receptor), both of which are continuous in the sequence. In the latter, the amino acid residues composing the message scatter in the whole sequence, and come together upon taking a defined structure.

On the basis of the message-address concept, Portoghese et al.[37] have synthesized interesting analogues combining opioid alkaloid pharmacophores

TABLE 4
Binding Assays of Opioid Peptide Analogues[a]

No.	Compound	[3H]DAGO		[3H]DSLET		K_i^δ/K_i^μ
		K_i^μ, nM	Relative potency[b]	K_i^δ, nM	Relative potency[b]	
1	H-Tyr-D-Orn-Phe-Asp-NH$_2$	10.4 ± 3.7	0.907 ± 0.032	2220 ± 65	0.00114 ± 0.00003	213
2	H-Tyr-Orn-Phe-Asp-NH$_2$	4830 ± 610	0.00195 ± 0.00025	>60000	<0.0000421	>12.4
3	H-Tyr-D-Orn-D-Phe-Asp-NH$_2$	3040 ± 470	0.00310 ± 0.00048	>380000	<0.0000067	>125
4	H-Tyr-D-Orn-Phe-D-Asp-NH$_2$	21.7 ± 3.2	0.435 ± 0.064	422 ± 17	0.00600 ± 0.00024	19.4
5	H-Tyr-D-Orn-Phe-(pNO$_2$)-Asp-NH$_2$	273 ± 21	0.0345 ± 0.0027	5060 ± 540	0.000500 ± 0.000053	18.5
6	H-Tyr-D-Orn-Phg-Asp-NH$_2$	445 ± 92	0.0212 ± 0.0044	3570 ± 370	0.000709 ± 0.000073	8.02
7	H-Tyr-D-Orn-Hfe-Asp-NH$_2$	11.9 ± 0.7	0.792 ± 0.047	849 ± 189	0.00298 ± 0.00066	71.3
8	H-Tyr-D-Orn-Phe(NMe)-Asp-NH$_2$	3570 ± 500	0.00264 ± 0.00037	28100 ± 2530	0.0000900 ± 0.0000081	7.87
9	H-Tyr-D-Asp-Phe-Orn-NH$_2$	9.55 ± 2.52	0.987 ± 0.261	1320 ± 150	0.00192 ± 0.00022	138
10	H-Tyr-D-Asp-Phe-A$_2$bu-NH$_2$	24.8 ± 1.1	0.380 ± 0.017	4170 ± 430	0.000607 ± 0.000062	168
11	H-Tyr-D-Cys-Phe-Cys-NH$_2$	11.0 ± 0.3	0.857 ± 0.023	373 ± 63	0.00678 ± 0.00115	33.9
12	[Leu5]enkephalin	9.43 ± 2.07	1	2.53 ± 0.35	1	0.268

a Mean of three determinations ± SEM.
b Potency relative to [Leu5]enkephalin.

From Schiller, P. W., Nguyen, T. M.-D., Maziak, L. A., Wilkes, B. C., and Lemieux, C., *J. Med. Chem.*, 30, 2094, 1987. With permission.

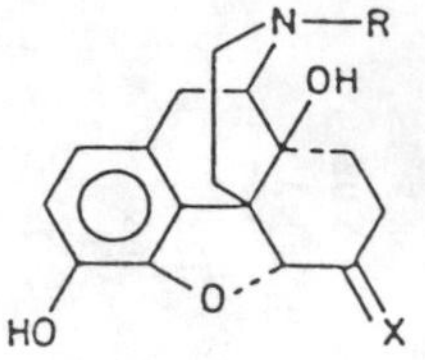

R	X
1. CH_3	O
2. $CH_2CH(CH_2)_2$	O
3. CH_3	$NNHCONH_2$
4. $CH_2CH(CH_2)_2$	$NNHCONH_2$
5. CH_3	NNHCO-Phe-Leu
6. $CH_2CH(CH_2)_2$	NNHCO-Phe-Leu
7. CH_3	NNHCO-Phe-Leu-Arg-Arg-Ile
8. $CH_2CH(CH_2)_2$	NNHCO-Phe-Leu-Arg-Arg-Ile

FIGURE 6. Analogues composed of opioid alkaloid pharmacophores and "address" segments of endogenous opioid peptides.

and address segments of endogenous opioid peptides. They have synthesized analogues of leucine-enkephalin and dynorphin (1—8) in which the N terminal dipeptide message sequence has been replaced by oxymorphone or naltrexone (Figure 6). As shown in Table 5, combination with the enkephalin address (5) increased the selectivity for δ-receptors, and attachment of the dynorphin address (7) increased the selectivity for κ-receptors when compared to oxymorphone 1 and its semicarbazone 3. It was successfully shown that peptide portions of the analogues can modulate the receptor selectivity of the attached alkaloid pharmacophores.

The message-address concept has been applied further to the development of a potent and selective δ-opioid receptor antagonist.[38] An indole system was fused to the C ring of naltrexone as a mimic of the address component (Figure 7). The benzene moiety of indole was viewed as the δ-address component, mimicking the phenyl group of Phe 4, and the pyrrole portion was used as a rigid spacer. Naltrindole (NTI) was the most potent member of the series, with K_e values of 0.1 nM at δ-receptors. The antagonism by NTI was about 220- and 350-fold greater at δ- than at μ- and κ-opioid receptors, respectively.

X. MEMBRANE COMPARTMENT CONCEPT

Another parameter, a peptide amphiphilic moment, for predicting receptor selection was presented by Schwyzer.[39] The amphiphilic moment represents the tendency of the peptide to orient on an aqueous-hydrophobic interphase boundary region on the basis of the segregation of hydrophobic and hydrophilic amino acid residues throughout the peptide molecule. The moment is calculated in analogy to the helical hydrophobic moment proposed by Eisen-

TABLE 5
Binding of Alkaloid-Peptide Hybrids to Opioid Receptor Types in Guinea Pig Membranes[a]

| Compound | Binding: K_d^b nM | | |
	μ	κ	δ
1	15 ± 1	725 ± 154	145 ± 17
3	14.5 ± 0.97	179 ± 22	69 ± 2.7
5	14.8 ± 1.8	96 ± 13	6.5 ± 0.18
7	15.5 ± 0.94	9.8 ± 0.21	24 ± 0.84
2	0.37 ± 0.03	4.8 ± 1.0	9.4 ± 1.0
4	0.48 ± 0.02	2.6 ± 0.24	3.7 ± 0.30
6	0.76 ± 0.08	2.8 ± 0.20	3.1 ± 0.10
8	2.3 ± 0.36	2.2 ± 0.41	5.7 ± 0.60
Leucine-enkephalin	380 ± 45	>10000	6.0 ± 0.5
Dynorphin(1-8)	340 ± 25	40 ± 14	50 ± 18

[a] Prepared by the method described by Tam: S. W. Tam, *Eur. J. Pharmacol.*, 1985, 109, 33.

[b] Values obtained from at least three replicate determinations. The following radioligands were employed: 0.5 nM of [³H]naloxone (μ binding); 1 nM of (−)-[³H]-ethylketazocine in the presence of DADLE (500 nM) and sufentanil (20 nM) (κ binding); 1 nM of [³H]DADLE in presence of sufentanil (4 nM) (δ binding). All binding was performed in 50 mM tris buffer, pH 7.4, containing 100 mM NaCl and bacitracin (50 μg/ml).

From Lipkowski, A. W., Tam, S. W., and Portoghese, P. S., *J. Med. Chem.*, 29, 1222, 1986. With permission.

berg et al.[40] In the helical hydrophobic moment, the direction vectors originate on the helix axis and are perpendicular to it, whereas in the amphiphilic moment the direction vectors originate from the center of the helix (Figure 8).

The concept of the amphiphilic moment arose from experimental discoveries of specific interactions between neuropeptides and lipid bilayer membranes.[41-47] These observations have led to the concept that the membrane may play an important role as catalyst for peptide-receptor interactions by serving as an antenna to capture peptides and as a device for the selection and proper orientation of the message segment ("membrane requirements"), thus facilitating productive receptor interactions.[48] Although it has been generally recognized that peptide structural features determine site specificity ("receptor requirements"), the concept of membrane requirements has been successfully applied to explain opioid receptor selection, in which opioid peptides have a common N terminal message segment and a different C terminal address segment, since the latter determines the receptor selection and the preferred conformation, orientation, and accumulation of the opioid peptide on the lipid membrane.

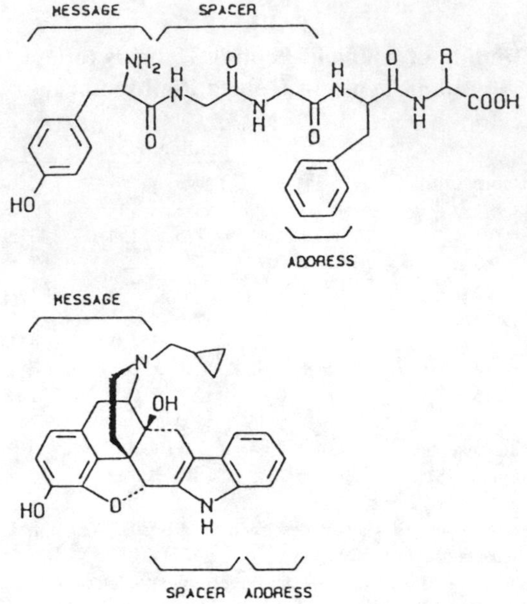

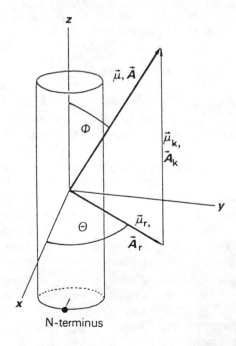

FIGURE 7. Antagonist having an indole ring fused to the C ring of naltrexone.

FIGURE 8. Schematic presentation of amphiphilic moment (A) of a helical peptide (cylinder). (From Schwyzer, R., *Biochemistry*, 25, 6335, 1986. With permission.)

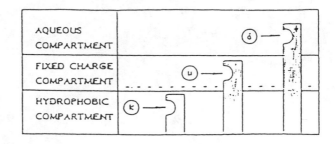

FIGURE 9. The location of binding sites of opioid receptors in lipid bilayers according to the membrane compartment concept. (From Schwyzer, R., *Biochemistry*, 25, 6335, 1986. With permission.)

According to the membrane compartment concept, the δ-site is exposed to the aqueous compartment surrounding the target cell at a distance comparable to or greater than the Debye-Huckel length and is in a cationic vicinity. The μ-site is exposed to the anionic fixed-charge compartment of the membrane in aqueous surroundings. The κ-site is buried in a more hydrophobic membrane compartment close to the fixed-charge compartment (Figure 9). The relative accumulation of the opioid message domains in these compartments is determined by the address domains and constitutes a major part of the receptor selection mechanism. Interestingly, the amphiphilic moment is a very useful parameter for explaining the orientation of peptide molecules in lipid membranes, thus being helpful for predicting opioid receptor selection.[49-52]

XI. ENKEPHALIN ANALOGUES CARRYING ARTIFICIAL ADDRESS PEPTIDES

Three different sequential peptides were synthesized as artificial address segments and connected to the C terminal of [Leu]enkephalinamide (Figure 10).[53] One of them, Tyr-Gly-Gly-Phe-Leu-Gly-Pro-(Lys-Aib-Leu-Aib)$_2$-OMe (ENK-A′8), was shown to be distributed to a lipid bilayer membrane upon taking an amphiphilic α-helical structure, while the other two analogues, Tyr-Gly-Gly-Phe-Leu-Gly-Pro(Lys-Sar-Sar-Sar)$_2$-OMe and Tyr-Gly-Gly-Phe-Leu-Gly-Pro-(Lys-Pro-Pro-Pro)$_2$-OMe, stayed in aqueous phase. Opioid receptor affinities of the latter two derivatives were similar to [Leu]enkephalinamide, indicating that the connection of hydrophilic peptide segments to the C terminal of [Leu]enkephalinamide did not affect the receptor affinity. On the other hand, δ- and μ-receptor affinities of ENK-A′8 were about 1/20 and 1/6 times, respectively, those of [Leu]enkephalinamide. Therefore, ENK-A′8 has higher selectivity for the μ-receptor than [Leu]enkephalinamide. The

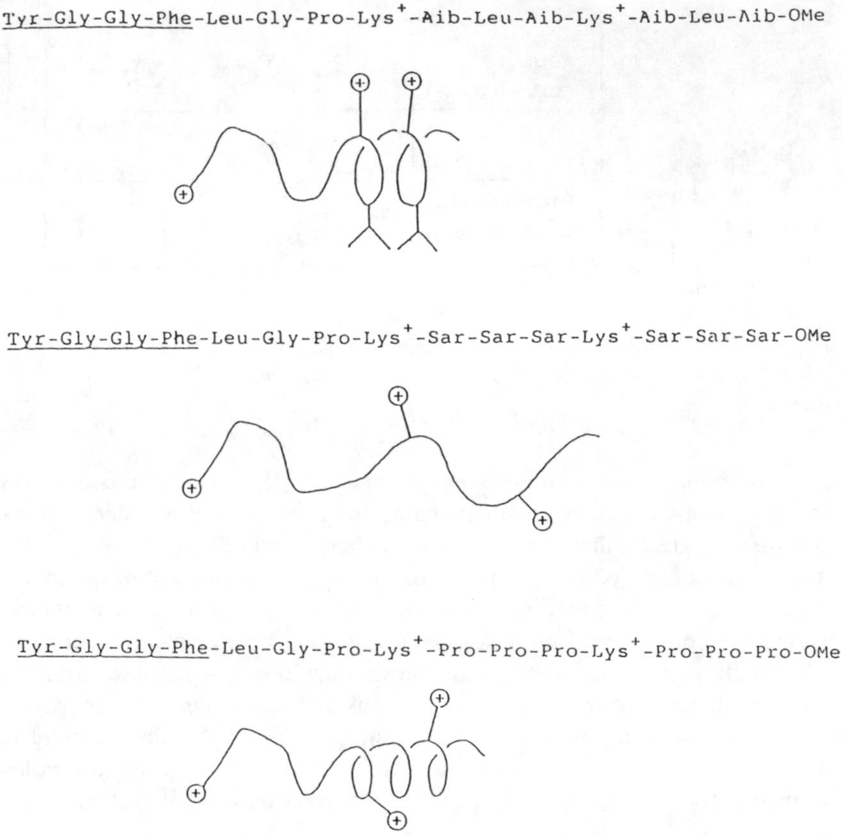

Tyr-Gly-Gly-Phe-Leu-Gly-Pro-Lys$^+$-Aib-Leu-Aib-Lys$^+$-Aib-Leu-Aib-OMe

Tyr-Gly-Gly-Phe-Leu-Gly-Pro-Lys$^+$-Sar-Sar-Sar-Lys$^+$-Sar-Sar-Sar-OMe

Tyr-Gly-Gly-Phe-Leu-Gly-Pro-Lys$^+$-Pro-Pro-Pro-Lys$^+$-Pro-Pro-Pro-OMe

FIGURE 10.   Enkephalin analogues carrying artificial address peptides.

result suggests that the derivative becomes a few times more selective for the μ receptor by accumulation on the lipid membrane, which is compatible with the membrane compartment concept.

XII. DERIVATIVE TAKING AN α-HELICAL CONFORMATION

Linear opioid peptides are so flexible that they take several conformations in solution, which do not necessarily include the active conformation to trigger the receptor. Since the molecular structure of opioid receptors has not been clarified so far, there is no way but synthesis of analogues having a defined conformation for obtaining information on the active conformation. Tyr-Gly-Gly-Phe-Leu-(Lys-Aib-Leu-Aib)$_2$-OH, in which a hinge sequence between the message and address sequences of ENK-A'8 was eliminated, was remarkably hampered to bind to δ- and μ-receptors.[54] CD measurements re-

receptors. The design is based on the following ideas: (1) the severe side effects of analgesic drugs are mediated through the central nervous system and (2) at least part of the analgesic action should be mediated in the periphery. Thus, polar derivatives will exhibit peripherally mediated analgesia being apart from centrally mediated side effects.[66,67] For example, Tyr-D-Arg-Gly-Phe(4-NO$_2$)-Pro-NH$_2$ showed analgesic effects predominantly mediated in the periphery, which was demonstrated by antagonism of antinociception by the peripheral opioid antagonist-*N*-methylanalorphine and by comparison of the activities in the writing (peripheral activity) and hot plate tests (central activity).

On the other hand, cyclic dynorphin A analogues, [Cys^5,Cys^{11}]Dyn A$^{1-11}$-NH$_2$ and [Cys^5,Cys^{11},D-Ala^8]Dyn A$^{1-11}$-NH$_2$, possess high κ- and μ-opioid receptor affinities centrally (guinea pig brain), but only weak activity at peripheral κ- and μ-receptors (GPI).[68] The results suggest the existence of distinct κ- and μ-opioid receptor types for the central and peripheral nervous systems. The existence of distinct δ-receptor subtypes for the central and peripheral nervous systems also has been proposed.[69]

Irreversible ligands have been prepared by introducing a photoaffinity group.[70] Since they do not dissociate from the receptor, they have advantages in receptor studies such as receptor isolation. *N,N*-Dialkylated leucine enkephalin was also shown to act as an irreversible antagonist of the δ-opioid receptor.[71]

REFERENCES

1. **Schofield, P. R., Shivers, B. D., and Seeburg, P. H.**, The role of receptor subtype diversity in the CNS, *TINS,* 13, 8, 1990.
2. **Martin, W. R., Eades, C. G., Thompson, J. A., Huppler, R. E., and Gilbert, P. E.**, The effects of morphine- and nalorphine-like drugs in the nondependent and morphine-dependent chronic spinal dog, *J. Pharmacol. Exp. Ther.,* 197, 517, 1976.
3. **Lord, J. A. H., Waterfield, A. A., Hughes, J., and Kosterlitz, H. W.**, Endogenous opioid peptides: multiple agonists and receptors, *Nature (London),* 267, 495, 1977.
4. **Ueda, H., Harada, H., Nozaki, M., Katada, T., Ui, M., Satoh, M., and Takagi, H.**, Reconstitution of rat brain μ opioid receptors with purified guanine nucleotide-binding regulatory proteins, G$_i$ and G$_o$, *Proc. Natl. Acad. Sci. U.S.A.,* 85, 7013, 1988.
5. **Millan, M. J.**, Multiple opioid systems and pain, *Pain,* 27, 303, 1986.
6. **Millan, M. J.**, κ-Opioid receptors and analgesia, *TIPS,* 11, 70, 1990.
7. **Hughes, J., Smith, T. W., Kosterlitz, H. W., Fothergill, L. A., Morgan, B. A., and Morris, H. R.**, Identification of two related pentapeptides from the brain with potent opiate agonist activity, *Nature (London),* 258, 577, 1975.

8. **Nakanishi, S., Inoue, A., Kita, T., Nakamura, M., Chnag, A. C. Y., Cohen, S. N., and Numa, S.,** Nucleotide sequence of cloned cDNA for bovine corticotropin-β-lipotropin precursor, *Nature (London)*, 278, 423, 1979.

9. **Noda, M., Furutani, Y., Takahashi, H., Toyosaato, M., Hirose, T., Inayama, S., Nakanishi, S., and Numa, S.,** Cloning and sequence analysis of cDNA for bovine adrenal preproenkephalin, *Nature (London)*, 295, 202, 1982.

10. **Kakidani, H., Furutani, Y., Takahashi, H., Noda, M., Morimoto, Y., Hirose, T., Asai, M., Inayama, S., Nakanishi, S., and Numa, S.,** Cloning and sequence analysis of cDNA for porcine β-neo-endorphin/dynorphin precursor, *Nature (London)*, 298, 245, 1985.

11. **Morley, J. S.,** Chemistry of opioid peptides, *Br. Med. Bull.*, 39, 5, 1983.

12. **Ramakrishnan, K. and Portoghese, P. S.,** Synthesis and biological evaluation of a metazocine-containing enkephalinamide. Evidence for nonidentical roles of the tyramine moiety in opiates and opioid peptides, *J. Med. Chem.*, 25, 1423, 1982.

13. **Sugg, E. E. and Portoghese, P. S.,** Investigation of 4-(3-hydroxyphenyl)-4-methylpipecolic acid as a conformationally restricted mimic of the tyrosyl residue of leucine-enkephalinamide, *J. Med. Chem.*, 29, 2028, 1986.

14. **Broccardo, M., Erspamer, V., Falconieri-Erspamer, G., Improta, G., Linari, G., Melchiorri, P., and Montecucchi, P. C.,** Pharmacological data on dermorphins, a new class of potent opioid peptides from amphibian skin, *Br. J. Pharmacol.*, 73, 625, 1981.

15. **Brantl, V., Teschemacher, H., Blasig, J., Henschen, A., and Lottspeich, F.,** Opioid activities of β-casomorphins, *Life Sci.*, 28, 1903, 1981.

16. **Roques, B. P., Gacel, G., Fournie-Zaluski, M.-C., Senault, B., and Lecomte, J. M.,** Demonstration of the crucial role of the phenylalanine moiety in enkephalin analogues for differential recognition of the μ- and δ-receptors, *Eur. J. Pharmacol.*, 60, 109, 1979.

17. **Gacel, G., Fournie-Zaluski, M.-C., and Roques, B. P.,** D-Tyr-Ser-Gly-Phe-Leu-Thr, a highly preferential ligand for δ-opiate receptors, *FEBS Lett.*, 118, 245, 1980.

18. **Zajac, J.-M., Gacel, G., Petit, F., Dodey, P., Rossignol, P., and Roques, B. P.,** Deltakephalin, Tyr-D-Thr-Gly-Phe-Leu-Thr: a new highly potent and fully specific agonist for opiate δ-receptors, *Biochem. Biophys. Res. Commun.*, 111, 390, 1983.

19. **Gacel, G., Zajac, J. M., Delay-Goyet, P., Dauge, V., and Roques, B. P.,** Investigation of the structural parameters involved in the δ and μ opioid receptor discrimination of linear enkephalin-related peptides, *J. Med. Chem.*, 31, 374, 1988.

20. **Do, K.-Q., Fauchere, J.-L., Schwyzer, R., Schiller, P. W., and Lemieux, C.,** Electronic, steric and hydrophobic factors influencing the action of enkephalin-like peptides on opiate receptors, *Hoppe-Seyler's Z. Physiol. Chem.*, 362, 601, 1981.

21. **Fauchere, J.-L.,** A quantitative structure-activity relationship study of the inhibitory action of aeries of enkephalin-like peptides in the guinea pig ileum and mouse vas deferens bioassays, *J. Med. Chem.*, 25, 1428, 1982.

22. **Schwyzer, R., Do, K.-Q., Eberle, A. N., and Fauchere, J.-L.,** Synthesis and biological properties of enkephalin-like peptides containing carboranylalanine in place of phenylalanine, *Helv. Chim. Acta*, 64, 2078, 1981.

23. **Schwyzer, R.,** Prediction of potency and receptor selectivity of regulatory peptides: the membrane compartment concept, in *Pept. Proc. Eur. Pept. Symp., 19th 1986*, Theodoropoulos, D., Ed., de Gruyter, Berlin, 1987, 7.

24. **Salvadori, S., Marastoni, M., Balboni, G., Sarto, G. P., and Tomatis, R.,** Synthesis and pharmacological activity of partially modified retro-inverso dermorphin tetrapeptides, *J. Med. Chem.*, 28, 769, 1985 (and references cited therein).

25. **Salvadori, S., Marastoni, M., Balboni, G., Sarto, G. P., and Tomatis, R.,** Synthesis and pharmacological activity of partially modified retro-inverso dermorphin tetrapeptides, *J. Med. Chem.*, 28, 769, 1985.

26. **Jacobson, A. R., Gintzler, A. R., and Sayere, L. M.,** Minimum-structure enkephalin analogues incorporating L-Tyrosine, D(or L)-phenylalanine, and a diamine spacer, *J. Med. Chem.*, 32, 1708, 1989.

27. **Gacel, G., Dauge, V., Breuze, P., Delay-Goyet, P., and Roques, B. P.,** Development of conformationally constrained linear peptides exhibiting a high affinity and pronounced selectivity for δ opioid receptors, *J. Med. Chem.,* 31, 1891, 1988.

28. **Imanishi, Y.,** Syntheses, conformation, and reactions of cyclic peptides, *Adv. Polym. Sci.,* 20, 1, 1976.

29. **Madison, V.,** Cyclic peptides revisited, *Biopolymers,* 24, 97, 1985.

30. **Schiller, P. W. and DiMaio, J.,** Opiate receptor subclasses differ in their conformational requirements, *Nature (London),* 297, 74, 1982.

31. **Schiller, P. W., Lipton, A., Horrobin, D. F., and Bodanszky, M.,** Unsulfated C-terminal 7-peptide of cholecystokinin: a new ligand of the opiate receptor, *Biochem. Biophys. Res. Commun.,* 85, 1332, 1978.

32. **Mosberg, H. I., Hurst, R., Hruby, V. J., Galligan, J. J., Burks, T. F., Gee, K., and Yamamura, H. I.,** Conformationally constrained cyclic enkephalin analogs with pronounced delta opioid receptor agonist selectivity, *Life Sci.,* 32, 2565, 1983.

33. **Schiller, P. W., Nguyen, T. M. D., Lemieux, L., and Maziak, L. A.,** Synthesis and activity profiles of novel cyclic opioid peptide monomers and dimers, *J. Med. Chem.,* 28, 1766, 1985.

34. **Schiller, P. W., Nguyen, T. M.-D., Maziak, L. A., Wilkes, B. C., and Lemieux, C.,** Structure-activity relationships of cyclic opioid peptide analogues containing a phenylalanine residue in the 3-position, *J. Med. Chem.,* 30, 2094, 1987.

35. **Mosberg, H. I., Hurst, R., Hruby, V. J., Gee, K., Yamamura, H. I., Galligan, J. J., and Burks, T. F.,** bis-Penicillamine enkephalins possess highly improved specificity toward δ opioid receptors, *Proc. Natl. Acad. Sci. U.S.A.,* 80, 5871, 1983.

36. **Schwyzer, R.,** ACTH: a short introductory review, *Ann. N.Y. Acad. Sci.,* 297, 3, 1977.

37. **Likowski, A. W., Tam, S. W., and Portoghese, P. S.,** Peptides as receptor selectivity modulators of opiate pharmacophores, *J. Med. Chem.,* 29, 1222, 1986.

38. **Portoghese, P. S., Sultana, M., and Takemori, A. E.,** Design of peptidomimetic δ opioid receptor antagonists using the message-address concept, *J. Med. Chem.,* 33, 1714, 1990.

39. **Schwyzer, R.,** Molecular mechanism of opioid receptor selection, *Biochemistry,* 25, 6335, 1986.

40. **Eisenberg, D., Weiss, R. M., and Terwilliger, T. C.,** The helical hydrophobic moment: a measure of the amphiphilicity of a helix, *Nature (London),* 299, 371, 1982.

41. **Schoch, P., Sargent, D. F., and Schwyzer, R.,** Hormone-receptor interactions: corticotropin-(1-24)-tetracosapeptide spans artificial lipid-bilayer membranes, *Biochem. Soc. Trans.,* 7, 846, 1979.

42. **Gysin, B. and Schwyzer, R.,** Head group and structure specific interactions of enkephalins and dynorphin with liposomes: investigation by hydrophobic photolabeling, *Arch. Biochem. Biophys.,* 225, 467, 1983.

43. **Gremlich, H.-U., Fringeli, H.-P., and Schwyzer, R.,** Conformational changes of adrenocorticotropin peptides upon interaction with lipid membranes revealed by infrared attenuated total reflection spectroscopy, *Biochemistry,* 22, 4257, 1984.

44. **Gysin, B. and Schwyzer, R.,** Hydrophobic and electrostatic interactions between adrenocorticotropin-(1-24)-tetracosapeptide and lipid vesicles. Amphiphilic primary structures, *Biochemistry,* 23, 1811, 1984.

45. **Gremlich, H.-U., Fringeli, H.-P., and Schwyzer, R.,** Interaction of adrenocorticotropin-(11-24)-tetradecapeptide with neutral lipid membranes revealed by infrared attenuated total reflection spectroscopy, *Biochemistry,* 23, 1808, 1984.

46. **Erne, D., Sargent, D. F., and Schwyzer, R.,** Preferred conformation, orientation, and accumulation of dynorphin A-(1-13)-tridecapeptide on the surface of neutral lipid membranes, *Biochemistry,* 24, 4261, 1985.

47. **Sargent, D. F., Bean, J. W., and Schwyzer, R.,** Reversible, binding of substance P to artificial lipid membranes studied by capacitance minimization techniques, *Biophys. Chem.,* 34, 103, 1989.

48. **Sargent, D. F. and Schwyzer, R.,** Membrane lipid phase as catalyst for peptide-receptor interaction, *Proc. Natl. Acad. Sci. U.S.A.,* 83, 5774, 1986.

49. **Schwyzer, R.,** Membrane structure and biologic activity of adrenocorticotropin (ACTH) and melanotropin (MSH) peptides. Estimation of structural parameters including the influence of the helix dipole moment, *Helv. Chim. Acta,* 69, 1685, 1986.

50. **Schwyzer, R.,** Estimated conformation, orientation, and accumulation of dynorphin A-(1-13)-tridecapeptide on the surface of neutral lipid membranes, *Biochemistry,* 25, 4281, 1986.

51. **Sargent, D. F., Bean, J. W., and Schwyzer, R.,** Conformation and orientation of regulatory peptides on lipid membranes. Key to the molecular mechanism of receptor selection, *Biophys. Chem.,* 31, 183, 1988.

52. **Schwyzer, R.,** Estimated membrane structure and receptor subtype selection of an opioid alkaloid-peptide hybrid, *Int. J. Pept. Protein Res.,* 32, 476, 1988.

53. **Kimura, S., Sasaki-Yagi, Y., and Imanishi, Y.,** Receptor selectivity of enkephalin analogs carrying artificial address peptides, *Int. J. Pept. Protein Res.,* 35, 550, 1990.

54. **Sasaki-Yagi, Y., Kimura, S., and Imanishi, Y.,** unpublished data.

55. **Shimohigashi, Y., Costa, T., Matsuura, S., Chen, H.-C., and Rodbard, D.,** Dimeric enkephalins display enhanced affinity and selectivity for the delta opiate receptor, *Mol. Pharmacol.,* 21, 558, 1982.

56. **Shimohigashi, Y., Costa, T., Chen, H.-C., and Rodbard, D.,** Dimeric tetrapeptide enkephalins display extraordinary selectivity for the δ opiate receptor, *Nature (London),* 297, 27, 333, 1982.

57. **Hazum, E., Chang, K.-J., and Cuatrecasas, P.,** Cluster formation of opiate (enkephalin) receptors in neuroblastoma cells: differences between agonists and antagonists and possible relationships to biological functions, *Proc. Natl. Acad. Sci. U.S.A.,* 77, 3038, 1980.

58. **Snyder, S. H.,** Brain peptides as neurotransmitters, *Science,* 209, 976, 1980.

59. **Lutz, R. A., Cruciani, R. A., Shimohigashi, Y., Costa, T., Kassis, S., Munson, P. J., and Rodbard, D.,** Increased affinity and selectivity of enkephalin tripeptide (Tyr-D-Ala-Gly) dimers, *Eur. J. Pharmacol.,* 111, 257, 1985.

60. **Kondo, M., Kitajima, H., Uchida, H., Kodama, H., Shimohigashi, Y., Herz, A., and Costa, T.,** Synthesis and biological activities of dimeric and tetrameric enkephalins cross-linked with cysteine derivative, in *Peptide Chemistry 1987,* Shiba, T. and Sakakibara, S., Eds., 1988, 565.

61. **Hazum, E., Chang, K. J., and Cuatrecasas, P.,** Receptor redistribution induced by hormones and neurotransmitters: possible relationships to biological functions, *Neuropeptides,* 1, 217, 1981.

62. **Smith, J. R. and Simon, E. J.,** Selective protection of stereospecific enkephalin and opiate binding against inactivation by *N*-ethylmaleimide: evidence for two classes of opiate receptors, *Proc. Natl. Acad. Sci. U.S.A.,* 77, 281, 1980.

63. **Simon, E. J., Dole, W. P., and Hiller, M. J.,** Coupling of a new, active morphine derivative to sepharose for affinity chromatography, *Proc. Natl. Acad. Sci. U.S.A.,* 69, 1835, 1972.

64. **Eberle, A. N., Kriwaczek, V. M., and Schulster, D.,** Corticotropin and melanotropin: organization and transmission of hormonal information, in *Perspectives in Peptide Chemistry,* Eberle, A. N., Geiger, R., and Wieland, T., Eds., de Gruyter, Berlin, 1981, 407.

65. **Sasaki-Yagi, Y., Kimura, S., and Imanishi, Y.,** unpublished data.

66. **Hardy, G. W., Lowe, L. A., Sang, P. Y., Simpkin, D. S. A., Wilkinson, S., Follenfant, R. L., and Smith, T. W.,** Peripherally acting enkephalin analogues. I. Polar pentapeptides, *J. Med. Chem.,* 31, 960, 1988.

67. **Hardy, G. W., Lowe, L. A., Mills, G., Sang, P. Y., Simpkin, D. S. A., Follenfant, R. L., Shankley, C., and Smith, T. W.,** Peripherally acting enkephalin analogues. II. Polar tri- and tetrapeptides, *J. Med. Chem.,* 32, 1108, 1989.

68. **Kawasaki, A. M., Knapp, R. J., Kramer, T. H., Wire, W. S., Vasquz, O. S., Yamamura, H. I., Burks, T. F., and Hruby, V. J.,** Design and synthesis of highly potent and selective cyclic dynorphin A analogues, *J. Med. Chem.,* 333, 1874, 1990.
69. **Vaughn, L. K., Wire, W. S., Davis, P., Shimohigashi, Y., Toth, G., Knapp, R. J., Hruby, V. J., Burks, T. F., and Yamamura, H. I.,** Differentiation between rat brain and mouse vas deferens δ opioid receptors, *Eur. J. Pharmacol.,* 177, 99, 1990.
70. **Landis, G., Lui, G., Shook, J. E., Yamamura, H. I., Burks, T. F., and Hruby, V. J.,** Synthesis of highly δ and μ opioid receptor selective peptides containing a photoaffinity group, *J. Med. Chem.,* 32, 638, 1989.
71. **Lovett, J. A. and Portoghese, P. S.,** Synthesis and evaluation of melphalan-containing *N,N*-dialkylenkephalin analogues as irreversible antagonists of the δ opioid receptor, *J. Med. Chem.,* 30, 1144, 1970.

3.2 ENZYME MODIFICATION FOR DRUG DELIVERY MOLECULAR DEVICES

Yoshihiro Ito

TABLE OF CONTENTS

I. INTRODUCTION

Enzymes are used in biomedical applications. As described in Chapter 2.2, some enzymes are themselves drugs. In this chapter, other uses of enzymes will be described, though the many ideas remain under investigation.

The purposes of drug delivery are to suppress decreases in drug effects due to instability and to avoid drug side effects. They are divided into two categories; time-controlled release and site-specific delivery as shown in Figure 1. For the drug to act at the desired site and time, drug delivery systems have been designed, a few of which have been reviewed.[1,2] In Chapter 3.1 site-specific delivery was featured. This chapter will describe time-controlled systems, especially external signal-responsive drug release.

The signals are classified as physical (e.g., pH, redox, photo, heat, ultrasonication, etc.) and chemical (e.g., glucose). To date, some systems responding to physical signals have been designed. However, it is necessary to use biological components to obtain a chemical signal response. Figure 2 shows an example of a glucose-sensitive insulin-releasing system. In this chapter, physical signal-responsive systems will be reviewed first. Second, combinations of enzymes with physical signal-responsive systems which obtain chemical signal responsivity will be mentioned.

II. ENVIRONMENTAL STIMULI-RESPONSIVE SYSTEMS

In this section, pH-responsive systems will be reviewed, followed by a description of other stimuli-responsive systems.

A. pH-RESPONSIVE RELEASING SYSTEMS
1. Polymer Systems

In 1971, Kopecek et al.[3] introduced ionogenic groups into a hydrogel composed of 2-hydroxyethyl methacrylate. The permeability of the hydrogel membrane was highly dependent on the degree of ionization as shown in Figure 3. It was found that the permeability for NaCl through a gel membrane containing acidic units of methacylic acid increased in the alkaline region, whereas the permeability through a gel membrane containing basic units of 2-(dimethylamino)ethyl methacrylate increased in the acidic region. Permeability through an ampholytic membrane — a gel containing approximately the same quantities of the methacrylic acid and 2-(dimethylamino)ethyl methacrylate units — passes through a minimum and increases in both directions from the isoelectric point.

Alhaique et al.[4] synthesized ethylene-co-vinyl-*N*,*N*-diethylglycinate copolymers as a pH-responsive reversible swelling and permeable film. They also synthesized another type of pH-responsive polymer membrane using a polysaccharide derivative.[5] On the other hand, Bala and Vasudevan[6] synthe-

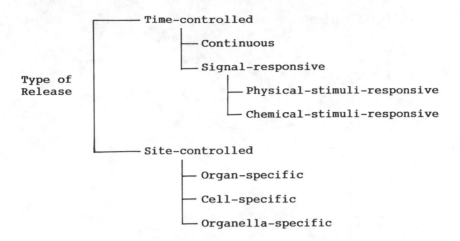

FIGURE 1.   Classification of the purposes of drug delivery systems.

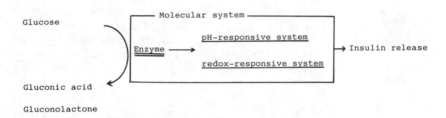

FIGURE 2.   An example of a drug delivery molecular device using enzyme modification. A glucose-sensitive insulin releasing system.

sized microcapsules consisting of a polyion complex.

This hydrogel system was investigated in more detail by Tanaka et al.[7-16] and Ilavsky et al.[17,18] As will be described in the next section, the hydrogel system has been used for various environmental stimuli-responsive release controls. They found that the variables which induced phase transition were temperature, solvent composition, pH, ionic composition, and a small electric field. Phase transitions accompanied by a reversible, discontinuous volume change as large as several hundred times, in response to infinitesimal changes in the environment, have been observed universally in various gels made of synthetic and natural polymers.

There are two methods of using this kind of hydrogel system for drug delivery, as shown in Figure 4. One permits release when the gel expands, and the other, when the gel shrinks.

Hydrogels have also been used in osmotic pumps.[19-21] An elementary osmotic pump can be used to achieve the zero order release of a drug.[19] When a highly soluble drug is used in the pump, the release rate is high and constant, as long as a saturated solution of the drug is present in the core. However,

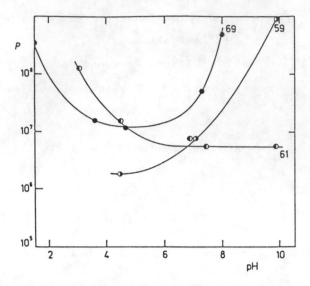

FIGURE 3. Dependence of the permeability coefficient P (cm^2/s) of modified hydrophilic membranes for NaCl on pH. 59, hydrogel containing methacrylic acid (MAA); 61, hydrogel containing 2-(diethylamino)ethyl methacrylate (DEAEMA); 69, hydrogel containing MAA and DEAEMA. (From Kopecek, J., Vacik, J., and Lim, D. J., *Polym. Sci., A-1*, 9, 2801, 1971. With permission.)

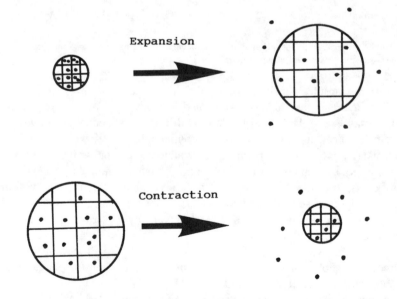

FIGURE 4. Responsive release mechanisms of hydrogels. One releases drugs when it expands, and the other, when it contracts.

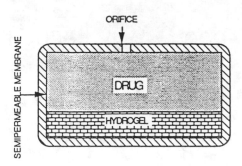

FIGURE 5. Hydrogel driven osmotic pump. (From Kabra, B. and Gehrke, S. H., *Polym. Prepr.*, 30, 490, 1989. With permission.)

when the solid is completely dissolved, a large fraction of the drug is still left solubilized in the device, and is then released at a diminishing rate. When an insoluble drug is used, the elementary osmotic pump will not work. Thus, hydrogels, whose osmotic properties can be altered easily, can be used to enhance the release from an osmotic pump (Figure 5). On the other hand, an osmotic pump consisting of a cylindrical reservoir containing the drug as a solution or paste, surrounded by a layer of an osmotic driving agent, which in turn is surrounded by a pH-sensitive membrane, has been synthesized. The rate of delivery of the drug from the osmotic pump was determined by the rate of the water influx into the device.

Another type of system using a polyelectrolyte-grafted porous membrane has been investigated.[22-29] Pefferkorn et al.,[22] and Idol and Anderson[23] demonstrated that solution pH could alter the solvent hydraulic permeability of porous membranes by adsorbing a polymer which undergoes a conformational transition as a threshold pH is crossed. Advantage could be taken of such systems to develop synthetic membranes whose selectivity can be altered simply by adjusting a physicochemical variable of the solution. On the other hand, Osada et al.[24,25] grafted poly(methacrylic acid) covalently onto a porous membrane. The water permeability reacted very sensitively to pH, in some cases with changes of as much as 1000-fold. Moreover, the pH dependence of the membrane permeability was the reverse of that of the hydrogel membrane. The phenomenon was explained as shown in Figure 6.

Iwata and Matsuda[26] prepared a porous poly(vinylidene fluoride) film graft-polymerized with polyacrylamide. A portion of the polyacrylamide was then hydrolyzed to poly(acrylic acid) and another portion converted to poly(vinyl amine) by a Hofmann rearrangement reaction. They demonstrated that the pH range of the permeation change could be controlled by adjusting the molar ratio of poly(acrylic acid) to poly(vinyl amine).

A useful and unique feature of these grafted membranes is that they are dimensionally stable because the basic substrate is a hydrophobic polymer and the grafting reaction is confined to the surface of this material, which essentially does not affect mechanical properties.

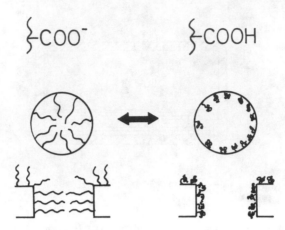

FIGURE 6.   Spreading and contraction of polyelectrolyte grafted on porous membrane. At high pH, poly(carbonic acid) is deprotonated and is spread by electrostatic repulsion. At low pH, poly(carbonic acid) is protonated and is shrunk by reduction of the repulsion.

On the other hand, Okahata's group[27,28] synthesized many kinds of grafted nylon microcapsules. The pH-sensitive polymers, poly(4-vinylpyridine) and poly(methyacrylic acid) were grafted. They explained the permeation control by a conformational change of the grafted chain. However, their proposed mechanism was ambiguous. The permeation control mechanism is considered to be based not upon conformational change, but upon swelling of the nylon capsule. In fact, their grafting method must have grafted the polymers not only onto the surface of the pore, but also onto the bulk. If the polymer was grafted only on the surface, the phenomenon observed by Osada et al. and Iwata et al. should have been observed in their investigation.

Ito et al.[29] revealed the mechanism of permeability control by pore size, as well as graft chain length and concentration. They grafted poly(acrylic acid) onto a straight-pored polycarbonate membrane. The density and the length of graft chains were controlled by the conditions of graft copolymerization, that is, the time of glow-discharge treatment and the monomer concentration in the polymerization, respectively. Graft chains with too high a density or excessive length restricted the mobility of the poly(acrylic acid) chains, and the pore size became pH-independent as shown in Figure 7.

Polypeptides have been used as the pH-responsive polymer, because of the transition of secondary structure accompanying ionic changes. Copolymer membrane containing glutamic acid,[30] cross-linked poly(glutamic acid),[31] and polyvinyl-polypeptide graft copolymers[32-34] were synthesized, as shown in Figure 8. In the graft copolymer,[32-34] it was suggested that continuous phases of the poly(L-aspargic acid) domain which function as the transmembrane permeating pathway ("channel") were formed. The channel-composing polypeptide segment in the membrane showed the pH-induced reversible conformational change which was considered to influence the permeation.

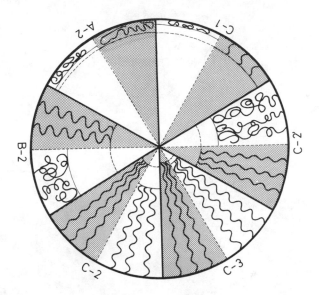

FIGURE 7.. Difference in the length of spreading and contraction of polyelectrolyte depends on the grafting density and chain length. Too dense and too long graft chains reduce the mobility of the chains. (From Ito, Y., Kotera, S., Inaba, M., Koni, K., and Imanishi, Y., *Polymer,* 31, 2157, 1990. With permission.)

However, these polypeptide derivatives are permeable at high pH. Therefore, polypeptide membranes composed of amino acids with amino groups in the side chains (i.e., methyl L-glutamate (MG)-N^5-(2-aminoethyl)-glutamine (AEG) copolymer were investigated in order to develop a membrane permeable at low pH.[35] MG/AEG copolymer cross-linked membranes were obtained by the aminolysis reaction of a poly(γ-methyl L-glutamate) (PMLG) solid membrane with ethylenediamine. The degree of hydration and the area of the MG/AEG copolymer membrane increased with the decreasing pH of the aqueous solution, reflecting ionization changes of the AEG moiety. The permeation coefficient of L-lysine through the membrane gradually increased with decreasing solution pH below 9.0, but steeply increased below 5.0. This behavior was explained in terms of the increase in the diffusibility of water and L-lysine in the membrane phase, based on the pH-induced conformational transition (swelling) of the copolymer.

A pH-sensitive polymer was introduced to control drug solubility.[36] Deprotonated macromolecules of partially quarternized poly(tertiary amine) of the poly[thio-1-(N,N-diethyl aminomethyl)ethylene] type with less than 20% quaternized repeating units, assumed globular conformations which form a molecularly dispersed organic microphase in water. Drugs reputedly insoluble in water can be dissolved and temporarily trapped in the lipophilic microphase. Because of the pH-dependent destabilization occurring through an all-or-none cooperative mechanism, instantaneous release can be obtained within a very narrow pH range.

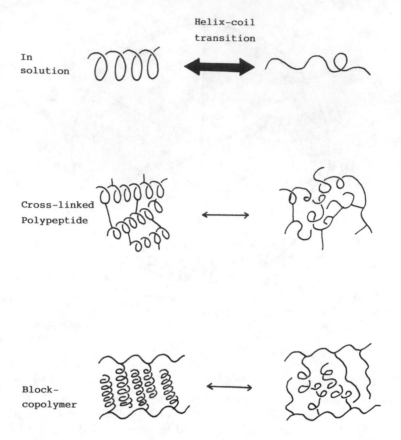

FIGURE 8.   Responsive polymer system utilizing the helix-coil polypeptide transition.

A pH-sensitive erosible polymer has been synthesized.[37-39] A partially esterified copolymer of methyl vinyl ether and maleic anhydride, underwent surface erosion at a rate that is extraordinarily pH dependent. The polymer dissolves by ionization of a carboxylic acid group. Another polymer system is poly(ortho esters) in which the erosion rate increases with decreasing pH. The polymer was prepared by adding diols to the diketone acetal 3,9-bis(ethylidene 2,4,8,10-tetraoxaspiro[5,5]undecane).

2. Polymer-Polymer Conjugate Systems

Seki and Kotaka[40] investigated the effect of hydrogen-bonding between polyoxyethylene and polyelectrolyte using the interpenetrating network system shown in Figure 9a. On the other hand, Osada et al.[24,25] reported that poly(methacrylic acid)-grafted porous membrane adsorbed with polyoxyethylene was a pH-responsive mechanochemical membrane as shown in Figure 9b. Poly(acrylic acid) or poly(methacrylic acid) complexed with pyrene end-labeled polyoxyethylene has been investigated by Hemker et al.[41] Since there

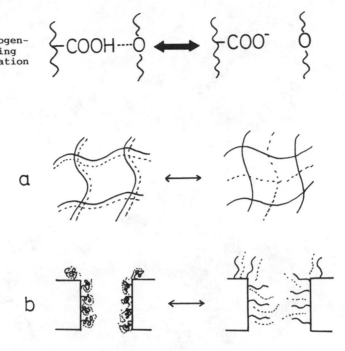

FIGURE 9. Responsive polymer system utilizing hydrogen bonding formation between poly(carbonic acid) and polyoxyethylene. (a) Interpenetrating networks of these polymers; (b) polyoxyethylene and grafted poly(carbonic acid).

are no hydrophobic regions in the poly(acrylic acid)-polyoxyethylene systems, none of the nonhydrogen-bonding phenomena were observed. All hydrophobically driven phenomena could be eliminated by destroying the consolidated hydrophobic regions in poly(methacrylic acid) systems. This was experimentally accomplished by the addition of methanol and by the ionization of the carboxy groups in poly(methacrylic acid).

3. Polymer-Liposome Conjugate Systems

The use of lipid interaction with a pH-sensitive polymer has been investigated. Tirrell et al.[42-46] utilized poly(ethacrylic acid) as a pH-sensitive polymer. Decreases in pH increased the hydrophobicity of the polymer and enhanced the interaction of the polymer with liposomes to destroy them (Figure 10). The method was based on the pH-induced solubility change of the polymer. Maeda et al.[47] combined the polymer and dimyristoylphosphatidylethanolamine, and investigated the H^+-induced release of vesicle contents containing the polymer derivative. Kitano et al.[48] synthesized a lipid with a polymerization initiator in the head group, and acrylic acid was polymerized with the initiator. The mixed liposome containing the polymer derivative was pH sensitive.

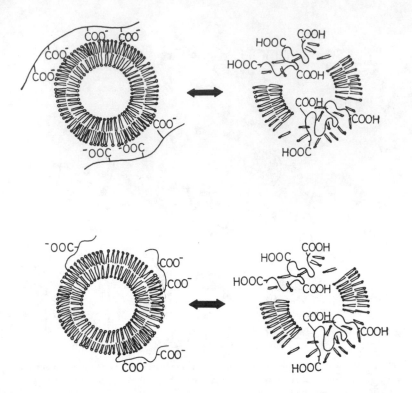

FIGURE 10.   Interaction between poly(ethacrylic acid) and liposome. Protonated poly(ethacrylic acid) "solves" the liposome. Lipid-containing poly(ethacrylic acid) was also synthesized.

Vesicle-polypeptide interactions have also been investigated by others.[49-51] Kono et al.[50,51] synthesized the amphiphilic sequential peptide, poly(Lys-Aib-Leu-Aib) in which Lys, Aib, and Leu represent lysine, 2-aminoisobutyric acid, and leucine as the sensitive polymer. The polypeptide bonded to a lipid bilayer membrane at high pH regions by assuming an amphiphilic α-helical conformation. At low pH regions, the structure was disturbed, resulting in the release of encapsulated substances as shown in Figure 11. Kono et al. used microcapsules coated with lipid as substance reservoirs. The microcapsule was used to enhance liposome stability.

4. Liposome Systems

Liposomes have been applied as safe and effective carriers for drug delivery purposes, responsiveness toward exogenous stimuli being a fundamental characteristic. The use of liposomes has been reviewed by Fendler[52] and Ringsdorf et al.[53]

Liposomes responsive to pH were synthesized by Yatvin et al.[54-56] and Ellens et al.[57] Palmitoyl homocysteine or cholesteryl hemisuccinate was used as a "pH-sensitive amphiphile". These liposomes were also used for fusion.[58-61] Figure 12 shows the examples of pH-sensitive lipids.

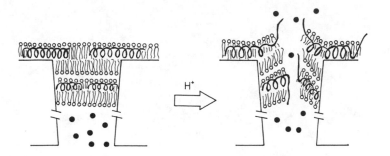

FIGURE 11. Interaction between amphiphilic polypeptide and molecular bilayer coated onto a porous microcapsule. The destruction of the amphiphilic structure of the polypeptide destroys the bilayer.

However, liposomes are not generally stable. The stability was improved by polymerization of the lipid,[52] coating with polysaccharide,[62] and poly-electrolyte.[63] Okahata et al.[64] coated a lipid layer on a microcapsule having a diameter of 2 to 5 mm. These multilayers effectively "corked" the porous nylon membrane, and diffusion of substances contained in the core of the capsules can be reversibly controlled by the external pH (Figure 13). Okahata et al.[65] designed other immobilization methods to stabilize the liposome systems as shown in Figure 14.

Takeoka et al.[66] mixed lipid liposomes composed of polymerizable and nonpolymerizable lipid and polymerized it to form a phase-separated membrane structure (Figure 15a). Phosphatidylethanolamine, phosphatidylserine, and phosphatidic acid were used as nonpolymerizable lipids. Their overall charge density and molecular packing arrangement was pH dependent.

On the other hand, Kitano et al.[67] synthesized homocysteine-carrying polymerized liposomes (Figure 15b). They concluded that the thiolactone formation, proposed by Yatvin et al., was not the reason for the pH sensitivity of the polymerized liposomes. Although the stabilization of liposomes is an important issue, the increase of stability accompanies sensitivity to environmental stimuli.

B. OTHER STIMULI-RESPONSIVE SYSTEMS

Molecular systems which respond to various stimuli, such as thermal change, photoirradiation, electrical current, etc., have been developed.

1. Thermosensitive Vinyl Polymer Systems

Thermally reversible hydrogels have been reviewed by Kaetsu et al.[68] Figure 16 shows the chemical structures. They described that the structure contained hydrophobic and hydrophilic moieties in one molecule.

Hoffman et al.[69,70] reported that thermally reversible hydrogels based on *N*-isopropylacrylamide (NiPAAm) cross-linked with methylene-bisacrylamide

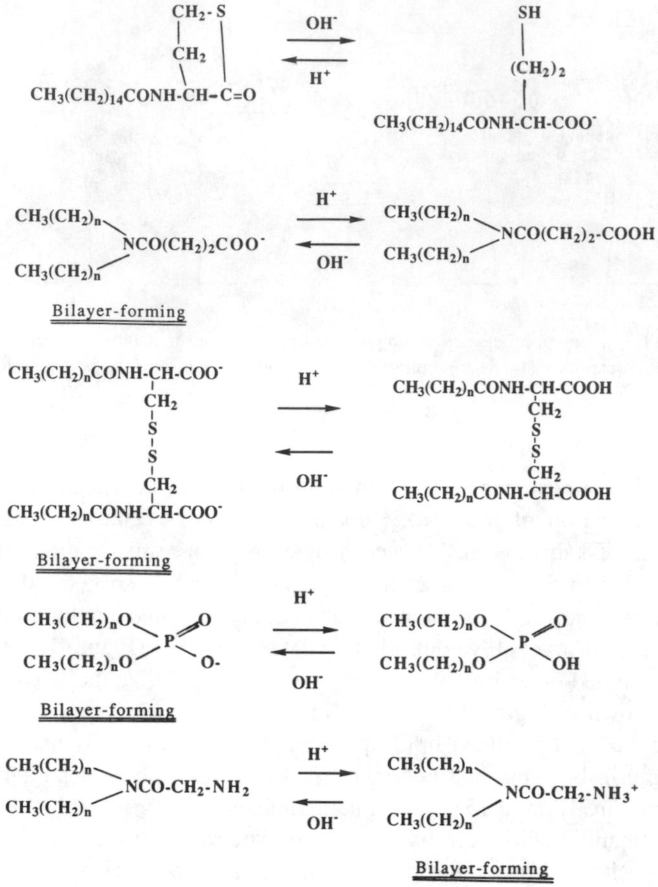

FIGURE 12.   pH-Sensitive, dissociative, bilayer-forming lipids.

(MBAAm) demix with water and shrink abruptly on heating just above 310°C. This collapse occurs as a result of the phase transition of poly(NiPAAm) at its lower critical solution temperature (LCST) of about 310°C. Freitas and Cussler[71] also synthesized temperature-sensitive gels.

Drug permeation through thermosensitive hydrogels was also investigated by Bae et al.[72-74] They considered that there are two classes of thermosensitive hydrogels, classified according to the origin of thermosensitivity in aqueous swelling. The first is derived from polymer-water interactions, especially specific hydrophobic/hydrophobic balancing effects and the configuration of side groups in cross-linked poly(*N,N'*-alkyl substituted acrylamides). The second is based on polymer-polymer interactions in addition to polymer-water interactions. An example is the interpenetrating polymer network (IPN) of poly(*N*-acryloyl pyrrolidine) (poly(APy)) and polyoxyethylene (POE). IPN

gel state
liquid crystal state
bilayer structure
pH7
Tc
(I)
(III)
OH⁻
H⁺
pH2
Tc
(II)
(IV)
destruction of
bilayer structure

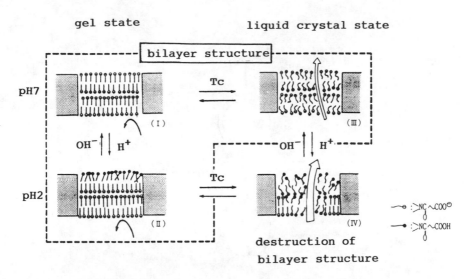

FIGURE 13. Structural changes in the molecular bilayer responds to environmental pH.

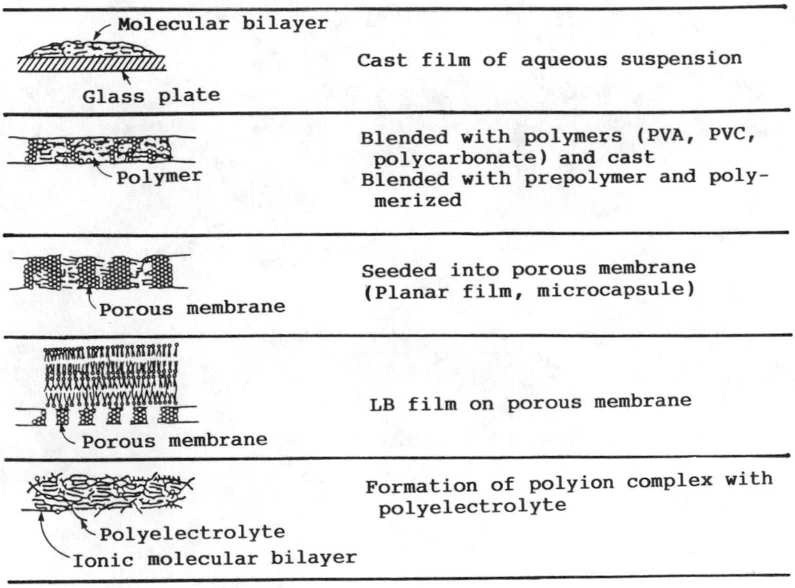

Molecular bilayer
Glass plate
Cast film of aqueous suspension
Polymer
Blended with polymers (PVA, PVC, polycarbonate) and cast
Blended with prepolymer and polymerized
Porous membrane
Seeded into porous membrane (Planar film, microcapsule)
Porous membrane
LB film on porous membrane
Polyelectrolyte
Ionic molecular bilayer
Formation of polyion complex with polyelectrolyte

FIGURE 14. Stabilization methods for supporting molecular bilayer structures.

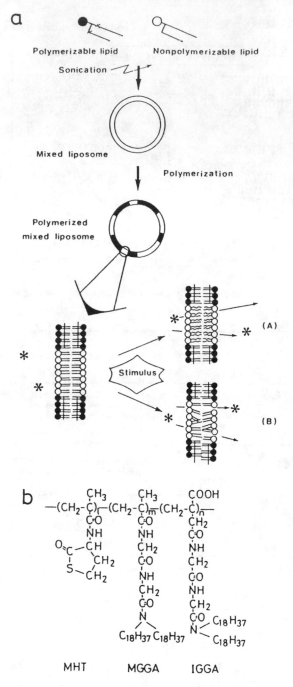

FIGURE 15.   Stabilization methods of responsive molecular bilayers. (a) Mixing polymerizable lipids with stimuli-responsive lipids. (From Takeoka, S., Ohno, H., Hayashi, N., and Tsuchida, E., *J. Controlled Release,* 9, 177, 1989, Figure 1. With permission.) (b) Polymerization of stimuli-sensitive polymerizable lipids. (From Kitano, H., Wolf, H., and Ise, N., *Macromolecules,* 23, 1958, 1990. With permission.)

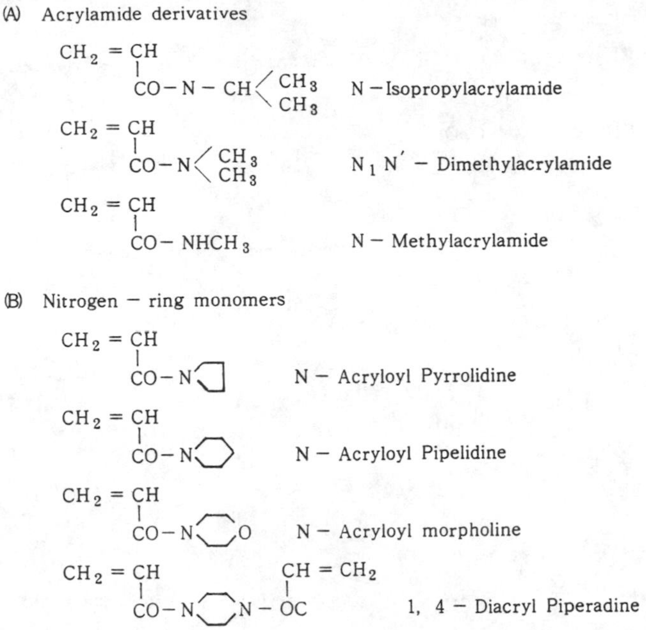

(A) Acrylamide derivatives

$$CH_2 = CH$$
$$CO-N-CH< \begin{matrix} CH_3 \\ CH_3 \end{matrix}$$ N −Isopropylacrylamide

$$CH_2 = CH$$
$$CO-N< \begin{matrix} CH_3 \\ CH_3 \end{matrix}$$ $N_1 N'$ − Dimethylacrylamide

$$CH_2 = CH$$
$$CO- NHCH_3$$ N − Methylacrylamide

(B) Nitrogen − ring monomers

$$CH_2 = CH$$
$$CO-N$$ N − Acryloyl Pyrrolidine

$$CH_2 = CH$$
$$CO-N$$ N − Acryloyl Pipelidine

$$CH_2 = CH$$
$$CO-N\quad O$$ N − Acryloyl morpholine

$$CH_2 = CH \qquad CH = CH_2$$
$$CO-N\quad N - OC$$ 1, 4 − Diacryl Piperadine

FIGURE 16.   Thermoresponsive polymerizable compounds for drug delivery systems.

exhibited enhanced thermosensitivity over both poly(APy) and POE networks due to incompatibility (repulsive interaction) between the two polymeric chains.

Bae et al. found that the deswelling and swelling rates were limited by the surface skin of the membrane as shown in Figure 17. When an amine-bearing membrane swollen at low pH is transferred into a higher pH, the amine groups near the surface of the membrane will deprotonate first, which forms a skin. However, behind the surface layer is a swollen gel containing free protons and protonated amine groups with concentrations determined by the equilibrium at low pH. Free protons in the gel will redistribute to the surface layer and reprotonate the amines there, thus preventing collapse of the surface into a skin. A skin forms in thermosensitive gels, and thermal equilibrium is established before significant gel deswelling occurs. The importance of the hydrogel surface layer has also been observed by other researchers.[75]

Ueda et al.[76] utilized polymers with a phospholipid polar group (desribed in Chapter 2.1), as thermally responsive hydrogels. Since these polymers are biocompatible, they are considered to be of use as drug delivery matrices.

Other types of thermal-sensitive permeable systems were investigated by Okahata et al.[77] and Iwata et al.[78] They grafted the poly(N-isopropylacrylamide) instead of poly(carboxylic acid) onto a nylon microcapsule or porous polytetrafluoroethylene membrane, as shown in Figure 6. Their systems responded to temperature changes differently. In the case of Iwata's study, water permeability increased when the temperature was below 35°C.

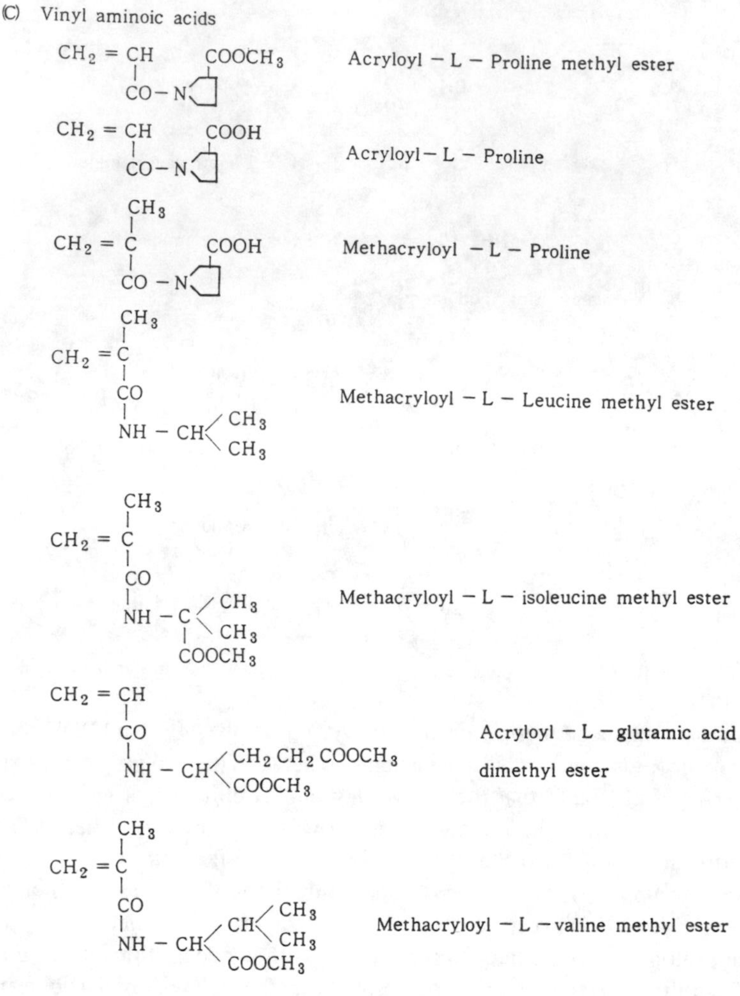

(C) Vinyl aminoic acids

$CH_2 = CH$ $COOCH_3$ Acryloyl $-$ L $-$ Proline methyl ester
        |
     $CO - N$

$CH_2 = CH$ $COOH$ Acryloyl$-$ L $-$ Proline
        |
     $CO - N$

        CH_3
         |
$CH_2 = C$ $COOH$ Methacryloyl $-$ L $-$ Proline
         |
      $CO - N$

        CH_3
         |
$CH_2 = C$
         |
        CO
         | Methacryloyl $-$ L $-$ Leucine methyl ester
     $NH - CH \big\langle \begin{smallmatrix} CH_3 \\ CH_3 \end{smallmatrix}$

          CH_3
           |
$CH_2 = C$
           |
          CO
           | CH_3
     $NH - C \big\langle \begin{smallmatrix} CH_3 \\ CH_3 \end{smallmatrix}$ Methacryloyl $-$ L $-$ isoleucine methyl ester
           |
          $COOCH_3$

$CH_2 = CH$
         |
        CO Acryloyl $-$ L $-$ glutamic acid
         | $CH_2 CH_2 COOCH_3$
     $NH - CH \big\langle \begin{smallmatrix} CH_2 CH_2 COOCH_3 \\ COOCH_3 \end{smallmatrix}$ dimethyl ester

        CH_3
         |
$CH_2 = C$
         |
        CO CH_3
         | $CH \big\langle \begin{smallmatrix} CH_3 \\ CH_3 \end{smallmatrix}$ Methacryloyl $-$ L $-$ valine methyl ester
     $NH - CH \big\langle$
                  $COOCH_3$

FIGURE 16 (continued).

2. Redox-Sensitive Vinyl Polymer Systems

Redox-sensitive or electrochemical control of drug delivery has been investigated.[79-84] Ion movement through membranes containing fixed ionic sites is dependent on the nature and number of such charged sites; changing these factors will affect the membrane ionic resistance. Poly(vinylferrocene), poly(tetrathiafulvalene), and poly(pyrrole) are examples of redox polymers with films, deposited on solid electrodes, that can be cycled electrochemically between charged-site and neutral states. Burgmayer and Murray[79,80] embedded a porous electrode inside a redox polymer membrane, and this membrane/ electrode was used to separate two pools of electrolyte solution. Control of the ion flow through the membrane was achieved by control of the polymer

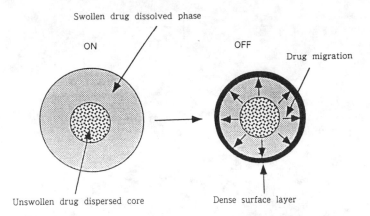

FIGURE 17.   Drug diffusion from a polymer device in the OFF state.

redox state. They termed this phenomenon an "ion gate". This idea was applied to the binding and release of anions and cations from conducting polymers.[81-83] These polymers are of particular interest because they have a high density of redox states, and their high conductivity permits relatively thick films to be promptly switched. These factors were considered to allow binding and prompt release of relatively large quantities of the ion of interest.

Figure 19 shows redox-sensitive polymers. Okahata et al.[84] grafted nylon-2,12 microcapsules with viologen-containing polymers. Poly [3-carbamoyl-1-(*p*-vinylbenzyl)pyridinium chloride] was grafted onto a porous membrane by Nishi et al.[85] The principle of controlling the water permeation was the same as shown in Figure 5. In the oxidized state, ammonium groups are formed in the PCVPC graft chains and repulsion between positive charges extended the polymer chain and closed the membrane pores. In the reduced state, the graft chains are de-ionized. Reduced electrostatic repulsion between the graft chains induces a coil-like conformation to open the pores.

Another unique system is the dihydropyridine-pyridium salt-type redox system, which has been applied to a number of pharmacologically active drugs for targeted and sustained delivery.[86] Several studies have shown that this chemical delivery system (CDS) can be successfully coupled with estradiol (E_2). The dihydropyridine derivative of *N*-methyl nicotinic acid is covalently attached to E_2 at the 17 position to result in E_2-CDS (Figure 20). The highly lipophilic moiety, E_2-CDS, can easily penetrate the blood-brain barrier (BBB) after systemic administration. However, oxidation of the labile carrier yields the corresponding quarternary salt (E_2-Q^+), which is retarded from efflux through the BBB because of its ionic hydrophilic character. The drug carrier moiety thus is "trapped" behind the BBB as evidenced by sustained elevated brain drug concentrations compared with those in peripheral tissues. The second step requires cleavage or hydrolysis of the E_2-Q^+, to release estradiol and the trigonelline carrier.

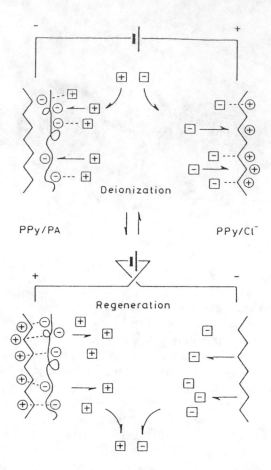

FIGURE 18.   Electrochemical deionization system utilizing a cation-exchangeable PPy/PVS⁻
electrode and an anion-exchangeable PPy/Cl⁻ electrode. (From Shimidzu, T., Ohtani, A., Iyoda,
T., and Honda, K., *J. Chem. Soc., Chem. Commun.*, 1415, 1986. With permission.)

3. Photoresponsive Vinyl Polymer Systems

Smets[87] synthesized various kinds of photoresponsive polymers. Irie and
Kungwatchakun[88] reported a synthesis of photosensitive gels by incorporating
photosensitive molecules, such as leukocyanide and leukohydroxide, into the
gel network. They observed the photoinduced phase transition of gels. Light
can be imposed instantly, which is in contrast to other variables. For example,
the temperature jump is limited by thermal diffusion, and pH change by ion
diffusion. The electric field phase transition is also limited by the accompanying ion diffusion. The phase transition induced by light provides a means
of performing high-speed experiments of volume changes in gels, where the
perturbation time is extremely fast. A permeation change resulting from photoirradiation was observed by Irie and Kungwatchakun.[88]

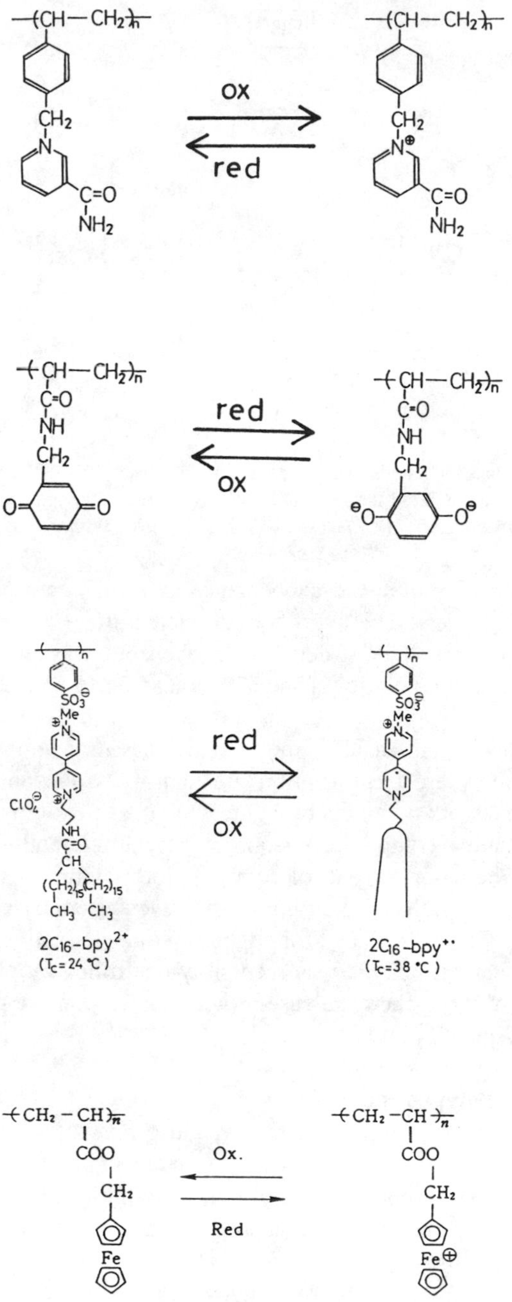

FIGURE 19.   Redox-sensitive polymers for permeation control.

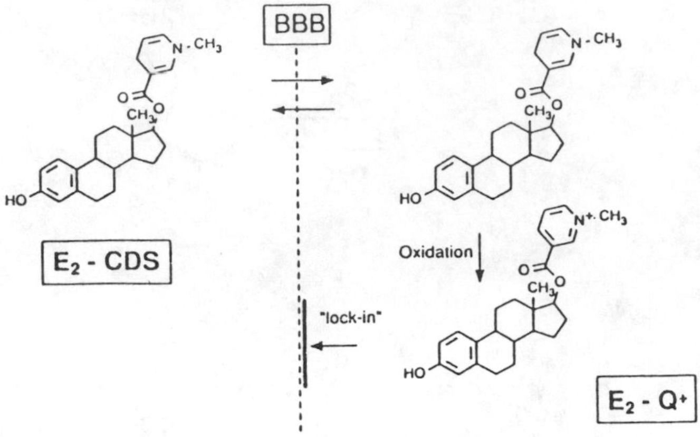

FIGURE 20. Mechanism of brain-targeted drug delivery of estradiol. BBB, blood-brain barrier. (From Dietzel, K., Keuth, V., Estes, K. S., Brewster, M. E., Clemmons, R. M., Vistelle, R., Bodor, N. S., and Derendorf, H., *Pharm. Res.,* 7, 879, 1990. With permission.)

Ishihara et al.[89] introduced azobenzene covalently into poly(2-hydroxy-ethylmethacrylate) gels. The hydrogel contracted after exposure to UV irradiation, since the *trans*-type structure of azobenzene was changed to the *cis*-type, which induced a polarity change, thus leading to an altered interaction with water.

Recently Suzuki and Tanaka[90] reported the phase transition of gels induced by visible light, where the transition mechanism is due only to the direct heating of the network polymers by light, which is an extremely fast process. The gel was mainly composed N-isopropylacrylamide and a light-sensitive chromophore, the trisodium salt of copper chlorophyllin.

Chung et al.[91] synthesized a photoresponsive system based on the mechanism shown in Figure 5. They grafted spiropyrane derivatives onto a porous membrane. Water permeation was reversibly controlled by photoirradiation. The structures of those chemicals used for a photoresponsive polymer system are shown in Figure 21.

4. Responsive Polypeptide Systems

Photoinduced permeability control was also investigated using polypeptides.[92-94] Poly(L-glutamic acid) containing pararosaniline groups in the side chains, azobenzene, or poly(L-asparic acid) containing azobenzene were synthesized. One was adsorbed onto a porous support membrane. The others were used to construct the membrane.

Inoue's group[95-99] has extensively investigated polyvinyl-polypeptide graft copolymers in which the permeability can be controlled by environmental stimuli, such as pH, light, divalent ions, urea, and ammonium salts. Another group has also investigated pH-, redox-, and photoinduced polypeptides. Okahata et al.[100] grafted ferrocene-bound polypeptide onto microcapsules.

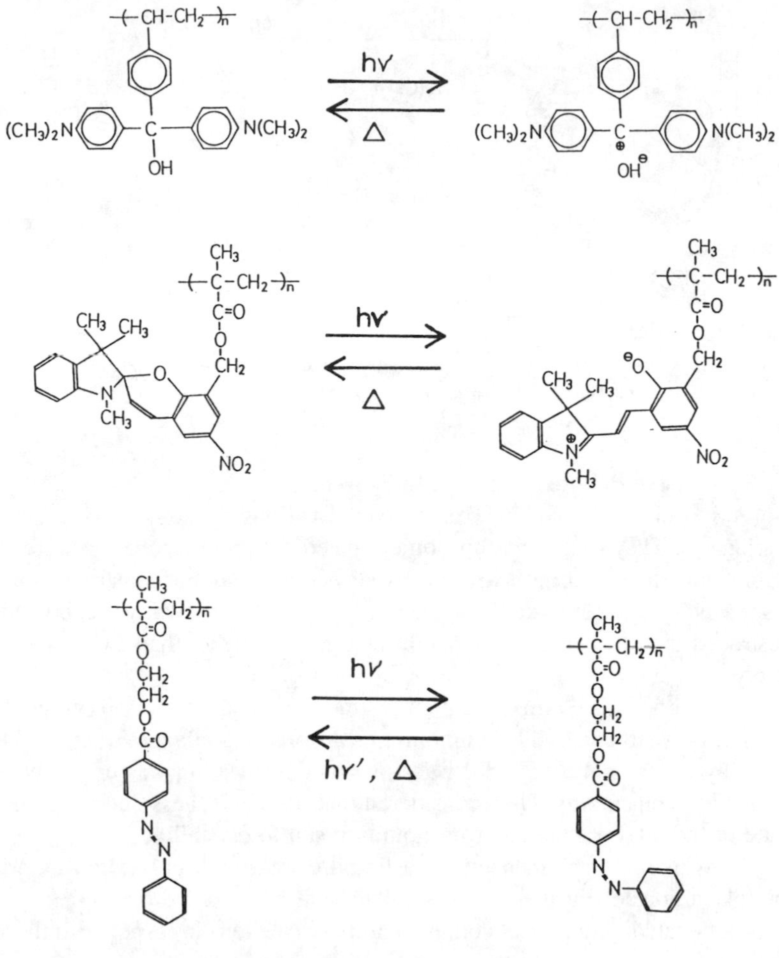

FIGURE 21.   Photosensitive polymers for permeation control.

5. Responsive Liposome Systems

Okahata et al.[101-117] have prepared numerous lipid-supporting systems, such as multibilayer-corked capsule membranes, monolayer-immobilized porous glass plates and multilayer-immobilized cast films as shown in Figure 14. Permeability across these artificial lipid assemblies supported by the physically strong membrane could be changed reversibly, responding to various outside effects such as temperature changes (phase transition), photoirradiation, ambient pH changes, electric field, and protein interactions. These permeability controls could be explained by orientation changes of the lipid matrices responding to external stimuli. The chemical structure of a photoresponsive lipid is exemplified in Figure 22.

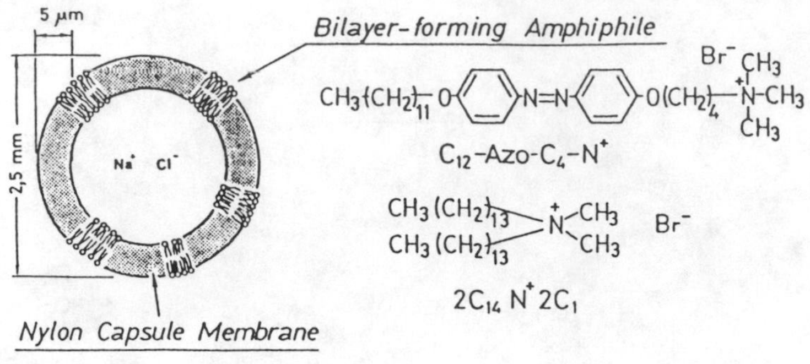

FIGURE 22. Photoresponsive microcapsule coated with photosensitive lipid. (From Okahata, Y., Lim, H.-J., and Hachiya, S., *Makromol. Chem. Rapid Commun.*, 4, 303, 1983. With permission.)

6. Responsive Polymer Conjugate Systems

Akiyoshi and Sunamoto[118] synthesized various liposome-polysaccharide conjugates. They found that liposomes coated with polysaccharide were very stable, and that a stimulus-responsive system could be constructed using responsive polysaccharide. For example, a pullulan derivative stabilized or destroyed the liposomes at temperatures lower or higher than LCST, respectively.

Kajiyama et al.[119] mixed a dialkyl ammonium salt with poly(vinyl chloride) or polycarbonate. The ammonium salt formed a bilayer structure, identified by X-ray analysis, and the gas or water permeation was critically controlled by temperature. They considered that responsiveness depended on the state of the bilayer, that is, from liquid crystal to crystalline.

Kajiyama et al.[120] then mixed a liquid crystal with poly(vinyl chloride) or polycarbonate. Furthermore, Shinkai et al.[121] mixed crown ether in the polycarbonate/liquid crystal conjugate to permeate ions in response to thermal change.

7. Other Stimuli-Responsive Systems
a. Charge Transfer (CT) Complex Formation

Electron donors and acceptors interact to form charge transfer (CT) complexes. Ishihara et al.[122] utilized this CT complex formation for control release. The water content of a hydrogel containing dinitrophenyl (DNP) group, which is an electron acceptor, was controlled by the addition of amino compounds (electron donors).

b. Ultrasonication

Okahata and Noguchi[123] reported that permeation of NaCl from the inner aqueous phase of a bilayer-coated capsule membrane was reversibly regulated by ultrasonic waves. Activation energy data showed that ultrasonic irradiation made coating bilayers fluid like a liquid crystalline state, which caused an

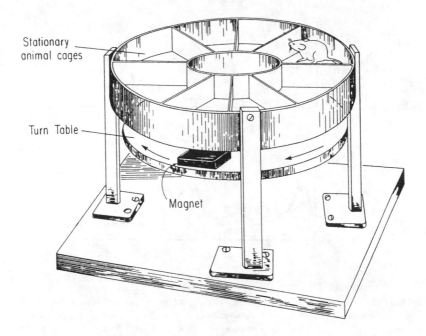

Stationary animal cages

Turn Table

Magnet

FIGURE 23. Magnetic triggering device. Up to eight rats can be placed on the top disk, each in a separate compartment. A motor regulator (not shown) allows for variation of field frequency by regulating the rotation of the bottom disk which contains two permanent magnets. (From Kost, J., Wolfrum, J., and Langer, R., *J. Biomed. Mater. Res.*, 21, 1367, 1987. With permission.)

increase in NaCl permeation. Kost et al.[124] investigated ultrasound-enhanced polymer degradation and release of incorporated substances. The release rate increased in proportion to the intensity of the ultrasound. Temperature and mixing were relatively unimportant in effecting enhanced polymer degradation, whereas cavitation appeared to play a significant role. Increased release rates were also observed when ultrasound was applied to biodegradable polymers implanted in rats.

c. Magnetic Field

Langer's group[125] investigated release control by magnet. They reported that an increase in drug release was directly proportional to field amplitude. The same pattern of results was observed *in vitro* and *in vivo*, though more powerful fields were required *in vivo* to achieve the same effect as that observed *in vitro*. Kost et al.[126] implanted polymer matrices containing insulin and embedded magnets subcutaneously in diabetic rats for 51 d. Figure 23 shows the method.

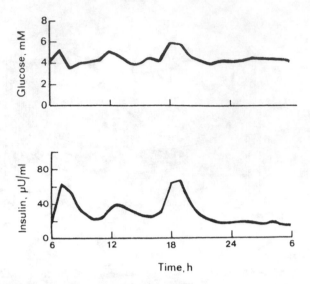

FIGURE 24.   Illustration of mean plasma glucose (upper panel) and insulin (lower panel) concentrations throughout the day in normal healthy subjects. (From Pickup, J. C. and Stevenson, R. W., *Rate-Controlled Drug Administration and Action*, Struyker-Boudier, Ed., CRC Press, Boca Raton, FL, 1986. With permission.)

III. ENZYMATICALLY CONTROLLED RELEASE SYSTEM

Enzymes as signal transducers have mostly been applied to the design of glucose-sensitive insulin releasing systems.

A. GLUCOSE-SENSITIVE INSULIN-RELEASING SYSTEMS

Insulin therapy is very important in the treatment of diabetes. Figure 24 shows the plasma glucose and insulin levels throughout the day in a group of nondiabetics.[127] There is an almost constant basal supply during the night and between meals, which maintains glucose levels by restraining and controlling hepatic glucose output. At mealtimes, glucose input from the gut increases and triggers a boost of insulin from the B cells of the islets of Langerhans. The augmented insulin level affects the dose-response curves of glucose disposal in the periphery, protein synthesis, lipogenesis, etc. as well as inhibiting liver glucose output and increasing glycogen deposition.

There are several insulin administration strategies. In conventional injection treatment, optimized injection regimens and self-monitoring of blood glucose are being investigated. Recent developments in electronics enables insulin infusion from a portable pump at a preprogrammed rate in the sense that the basal and prandial rates are decided by the operator and set for the individual patient. This is called the open-loop system. Another strategy under

Artificial Pancreas

1. **Immuno-isolation of pancreas cell**

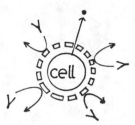

2. **Combination of glucose sensor and insulin injector**

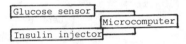

3. **Molecular system**

   Exchange-type

   Transform-type

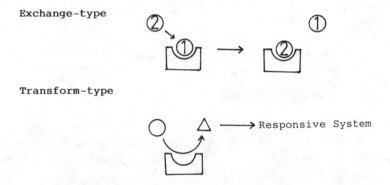

FIGURE 25.   Ideas for synthesis of an artificial pancreas.

investigation is utilization of magnetically and ultrasonically stimulated, or biodegradable polymer matrices. The advantage of implantable polymer pellets compared with insulin infusion pumps is considered to be that a large amount of insulin can be concentrated into a small-sized polymer as opposed to a mechanized pumping system which must be refilled daily.[128,129]

Still another strategy is the feedback system (closed system), which, in response to blood glucose levels, modulates or triggers the release of insulin. The approaches are further segmented into three, as shown in Figure 25.

One is encapsulation of pancreas islet cells in an inert matrix whose purpose is immunoisolation.[130-133] The second is an electromechanical system consisting of a glucose sensor and an insulin infusion pump.[134-139] Some

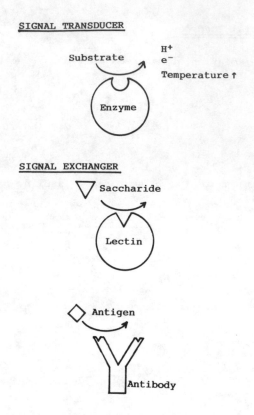

FIGURE 26.   Biospecific interaction for chemical responsiveness.

success has been achieved with wearable and implantable open-loop systems. The third is a molecular closed-loop system which was reviewed by Pitt et al. in 1985[140] and by Heller in 1988.[141] Recent development in the molecular system will be described in this section.

To obtain chemical-stimuli responsiveness, the use of biological interactions (such as the enzyme-substrate, the antigen-antibody, and the saccharide-lectin [Figure 26]) are very useful. Enzyme-substrate interactions are used as signal transducers. A typical example is the biosensor. The other interactions are used as signal substitutions for biomedical and clinical analyses, as well as in glucose-sensitive insulin-releasing systems.

A molecular system using the lectin-saccharide interaction was first designed by Brownlee and Cerami,[142,143] and then developed by Kim's group.[144-150] The system is composed of glycosylated insulin bound to concanavalin A, which is displaced by the competitive binding of glucose as shown in Figure 27a. Recently, new ideas involving boronic acid derivatives instead of lectin were reported by a Japanese group (Figure 27b,c).[151,152] The advantage of the new system is that there is no need of immunoisolation,

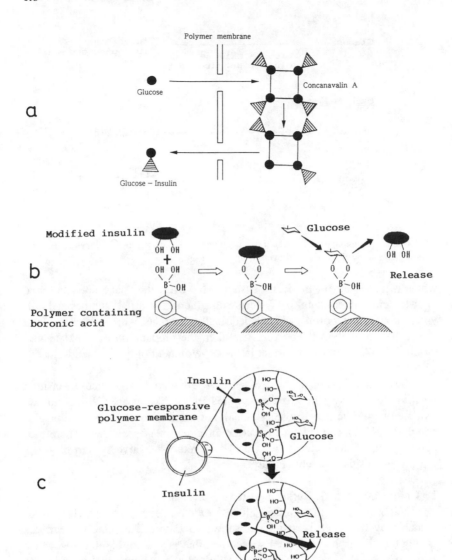

FIGURE 27.   Glucose-sensitive insulin-releasing systems using the mechanism of signal exchange. (a) The biospecific interaction between lectin (concanavalin A) and glucose. Glucose-
bound insulin was synthesized. (b) The interaction between diol and boronic acid. Insulin was
modified to contain a diol group. (c) A complex composed of polymers containing boronic acid
and polymers having diol groups was used as the glucose-responsive polymer membrane. Glucose
destroys the polymer complex to release insulin.

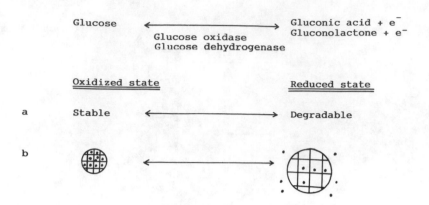

FIGURE 28.   Coupling of redox-responsive polymers with glucose-sensitive enzymes. (a) Polymer degradable by H_2O_2. (b) Hydrogel swollen by H_2O_2.

which is necessary for the lectin system. However, they have not succeeded in specifically interacting the system with glucose. Wulff[153] synthesized polymers containing boron, which specifically interacts with saccharide, and found that the polymer was very useful in the preparation of template materials. His idea will be important in the development of boronic acid-glucose systems.

The idea of using a glucose-sensitive enzyme for a glucose-sensitive insulin-releasing system was first applied by Updike et al.[54] When glucose oxidase or glucose dehydrogenase catalyzes glucose to gluconic acid or gluconolactone, respectively, a proton or an electron is produced. Therefore, combination of the enzymes with a pH- or redox-sensitive system becomes a glucose-sensitive system (Figure 2).

1. Combination with Redox-Sensitive Systems

Updike et al.[154] immobilized glucose oxidase in polymer which was degradable by hydrogen peroxide, as shown in Figure 28a. One system was insulin covalently bonded via a spacer arm that is susceptible to cleavage by H_2O_2 to a biodegradable, tissue-injectable polymer support matrix. Glucose oxidase attached to the support matrix. H_2O_2 was generated enzymatically as a function of the glucose concentration and, in turn, released insulin into the circulation as required. The other type of negative feedback control system was achieved by entrapping glucose oxidase and insulin molecules in highly cross-linked polymer molecules, providing the cross-linker was cleavaged by H_2O_2. When glucose entered the polymer matrix, H_2O_2 was generated, which released insulin. Updike et al.[154] considered that no modification of the insulin would be required due to the use of polymer gel entrapment rather than covalent bonding of the insulin. In their laboratory, they studied the ability of H_2O_2 to degrade such polymers as starch, polyacrylamide, agar, and cel-

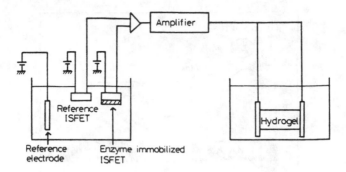

FIGURE 29. Model scheme for a glucose-responsive insulin-releasing system consisting of an ISFET glucose sensor and an electrically responsive hydrogel. (From Kaetsu, I., Morita, Y., Otori, A., and Naka, Y., *Artif. Organs*, 14, 237, 1990. With permission.)

lulose and its derivatives. They found that only starch was degraded at relatively low H_2O_2 concentrations. However, the rate of degradation was still too slow. Therefore, the use of hydrazides or boron derivatives to form a redox-sensitive degradable polymer was considered.

Ishihara et al.[155] immobilized glucose oxidase onto a redox-sensitive hydrogel (Figure 28b) containing nicotinamide groups. The membrane responded to hydrogen peroxide formed during the oxidation of glucose to gluconic acid. The oxidation of the dihydronicotinamide portion of the molecule by hydrogen peroxide increases hydrophilicity due to generation of a positive charge. This increase in hydrophilicity should enhance permeability to insulin. The permeation coefficient of the complex membrane increased about 1.5 times by addition of glucose. However, the permeation rate changes were not reversible.

Kaetsu et al.[156] designed a new device composed of a biochips sensor and an electrically responsive hydrogel. A mixture of a photocurable prepolymer and glucose oxidase was coated onto the surface of the gate portion of ISFET and polymerized into the enzyme membrane by UV irradiation. Mixtures of isopropyl acrylamide, acrylic acid, or methylacrylic acid and polyfunctional monomers (such as polyethyleneglycol dimethyacrylate) were polymerized into a hydrogel by γ-ray irradiation. The formed gel was further hydrolyzed with NaOH. Insulin was entrapped into the gel. Finally, the ISFET glucose sensor, the insulin entrapping hydrogel, and an amplifier were connected into the whole system as shown in Figure 29.

Chung et al.[157] immobilized insulin through a disulfide bond to a polymer membrane. The disulfide bone was cleaved upon oxidation of glucose with glucose dehydrogenase (GDH), releasing insulin from the membrane, as shown in Figure 30. Sensitivity to glucose concentration was enhanced by the presence of enzyme cofactors acting as electron mediators, and further enhanced by the coimmobilization of GDH (Figure 30B). The membrane system was specific for glucose, and the released insulin was indistinguishable from native insulin in HPLC elution patterns.

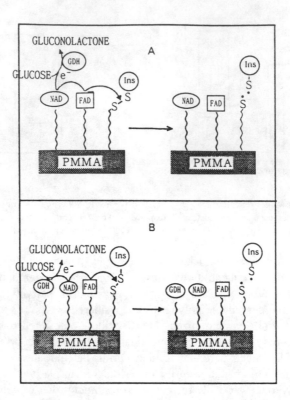

FIGURE 30.   A composite membrane, on which insulin was immobilized through a disulfide linkage. Free (A) or immobilized (B) glucose dehydrogenase oxidizes glucose and the produced electron cleaves the disulfide bond to release the immobilized insulin.

2. Combination with Redox-(Electrically) Sensitive Systems

The pH-sensitive systems used for glucose-sensitive insulin-release are summarized in Figure 31. Kost et al.[158] used a cross-linked hydrogel membrane having a *N,N*-dimethylaminoethyl methacrylate (DEA) moiety and a glucose oxidase molecule. The electrostatic repulsion caused by protonation of the amino groups led to expansion of the network structure of the hydrogel membrane. Faster delivery of insulin through the swollen membrane in the presence of glucose was realized. Ishihara et al.[159] also reported that there was a particularly large increase in the equilibrium water content and insulin permeability of the poly[DEA-co-poly-(2-hydroxypropyl methacrylate)] membrane in the presence of glucose.

After the establishment of the design concept, groups have been focusing upon increasing the sensitivity. Albin et al.[160,161] developed a macroporous hydrogel system to achieve the purpose. They reported that gels are 8 to 12 times more permeable to insulin at pH 4.0 than at pH 7.4, and gels containing glucose oxidase are 2.4 to 5.5 more permeable to insulin when 400 mg% glucose is introduced into an initially glucose-free environment.

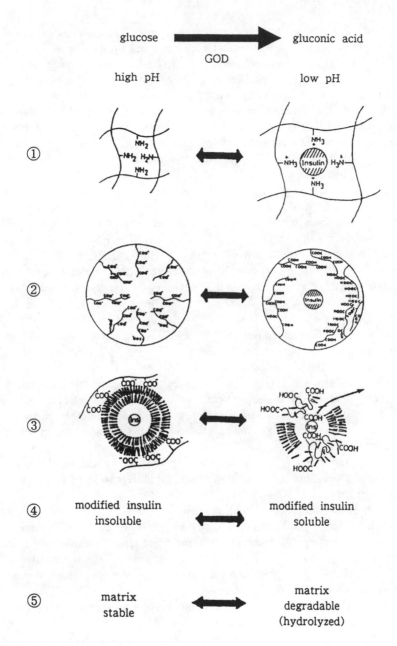

FIGURE 31.   Glucose-sensitive insulin-releasing strategies composed of a glucose-sensitive enzyme and a pH-responsive system.

On the other hand, Ishihara et al.[162,163] synthesized capsules composed of an amphiphilic electrolyte hydrogel, containing insulin and glucose oxidase. The polymer was poly(hydroxyethyl acrylate-co-DEA-co-trimethylsilylstyrene) (HEA-DEA-TMS). The HEA is hydrophilic, DEA is pH sensitive, and TMS is the hydrophobic and capsule-forming moiety. The permeation coefficient change before and after glucose addition was improved by encapsulation, about 1.5 mm in diameter. These approaches were aimed at increasing the surface area in contact with external stimuli.

Siegel and Firestone[164,165] proposed two implantable self-regulating insulin pumps with action dependent on the expansion and contraction of certain pH-sensitive hydrogel membranes that respond to changes in glucose concentration via enzymatic conversion of glucose to gluconic acid, as shown in Figure 32. One was based on the design of an osmotic pump. A potential advantage of the osmotic pump over the other mechanochemically controlled systems is that the insulin and the rate-controlling membrane are spatially separated. Thus insulin will not "clog" the membrane, which is a distinct possibility in those systems where the polymers control the permeability of insulin. Moreover, insulin can be formulated in a variety of ways in the osmotic pump.

Ito et al.[166] synthesized poly(acrylic acid)-grafted porous membrane immobilized with glucose oxidase as shown in Figure 31-2. The principle of the system was based on the conformational change of the graft chains as described in Chapter 2.1. Glucose oxidase catalyzed the oxidation of glucose to gluconic acid, and protonated the carboxylate groups. To control the release rate of insulin near at pH 7, they are attempting to graft poly(ethacrylic acid). Iwata et al.[167] also grafted an amphiphilic polyelectrolyte onto a porous membrane for this purpose.

Devlin and Tirrell[168] demonstrated the possibility of glucose-dependent disruption of phospholipid vesicles, though insulin release experiments were not performed.

Fischel-Ghodsian et al.[169] modified insulin by adding three lysine groups, causing its solubility to be minimal at physiological pH (7.4), and to increase substantially as the pH is decreased. The modified insulin was dispersed into a porous ethylene vinyl acetate copolymer matrix immobilized with glucose oxidase. A rise in blood glucose level triggered an increased insulin release, because of the increase in the solubility.

Heller et al.[170,171] synthesized erodible poly(ortho esters) with erosion rates that increase sharply as the pH is lowered. Siegel and Firestone[164] considered that the combination of this pH-sensitive degradable polymer with glucose oxidase will be a glucose-sensitive insulin-releasing system.

3. Protein Devices

Recently, Ito et al.[172] designed a new device, which is composed of protein for a glucose-sensitive insulin-releasing system (Figure 33). After the

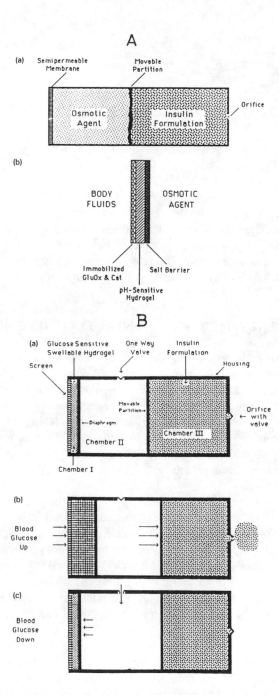

FIGURE 32. (A) (a) Schematic of a proposed self-regulating osmotic insulin pump; (b) detail of glucose-sensitive semipermeable membrane. (B) (a) Schematic of proposed self-regulating mechanochemical insulin pump; (b),(c) expected operation of mechanochemical pump. Arrows indicate water flux. (From Siegel, R. A. and Firestone, B. A., *J. Controlled Release,* 11, 181, 1190. With permission.)

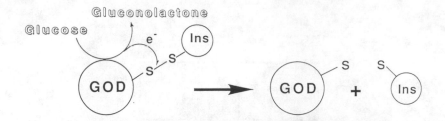

FIGURE 33.   Principle of glucose-sensitive insulin-releasing protein device.

carboxyl groups of insulin were blocked, 5,5'-dithiobis(2-nitrobenzoic acid) was coupled to it. The insulin derivative was then coupled to glucose oxidase using water-soluble carbodiimide. The protein device was purified by column chromatography. Upon adding glucose, a new peak corresponding to the released insulin was observed. The released insulin had about 80% activity of the native insulin. This device will be a new application of protein engineering.

B. OTHER ENZYMATICALLY CONTROLLED RELEASE SYSTEMS

Though there is no current therapeutic relevance, some enzymatically controlled systems have been propounded. A hydrocortisone release system which responds to the concentration of external urea has been designed by Heller et al.[141] The pH-sensitive erodible copolymer was combined with urease. The principle was based on the conversion of urea to NH_4HCO_3 and NH_4OH by the action of urease. While this ingenious system has no foreseeable therapeutic application, it established the feasibility of using immobilized enzymes to induce a local pH change, and was the first example of a synthetic self-regulated system.

The preceding system was modified to use a hydrogel containing an immobilized enzyme which, when activated, is capable of degrading the hydrogel (Figure 34a). The enzyme is covalently bound to a trigger molecule which is also the hapten for an antibody. While the enzyme-linked hapten is bound by the antibody, the enzymatic activity is suppressed by steric hindrance of the active site. The appearance of the trigger molecule in the circulatory system results in competitive binding. The uncovering of the active site of the enzyme initiates the degradation of the hydrogel and the surface erosion of the inner core.

Other uses of biological molecular recognition have included an antibody-based drug delivery system. The strategy being explored is illustrated conceptually in Figure 34b in the form of a delivery system for the narcotic antagonist naltrexone. The system is designed to remain dormant after subdermal implantation, until triggered by the appearance of morphine in the circulatory system.

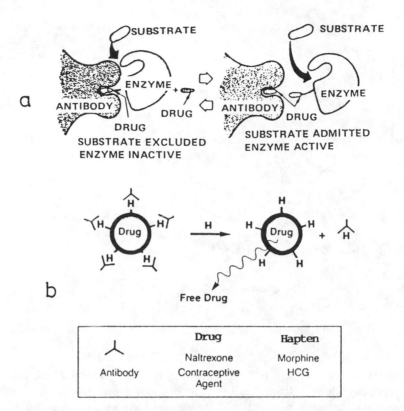

FIGURE 34.   (a) Reversible enzyme inactivation by hapten-antibody interactions. (b) Triggered release based on reversible antibody binding to haptens attached to the surface of permeable or biodegradable polymers. (From Heller, J. J., *J. Controlled Release*, 8, 111, 1988. With permission.)

Lectin-saccharide interaction has been used by Okahata and Noguchi.[106] They grafted polymers containing a saccharide group in the side chain onto microcapsules. Upon adding concanavalin A, permeation of the microcapsule was enhanced. They considered that the increase was based on aggregation of the polymers as shown in Figure 35.

IV. FUTURE OUTLOOK

The molecular devices described in this chapter have two main problems. One is immunoisolation. The other is "completeness" of the device. The former is a perpetual function of any device used in the body. To overcome this problem, modification of the system will be useful as described in Chapter 2.2.

The latter problem is interesting from the standpoint of designing intelligent materials. For example, a method to mimic real pancreatic release

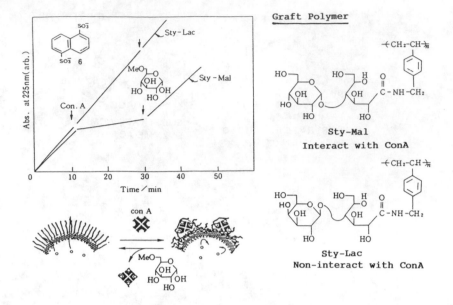

FIGURE 35.   Permeability control utilizing the interaction between concanavalin A and saccharide.

presents a challenge. Devices that exhibit proportional-derivative control, or that simply release a "burst" of insulin followed by a secondary slower release phase will probably be superior. It is necessary to design a programmable material or device.

REFERENCES

1. **Poznansky, M. J. and Juliano, R. L.,** Biological approaches to the controlled delivery of drugs: a critical review, *Pharmacol. Rev.,* 36, 277, 1984.
2. **Langer, R.,** New methods of drug delivery, *Science,* 249, 1527, 1990.
3. **Kopecek, J. Vacik, J., and Lim, D.,** Permeability of membranes containing ionogenic groups, *J. Polym. Sci., Part A-1,* 9, 2801, 1971.
4. **Alhaique, F., Marchetti, M., Riccieri, F. M., and Santucci, A.,** Polymeric film responding in diffusion properties to environmental pH stimuli: a model for a self-regulating drug delivery system, *J. Pharmacol.,* 33, 413, 1981.
5. **Alhaique, F., Riccieri, F. M., Santucci, E., Crescezi, V., and Gamini, A.,** A possible pH-controlled drug delivery system based on a derivative of the polysaccharide sclero-glucan, *J. Pharm. Pharmacol.,* 37, 310, 1984.
6. **Bala, K. and Vasudevan, P.,** pH-Sensitive microcapsules for drug release, *J. Pharm Sci.,* 71, 960, 1982.
7. **Tanaka, T.,** Collapse of gels and the critical end point, *Phys. Rev. Lett.,* 40, 820, 1978.
8. **Tanaka, T., Fillmore, D. J., Sun, S.-T., Nishio, I., Swislow, G., and Shah, S.,** *Phys. Rev. Lett.,* 45, 1636, 1980.

9. **Tanaka, T., Nishio, I., Sun, S., and Ueno-Nishio, S.,** Collapse of gels in an electrical field, *Science,* 218, 467, 1982.
10. **Tanaka, T.,** Kinetics of phase transition in polymer gels, *Physica,* 140A, 261, 1986.
11. **Matsumoto, E. S. and Tanaka, T.,** Kinetics of discontinuous volume-phase transition of gels, *J. Chem. Phys.,* 86, 1695, 1988.
12. **Hirokawa, Y. and Tanaka, T.,** Volume phase transition in a nonionic gel, *J. Chem. Phys.,* 81, 6379, 1984.
13. **Ohmine, I. and Tanaka, T.,** Salt effects on the phase transition of ionic gels, *J. Chem. Phys.,* 11, 5725, 1982.
14. **Amiya, T. and Tanaka, T.,** Phase transition in cross-linked gels of natural polymers, *Macromolecules,* 20, 1162, 1987.
15. **Matsuo, E. S. and Tanaka, T.,** Kinetics of discontinuous volume-phase transition of gels, *J. Chem. Phys.,* 89, 1695, 1988.
16. **Hirotsu, S., Hirokawa, Y., and Tanaka, T.,** Volume-phase transitions of ionized *N*-isopropylacrylamide gels, *J. Chem. Phys.,* 87, 1392, 1987.
17. **Ilavsky, M.,** Phase transition in swollen gels. II. Effect of charge concentration on the collapse and mechanical behavior of polyacrylamide networks, *Macromolecules,* 15, 782, 1982.
18. **Horouz, J., Ilavsky, M., Ulbrick, and Kopecek, K.,** The photoelastic behavior of dry and swollen networks of poly(*N,N*-diethylacrylamide) and of its copolymer with *N-tert*-butylacrylamide, *Eur. Polym. J.,* 17, 361, 1981.
19. **Theeuwes, I. F. and Yum, S. I.,** Principles of the design and operation of generic osmotic pumps for the delivery of semisolid or liquid drug formations, *Ann. Biomed. Eng.,* 4, 353, 1976.
20. **Kabra, B. and Gehrke, S. H.,** Hydrogels for driving an osmotic pump, *Polym. Prepr.,* 30, 490, 1989.
21. **Firestone, B. A. and Siegel, R. A.,** Swelling equilibria and kinetics in pH-sensitive hydrophobic membranes, *Proc. Int. Symp. Controlled Release Bioact. Mater.,* 14, 81, 1987.
22. **Pefferkorn, E., Schmitt, A., and Varoqui, R.,** Interfacial properties of poly(a-glutamic acid). Correlation with static and dynamic membrane properties, *Biopolymers,* 21, 1451, 1982.
23. **Idol, W. K. and Anderson, J. L.,** Effects of adsorbed polyelectrolytes on convective flow and diffusion in porous membranes, *J. Membr. Sci.,* 28, 269, 1986.
24. **Osada, Y. and Takeuchi, Y.,** Water and protein permeation through polymer membrane having mechanochemically expanding and contracting pores. Function of chemical valves. I., *J. Polym. Sci., Polym. Lett. Ed.,* 19, 303, 1981.
25. **Osada, Y., Hondo, K., and Ohta, M.,** Control of water permeability by mechano-chemical contraction of poly(methacrylic acid)-grafted membranes, *J. Membr. Sci.,* 27, 327, 1986.
26. **Iwata, H. and Matsuda, T.,** Preparation and properties of novel environment-sensitive membranes prepared by graft polymerization onto a porous membrane, *J. Membr. Sci.,* 38, 185, 1988.
27. **Okahata, Y., Ozaki, K., and Seki, T.,** pH-Sensitive permeability control of polymer grafted nylon capsule membrane, *J. Chem. Soc., Chem. Commun.,* 519, 1984.
28. **Okahata, Y., Noguchi, H., and Seki, T.,** Functional capsule membranes. 26. Permeability control of polymer grafted capsule membranes responding to ambient pH changes, *Macromolecules,* 20, 15, 1987.
29. **Ito, Y., Kotera, S., Inaba, M., Kono, K., and Imanishi, Y.,** Control of poly(acrylic acid) grafts of pore size of polycarbonate membrane with straight pores, *Polymer,* 31, 2157, 1990.
30. **Kinoshita, T., Iwata, T., Takizawa, A., and Tsujita, Y.,** Ionic salt permeabilities and membrane potentials of asymmetric polypeptide membranes, *Colloid Polym. Sci.,* 261, 933, 1983.

31. **Kinoshita, T., Yamashita, I., Iwata, T., Takizawa, A., and Tsujita, Y.,** Ionic salt permeabilities and membrane potentials of glutamic acid-methyl-L-glutamate copolymer membranes, *J. Macromol. Sci., Phys.,* B22, 1, 1983.

32. **Maeda, M., Kimura, M., Hareyama, Y., and Inoue, S.,** pH-Dependent ion transport across polymer membrane. pH-Induced reversible conformational change of transmembrane poly(L-aspartic acid) domain in polymer membrane, *J. Am. Chem. Soc.,* 106, 250, 1984.

33. **Chung, D.-W., Higuchi, S., Maeda, M., and Inoue, S.,** pH-Induced regulation of permselectivity of sugars by polymer membrane from polyvinyl-polypeptide graft copolymer, *J. Am. Chem. Soc.,* 108, 5823, 1986.

34. **Higuchi, S., Mozawa, T., Maeda, M., and Inoue, S.,** pH-Induced regulation of the permeability of a polymer membrane with a transmembrane pathway prepared from a synthetic polypeptide, *Macromolecules,* 19, 2263, 1986.

35. **Kinoshita, T., Takizawa, A., Tsujita, Y., and Ishikawa, M.,** The acceleration of L-lysine transport through poly(α-amino acid) membranes in acid water, *Kobunski Ronbunshu,* 43, 827, 1986.

36. **Huguet, J. and Vert, M.,** Partially quarternized poly(tertiary amine) as pH-dependent drug carrier for solubilization and temporary trapping of lipophilic drugs in aqueous media, *J. Controlled Release,* 1, 217, 1985.

37. **Heller, J., Baker, R. W., Gale, R. M., and Rodin, J. O.,** Controlled drug release by polymer dissolution. I. Partial esters of maleic anhydride copolymers. Properties and theory, *J. Appl. Polym. Sci.,* 22, 1991, 1978.

38. **Heller, J., Penhale, D. W. H., Fritzinger, B. K., Rose, J. E., and Helwing, R. F.,** Controlled release of contraceptive steroids from biodegradable poly(ortho esters), *Contracep. Delv. Syst.,* 4, 43, 1983.

39. **Heller, J., Penhale, D. W. H., and Helwing, R. F.,** Preparation of poly(ortho esters) by the reaction of ketene acetal and diol, *J. Polym. Sci., Polym. Lett. Ed.,* 18, 691, 1980.

40. **Seki, S. and Kotaka, T.,** Complex-forming poly(oxyethylene)/poly(acrylic acid) interpenetrating polymer networks. II. Function as chemical valve, *Macromolecules,* 17, 978, 1986.

41. **Hemker, D. J., Garza, V., and Frank, C. W.,** Complexation of poly(acrylic acid) and poly(methacrylic acid) with pyrene-end-labeled poly(ethylene glycol). pH and fluorescence measurements, *Macromolecules,* 23, 4411, 1990.

42. **Seki, K. and Tirrell, D. A.,** pH-Dependent complexation of poly(acrylic acid) derivatives with phospholipid vesicle membranes, *Macromolecules,* 17, 1692, 1982.

43. **Petrukhina, O. O., Ivanov, N. N., Feldstein, M. M., Vasilev, A. E., Plate, N. A., and Torchilin, V. P.,** The regulation of liposome permeability by polyelectrolyte, *J. Controlled Release,* 3, 137, 1986.

44. **Tirrell, D. A., Takigawa, D. Y., and Seki, K.,** pH Sensitization of phospholipid, vesicles via complexation with synthetic poly(carboxylic acid)s, *Ann. N.Y. Acad. Sci.,* 446, 237, 1985.

45. **Tirrell, D. A.,** Macromolecular switches for bilayer membranes, *J. Controlled Release,* 6, 15, 1987.

46. **Takigawa, D. Y. and Tirrell, D. A.,** Reversal of Mg^{2+} induced structural changes in dipalmitoylphosphatidylglycerol bilayers by adsorbed poly(ethyleneimine)s, *Makromol. Chem. Rapid Commun.,* 6, 653, 1985.

47. **Maeda, M., Kumano, A., and Tirrell, D. A.,** H^+-induced release of calcine from phosphatidylcholine vesicles bearing surface-bound polyelectrolyte chains, *J. Am. Chem. Soc.,* 110, 7455, 1988.

48. **Kitano, H., Akatsuka, Y., and Ise, N.,** pH-Responsive liposome, *Macromolecule,* in press.

49. **Subbarao, N. K., Parente, R. A., Szoka, F. C., Jr., Nadasdi, L., and Pongracz, K.,** pH-Dependent bilayer destabilization by an amphipathic peptide, *Biochemistry,* 26, 2964, 1987.

50. **Kono, K., Kimura, S., and Imanishi, Y.,** pH-Dependent interaction of amphiphilic polypeptide, poly(Lys-Aib-Leu-Aib), with lipid bilayer membrane, *Biochemistry,* 29, 3631, 1990.

51. **Kono, K., Kimura, S., and Imanishi, Y.,** Preparation and characterization of lipid membranes coated on polyamide microcapsules, *J. Membr. Sci.,* 50, 85, 1990.

52. **Fendler, J. H.,** Potential membrane-mimetic polymers in membrane technology, *J. Membr. Sci.,* 30, 323, 1987.

53. **Ringsdorf, H., Scharb, B., and Venzmer, J.,** Molecular architecture and function of polymeric orientated systems: models for the study of organization, surface recognition and dynamics of biomembranes, *Angew. Chem., Int. Ed. Engl.,* 27, 113, 1988.

54. **Yatvin, M. B., Kreutz, W., Horwitz, B. A., and Shinitzky, M.,** pH-Sensitive liposomes: possible clinical implications, *Science,* 210, 1253, 1980.

55. **Yatvin, M. B., Weinstein, J. N., Dennis, W. H., and Blumenthal, R.,** Design of liposome for enhanced local release of drugs by hyperthermia, *Science,* 202, 1290, 1978.

56. **Connor, J., Yatvin, M. B., and Huang, L.,** pH-Sensitive immunoliposomes: acid-induced liposome fusion, *Proc. Natl. Acad. Sci. U.S.A.,* 81, 1715, 1984.

57. **Ellens, H., Bentz, J., and Szoka, F. C.,** pH-Induced destabilization of phosphatidylethanolamine-containing liposomes: role of bilayer contact, *Biochemistry,* 23, 1532, 1984.

58. **Tsusaki, B. E., Kanda, S., and Huang, L.,** Stimulation of superoxide release in neutrophiles by oleoyl-2-acetylglycerol incorporated into pH-sensitive liposomes, *Biochem. Biophys. Res. Commun.,* 136, 242, 1986.

59. **Dugzunes, N., Straubinger, R. M., Baldwin, P. A., Friend, D. S., and Papahadjapoulos, D.,** Proton-induced fusion of oleic acids-phosphatidylethanolamine liposomes, *Biochemistry,* 24, 3091, 1985.

60. **Huang, L. and Lui, S. S.,** Acid-induced fusion of liposomes with inner membranes of mitochondria, *Biophys. J.,* 45, 72, 1984.

61. **Straubinger, R. M., Duzguners, N., and Papahadjoulos, D.,** pH-Sensitive liposome mediated cytoplasmic delivery of encapsuled macromolecules, *FEBS Lett.,* 179, 148, 1985.

62. **Iwamoto, K. and Sunamoto, J.,** Liposomal membrane. XII. Adsorption of polysaccharide on liposomal membrane as monitored by fluorescence depolarization, *J. Biochem.,* 91, 975, 1982.

63. **Regen, S. L., Shin, J.-C., Yamaguchi, K.,** Polymer-encased vesicles, *J. Am. Chem. Soc.,* 106, 2446, 1984.

64. **Okahata, Y. and Seki, T.,** pH-Responsive permeation of bilayer-coated capsule membranes by ambient pH changes, *Chem. Lett.,* 1251, 1984.

65. **Okahata, Y. and Seki, T.,** Permeability control of bilayer-immobilized films, *Oil Chem.,* 34, 820, 1985.

66. **Takeoka, S., Ohno, H., Hayashi, N., and Tsuchida, E.,** Control of release of encapsuled molecules from polymerized mixed liposomes induced physical or chemical stimuli, *J. Controlled Release,* 9, 177, 1989.

67. **Kitano, H., Wolf, H., and Ise, N.,** pH-Responsive release of fluorophore from homocysteine-carrying polymerized liposomes, *Macromolecules,* 23, 1958, 1990.

68. **Kaetsu, I., Naka, Y., Yoshida, M., and Asano, M.,** Drug delivery system using polymer, *Rept. Poval Commit.,* 92, 108, 1988.

69. **Hoffman, A. S., Affrassiabi, A., and Dong, L. C.,** Thermally reversible hydrogels. II. Delivery and selective removal of substances from aqueous solutions, *J. Controlled Release,* 4, 213, 1986.

70. **Afrassiabi, A., Hoffman, A. S., and Cadwell,** Effect of temperature on the release rate of biomolecules from thermally reversible hydrogels, *J. Membr. Sci.,* 33, 191, 1987.

71. **Freitas, R. F. S. and Cussler, E. L.,** Temperature sensitive gels as size selective absorbants, *Sep. Sci. Technol.,* 22, 911, 1987.

72. **Bae, Y. H., Okano, T., Hsu, R., and Kim, S. W.,** Thermosensitive polymers as on-off switches for drug release, *Makromol. Chem., Rapid Commun.*, 8, 481, 1987.

73. **Bae, Y. H., Okano, T., and Kim, S. W.,** Insulin permeation through thermo-sensitive hydrogels, *J. Controlled Release*, 9, 271, 1989.

74. **Okano, T., Bae, Y. H., and Kim, S. W.,** Thermally on-off switching polymers for drug permeation and release, *J. Controlled Release*, 11, 255, 1990.

75. **Drummond, W. R., Knight, M. L., Brannon, M. L., and Peppas, N. A.,** Surface instabilities during swelling of pH-sensitive hydrogels, *J. Controlled Release*, 7, 181, 1988.

76. **Ueda, T., Ishihara, K., and Nakabayashi, N.,** Thermally responsive release of 5-fluorouracil from a biocompatible hydrogel membrane with phospholipid structure, *Makromol. Chem., Rapid Commun.*, 11, 345, 1990.

77. **Okahata, Y., Noguchi, H., and Seki, T.,** Thermoselective permeation from a polymer-grafted capsule membrane, *Macromolecules*, 19, 493, 1986.

78. **Iwata, H., Oodate, M., Uyama, Y., Amemiya, H., and Ikada, Y.,** Preparation of temperature-sensitive membranes by graft polymerization onto a porous membrane, *J. Membr. Sci.*, in press.

79. **Burgmayer, P. and Murray, R. W.,** An ion gate membrane: electrochemical control of ion permeability through a membrane with an embedded electrode, *J. Am. Chem. Soc.*, 104, 6139, 1982.

80. **Burgmayer, P. and Murray, R. W.,** Ion gate electrodes. Polypyrrole as a switchable ion conductor membrane, *J. Phys. Chem.*, 88, 2515, 1984.

81. **Zinger, B. and Miller, L. L.,** Timed release of chemicals from polypyrrole films, *J. Am. Chem. Soc.*, 106, 6861, 1984.

82. **Shimidzu, T., Ohtani, A., Iyoda, T., and Honda, K.,** Effective adsorption-desorption of cations on a polypyrrole-polymer anion composite electrode, *J. Chem. Soc., Chem. Commun.*, 1415, 1986.

83. **Miller, L. L. and Zhou, Q. X.,** Poly (*N*-methylpyrrolylium) poly(styrenesulfonate). A conductive, electrically switchable cation exchanger that cathodically binds and anodically releases dopamine, *Macromolecules*, 20, 1594, 1987.

84. **Okahata, Y., Ariga, K., and Seki, T.,** Redox-sensitive permeation from a capsule membrane grafted with viologen-containing polymers, *J. Chem. Soc., Chem. Commun.*, 1986, 73.

85. **Nishi, S., Ito, Y., and Imanishi, Y.,** Graft-polymerization of 3-carbamoyl-1-(*p*-vinyl-benzyl)pyridinium chloride onto porous membrane and redox-sensitive solute-permeation of the grafted membrane, 36th Polym. Symp. Kobe, Japan, June 6, 1990, 56.

86. **Dietzel, K., Keuth, V., Estes, K. S., Brewster, M. E., Clemmons, R. M., Vistelle, R., Bodor, N. S., and Derendorf, H.,** A redox-based system that enhances delivery of estradiol to the brain: pharmacokinetic evaluation in the dog, *Pharm. Res.*, 7, 879, 1990.

87. **Smets, G.,** New developments in photochromic polymers, *J. Polym. Sci., Polym. Chem. Ed.*, 13, 2223, 1975.

88. **Irie, M. and Kungwatchakun, D.,** Photoresponsive polymers. IX. Photostimulated dilation of polyacrylamide gels having triphenylmethane leuco derivatives, *Macromolecules*, 19, 2476, 1986.

89. **Ishihara, K., Hamada, N., Sato, S., and Shinohara, I.,** Photo induced swelling control of amphiphilic azoaromatic polymer membrane, *J. Polym. Sci., Polym. Lett. Ed.*, 22, 121, 1984.

90. **Suzuki, A. and Tanaka, T.,** Phase transition in polymer gels induced by visible light, *Nature (London)*, 346, 345, 1990.

91. **Chung, D. J., Ito, Y., and Imanishi, Y.,** Permeability control of photo-responsive porous polymer membrane grafted with vinyl polymers containing spiropyran groups, *Polym. Prepr. Jpn.*, 40, 740, 1991.

92. **Sato, M., Kinoshita, T., Takizawa, A., Tsujita, Y., and Ito, R.,** Permeability changes of a porous membrane by an adsorbed photoresponsive polypeptide, *Polym. J.,* 20, 761, 1988.

93. **Sato, M., Kinoshita, T., Takizawa, A., and Tsujita, Y.,** Photoinduced conformational transition of polypeptide membrane composed of poly(L-glutamic acid) containing pararosaniline groups in the side chains, *Macromolecules,* 21, 3419, 1988.

94. **Sato, M., Kinoshita, T., Takizawa, A., and Tsujita, Y.,** Photocontrol of polypeptide membrane functions by photo-dissociation of pararosaniline side-chain groups, *Polym. J.,* 20, 729, 1988.

95. **Chung, D.-W., Tanaka, K., Maeda, M., and Inoue, S.,** Effects of cationic surfactants with an alkyl group of different chain lengths on permeability across polymer membranes having poly(L-glutamic acid) domains, *Polym. J.,* 20, 933, 1988.

96. **Maeda, M., Aoyama, M., and Inoue, S.,** Divalent cation-induced permeability regulation of polymer membrane with permeating pathway of a synthetic polypeptide, *Macromol. Chem.,* 187, 2137, 1986.

97. **Chung, D.-W., Maeda, M., and Inoue, S.,** Urea-induced "denaturation" of a polymer membrane with poly(L-glutamic acid) domains for permeation, *Macromol. Chem.,* 189, 1635, 1988.

98. **Chung, D.-W., Kurosawa, K., Maeda, M., and Inoue, S.,** Effect of cationic surfactants on the permeability across a membrane of poly[butyl methacrylate-co-N-methyl-N-(4-vinylphenethyl)ethylenediamine]-graft-poly(L-glutamic acid), *Makromol. Chem.,* 189, 2731, 1988.

99. **Aoyama, M., Youda, A., Watanabe, J., and Inoue, S.,** Synthesis and circular dichroic and photoresponsive properties of a graft copolymer containing an azoaromatic polypeptide branch and its membrane, *Macromolecules,* 23, 1458, 1990.

100. **Okahata, Y. and Takenouchi, K.,** Permeability-controllable membranes. IX. Electrochemical redox-sensitive gate membranes of polypeptide films having ferrocene groups in the side chains, *Macromolecules,* 22, 308, 1989.

101. **Okahata, Y., Enna, G., Taghuchi, K., and Seki, T.,** Electrochemical permeability control through a bilayer-immobilized film containg redox sites, *J. Am. Chem. Soc.,* 107, 5300, 1985.

102. **Okahata, Y., Iizuka, N., Nakamura, G., and Seki, T.,** Permeability of fluorescent probes at phase transitions from bilayer-coated capsule membranes, *J. Chem. Soc., Perkin Trans. 2,* 1592, 1985.

103. **Okahata, Y.,** Lipid bilayer-corked capsule membranes. Reversible, signal-responsive permeation control, *Acc. Chem. Res.,* 19, 57, 1986.

104. **Okahata, Y., Lim, H.-J., Nakamura, G., and Hachiya, S.,** A large nylon capsule coated with a synthetic bilayer membrane. Permeability control of NaCl by phase transition of the dialkylammonium bilayer coating, *J. Am. Chem. Soc.,* 105, 4855, 1983.

105. **Okahata, Y., Lim, H.-J., Hachiya, S., and Nakamura, G.,** Bilayer-coated capsule membranes. IV. Control of NaCl permeability by phase transition of synthetic bilayer coatings, depending on their hydrophilic head groups, *J. Membr. Sci.,* 19, 237, 1984.

106. **Okahata, Y. and Noguchi, H.,** Ultrasound-responsive permeability control of bilayer-coated capsule membranes, *Chem. Lett.,* 1517, 1983.

107. **Okahata, Y., Lim, H.-J., and Hachiya, S.,** Photoresponsible capsule membranes. Permeability control of NaCl from a nylon capsule coated with a synthetic bilayer membrane containing an azo-chromophore, *Makromol. Chem. Rapid Commun.,* 4, 303, 1983.

108. **Okahata, Y., Lim, H.-J., and Hachiya, S.,** Bilayer coated capsule membranes. II. Photoresponsive permeability control of sodium chloride across a capsule membrane, *J. Chem. Soc., Perkin Trans.,* 2, 989, 1984.

109. **Okahata, Y., Lim, H.-J., and Nakamura, G.,** Signal receptive capsule membrane. Ca^{2+}-induced permeability control of large nylon capsule coated with synthetic bilayer membrane, *Chem. Lett.,* 755, 1983.

110. **Okahata, Y. and Lim, H.-J.,** Functional capsule membranes. VIII. Signal-receptive permeability control of NaCl from a large nylon capsule coated with phospholipid bilayers, *J. Am. Chem. Soc.,* 106, 4696, 1984.
111. **Okahata, Y., Nakamura, G., Hachiya, S., Noguchi, H., and Lim, H.-J.,** Selective storage and permeation of NaOH and HCL across a capsule membrane coated with charged synthetic bilayers, *J. Chem. Soc., Chem. Commun.,* 1206, 1983.
112. **Okahata, Y., Noguchi, H., and Seki, T.,** Selective pH gradient formation across a capsule membrane corked with charged synthetic bilayers, *J. Membr. Sci.,* 24, 169, 1985.
113. **Seki, T. and Okahata, Y.,** pH-Sensitive permeation of ionic fluorescent probes from nylon capsule membrane, *Macromolecules,* 17, 1880, 1984.
114. **Okahata, Y., Hachiya, S., and Seki, T.,** Activity control of an enzyme immobilized on a capsule membrane with synthetic bilayers, *J. Polym. Sci., Polym. Lett. Ed.,* 22, 595, 1984.
115. **Okahata, Y., Hachiya, S., and Seki, T.,** The formation of stable transient pores across bilayer-coated capsule membranes by external electric fields, *J. Chem. Soc., Chem. Commun.,* 1373, 1984.
116. **Okahata, Y., Hachiya, S., Ariga, K., and Seki, T.,** The electric breakdown and permeability control of a bilayer-corked capsule membrane in an external electric field, *J. Am. Chem. Soc.,* 108, 2863, 1986.
117. **Okahata, Y. and Seki, T.,** pH-Sensitive capsule membranes. Reversible permeability control from the dissociative bilayer-coated capsule membrane by an ambient pH change, *J. Am. Chem. Soc.,* 106, 8065, 1984.
118. **Akiyoshi, K. and Sunamoto, J.,** Synthesis and characterization of stimuli-responsive polysaccharides, *Polym. Prepr., Jpn.,* 38, 3020, 1989.
119. **Kajiyama, T.,** Gas and liquid permeabilities through polymer/liquid crystal blend, *Membrane,* 4, 229, 1979.
120. **Kajiyama, T., Washizu, M., and Takayanagi, M.,** Membrane structure and permeation properties of poly(vinyl chloride)/liquid crystal composite membrane, *J. Appl. Polym. Sci.,* 29, 3955, 1984.
121. **Shinkai, S., Nakamura, S., Ohara, K., Tachiki, S., Manabe, O., and Kajiyama, T.,** "Complete" thermocontrol of ion permeation through ternary composite membranes composed of polymer/liquid crystal/amphiphilic crown ethers, *Macromolecules,* 20, 21, 1987.
122. **Ishihara, K., Muramoto, N., Iida, T., and Shinohara, I.,** Preparation of polymer membrane with responsive function for amino compounds, *Polym. Bull.,* 7, 457, 1982.
123. **Okahata, Y. and Noguchi, H.,** Ultrasound-responsive permeability control of bilayer-coated capsule membranes, *Chem. Lett.,* 1983, 1517.
124. **Kost, J., Leong, K., and Langer, R.,** Ultrasound enhanced polymer degradation and release of incorporated substances, *Proc. Natl. Acad. Sci. U.S.A.,* 86, 7663, 1989.
125. **Edelman, E. R., Brown, L., Taylor, J., and Langer, R.,** *In vitro* and *in vivo* kinetics of regulated drug release from polymer matrices by oscillating magnetic fields, *J. Biomed. Mater. Res.,* 21, 339, 1987.
126. **Kost, J., Wolfrum, J., and Langer, R.,** Magnetically enhanced insulin release in diabetic rats, *J. Biomed. Mater. Res.,* 21, 1367, 1987.
127. **Pickup, J. C. and Stevenson, R. W.,** Rate-controlled insulin delivery in normal physiology and in the treatment of insulin-dependent diabetes, in *Rate-Controlled Drug Administration and Action,* Struyker-Boudier, Ed., CRC Press, Boca Raton, FL, 1986, 143.
128. **Sefton, M. V.,** Implantable pumps, *CRC Crit. Rev. Biomed. Eng.,* 14, 201, 1986.
129. **Sefton, M. V., Horvath, Y., and Zingg, W.,** Insulin delivery by a diffusion-controlled micropump in pancreactomized dogs: phase 1, *J. Controlled Release,* 12, 1, 1990.
130. **Goosen, M. F. A.,** Insulin delivery systems and the encapsulation of cells for medical and industrial use, *CRC Crit. Rev. Biocompatibility,* 3, 1, 1987.
131. **Lim, F. and Sun, A. M.,** Microencapsulated islets as bioartificial endocrine pancreas, *Science,* 210, 908, 1980.

132. **Sefton, M. C., Broughton, R. L., Sugamori, M. E., and Mallabone, C. L.,** Hydrophilic polyacrylates for the microencapsulation of fibroblasts or pancreatic islets, *J. Controlled Release,* 6, 177, 1987.

133. **Iwata, H., Amemiya, H., and Akutsu, T.,** Hybrid artificial pancreas: islets immobilized in a hydrogel tablet, *Jpn. J. Artif. Organs,* 18, 1324, 1989.

134. **Schade, D. S., Eaton, R. P., and Spencer, W.,** Normalization of plasma insulin profiles in diabetic subjects with programmed insulin delivery, *Diabetes Care,* 3, 9, 1980.

135. **Franetzki, M.,** Drug delivery by program or sensor controlled infusion devices, *Pharm. Res.,* 1, 237, 1984.

136. **Rupp, V. M., Barbosa, J. J., Blackshear, P. J., McCarthy, H. B., Rohde, T. D., Goldenberg, F. J., Rublein, T. G., Dorman, F. D., and Buchwald, H.,** The use of an implantable insulin pump in the treatment of Type II diabetes, *N. Engl. J. Med.,* 307, 265, 1982.

137. **Spencer, W. J.,** A review of programmed insulin delivery systems, *Trans. Biomed. Eng.,* 28, 237, 1981.

138. **Fischer, U., Schenk, W., Salzider, E., Albrecht, G., Abel, P., and Freyse, E.-J.,** Does physical blood glucose control require an adaptive control strategy?, *IEEE Trans. Biomed. Eng.,* 34, 575, 1987.

139. **Shichiri, M., Kawamori, R., Hakin, N., Asakawa, N., Yamasaki, Y., and Abe, H.,** The development of wearable-type artificial endocrine pancreas and its usefulness in glycemic control of human diabetes mellitus, *Biomed. Biochim. Acta,* 43, 561, 1984.

140. **Pitt, C. G., Gu, Z.-W., Hendren, R. W., Thompson, J., and Wani, M. C.,** Triggered drug delivery systems, *J. Controlled Release,* 2, 363, 1985.

141. **Heller, J.,** Chemically self-regulated drug delivery systems, *J. Controlled Release,* 8, 111, 1988.

142. **Brownlee, M. and Cerami, A.,** A glucose-controlled insulin delivery system: semisynthetic insulin bound lectin, *Science,* 206, 1190, 1979.

143. **Brownlee, M. and Cerami, A.,** Glycosylated insulin complexed to Con A, *Diabetes,* 32, 499, 1983.

144. **Jeong, S. Y., Kim, S. W., Eenink, M. J. D., and Feijen, J.,** Self-regulating insulin delivery systems. I. Synthesis and characterization of glycosylated insulin, *J. Controlled Release,* 1, 57, 1984.

145. **Sato, S., Jeong, S. Y., McRea, J. C., and Kim, S. W.,** Self-regulating insulin delivery systems. II. *In vivo* studies, *J. Controlled Release,* 1, 67, 1984.

146. **Jeong, S. Y., Kim, S. W., Holmberg, D. L., and McRea, J. C.,** Self-regulating insulin delivery systems. III. *In vivo* studies, *J. Controlled Release,* 2, 143, 1985.

147. **Seminoff, L. A., Olsen, G. B., and Kim, S. W.,** A self-regulating insulin delivery system. I. Characterization of a synthetic glycosylated derivative, *Int. J. Pharm.,* 54, 241, 1989.

148. **Seminoff, L. A., Gleeson, J. M., Zheng, J., Olsen, G. B., Holmberg, D. L., Mohammad, S. F., Wilson, D. E., and Kim, S. W.,** A self-regulating insulin delivery system. II. *In vivo* characteristics of a synthetic glycosylated insulin, *Int. J. Pharm.,* 54, 251, 1989.

149. **Kim, S. W., Pai, C. M., Makino, K., Seminoff, L. A., Holmberg, D. L., Gleeson, J. M., Wilson, D. E., and Mack, E. J.,** Self-regulated glycosylated insulin delivery, *J. Controlled Release,* 11, 193, 1990.

150. **Makino, K., Mack, E. J., Okano, T., and Kim, S. W.,** A microcapsule self-regulating delivery system for insulin, *J. Controlled Release,* 12, 235, 1990.

151. **Shiino, D., Waki, K., Murata, Y., Koyama, Y., Kataoka, K., Yokoyama, M., Okano, T., Sakurai, Y.,** Preparation of glucose sensitive release system and its characterization, *Polym. Prepr. Jpn.,* 39, 2604, 1990.

152. **Kitano, S., Kataoka, K., Koyama, Y., Yokoyama, M., Okano, T., and Sakurai, Y.,** Novel drug delivery system based on glucose sensitive polymer complex, *Polym. Prepr., Jpn.,* 39, 2607, 1990.

153. **Wulff, G.,** Main-chain chirality and optical activity in polymers consisting of C-C chains, *Angew. Chem., Int. Ed. Engl.,* 28, 21, 1989.
154. **Updike, S. J., Schuls, M., and Ekman, B.,** Implanting the glucose enzyme electrode-problem, progress, and alternative solutions, *Diabetes Care,* 5, 207, 1982.
155. **Ishihara, K., Kobayashi, M., and Shinohara, I.,** Control of insulin permeation through a polymer membrane with responsive function for glucose, *Makromol. Chem. Rapid Commun.,* 4, 327, 1983.
156. **Kaetsu, I., Morita, Y., Otori, A., and Naka, Y.,** Drug release systems controlled by biochip sensor and electrically responsive hydrogel, *Artif. Organs,* 14, 237, 1990.
157. **Chung, D.-J., Ito, Y., and Imanishi, Y.,** An insulin-releasing membrane system on the basis of redox reaction of glucose, *J. Controlled Release,* in press.
158. **Kost, J., Horbett, T. A., Ratner, B. D., and Singh, M.,** Glucose-sensitive membranes containing glucose oxidase: activity, swelling, and permeability studies, *J. Biomed. Mater. Res.,* 19, 1117, 1984.
159. **Ishihara, K., Kobayashi, M., Ishimura, N., and Shinohara, I.,** Glucose induced permeation control of insulin through a complex membrane consisting of immobilized glucose oxidase and a polyamine, *Polym. J.,* 16, 625, 1984.
160. **Albin, G., Horbett, T. A., and Ratner, B. D.,** Glucose sensitive membranes for controlled delivery of insulin: insulin transport studies, *J. Controlled Release,* 2, 153, 1985.
161. **Albin, G., Horbett, T. A., Miller, S. R., and Ricker, N. L.,** Theoretical and experimental studies of glucose sensitive membranes, *J. Controlled Release,* 6, 267, 1987.
162. **Ishihara, K. and Matsui, K.,** Glucose-responsive insulin release from a polymer capsule, *J. Polym. Sci., Polym. Lett. Ed.,* 24, 413, 1986.
163. **Ishihara, K.,** The use of bioresponsive membranes in the treatment of diabetes, in *Topics of Pharmaceutical Sciences 1989* Breimer, D. D., Crommelin, D. J. A., and Midha, K. K., Eds., FIP, 1989, 13.
164. **Siegel, R. A. and Firestone, B. A.,** Mechanochemical approaches to self-regulating insulin pump design, *J. Controlled Release,* 11, 181, 1990.
165. **Siegel, R. A. and Firestone, B. A.,** pH-Dependent equilibrium swelling properties of hydrophilic polyelectrolyte copolymer gels, *Macromolecules,* 21, 3254, 1988.
166. **Ito, Y., Casolaro, M., Kono, K., and Imanishi, Y.,** An insulin-releasing system that is responsive to glucose, *J. Controlled Release,* 10, 195, 1989.
167. **Iwata, H., Hata, Y., Amemiya, H., Matsuda, T., Takano, H., and Akustu, T.,** Self regulated insulin delivery system composed of pH sensitive membranes and glucose oxidase. VI. Preparation of membranes having various pH sensitive regions and development of immobilization method GOD, *Polym. Prepr., Jpn.,* 37, 726, 1988.
168. **Devlin, B. P. and Tirrell, D. A.,** Glucose-dependent disruption of phospholipid vesicle membranes, *Macromolecules,* 19, 2465, 1986.
169. **Fischel-Ghodsian, F., Brown, L., Mathiowitz, E., Brandenberg, D., and Langer, R.,** Enzymatically controlled drug delivery, *Proc. Natl. Acad. Sci. U.S.A.,* 85, 2403, 1988.
170. **Heller, J. and Trescony, P. V.,** Controlled drug release by polymer dissolution. II. An enzyme mediated delivery system, *J. Pharm. Sci.,* 68, 919, 1979.
171. **Pangburn, S. H., Trescony, P. V., and Heller, J.,** Partially deacetylated chitin: its use in self-regulating drug delivery systems, in *Chitin, Chitosan, and Related Enzymes,* Academic Press, New York, 1984, 3.
172. **Ito, Y., Chung, D. J., and Imanishi, Y.,** Glucose-sensitive insulin-releasing system based on redox reaction, *Artif. Organs,* 14, 234, 1990.
173. **Okahata, Y., Nakamura, G., and Noguchi, H.,** Functional capsule membranes. XIX. Concanavalin A-induced permeability control of capsule membranes corked with synthetic glycolipid bilayers or grafted with synthetic glycopolymers, *J. Chem. Soc., Perkin Trans.,* 2, 1317, 1987.

Modification of Proteins for Synthesizing Biofunctional Materials

4.1 GRAFTING OF ENZYMES WITH SYNTHETIC POLYMERS FOR USE IN ORGANIC SOLVENTS

Hideo Kise

TABLE OF CONTENTS

I. INTRODUCTION

Although enzymes have been known for many years as highly efficient and specific catalysts, only recently have they become accepted as useful catalysts for synthetic reactions.[1,2] In general, there are several disadvantages in using enzymes as synthetic catalysts, such as lack of stability and restricted reaction conditions. For example, it has been believed that enzymes work only in aqueous solutions. This has prevented many of organic chemists from employing enzymes as synthetic catalysts, because various organic solvents are generally used as reaction media.

One of the main breakthroughs in biotechology is biocatalysis in media mainly composed of organic solvents. It is not surprising that enzymes work in organic solvents, since, in living cells, many enzymes are located either within or at the surface of biological membranes. As will be described later, there are many advantages in employing enzymes in organic solvents. However, there are still many problems to be solved for the establishment of efficient enzyme reaction systems for organic synthesis. The major problem is how to stabilize and maintain catalytic activity of enzymes in organic solvents, since organic solvents often unfold peptide chains of enzymes. Although recent studies have revealed that enzymes can exhibit catalytic activities in selected organic solvents, the magnitude of activity is generally much lower than in aqueous solutions. It is known that activity and stability of enzymes are highly sensitive to the nature of the reaction media. Therefore, modification of enzymes, including complexation or immobilization, would change catalytic properties of enzymes. In this chapter chemical or physical modifications of enzymes as the most promising method for stabilization and activation of enzymes in organic solvents will be described.

II. GENERAL ASPECTS OF ENZYMATIC REACTIONS IN ORGANIC SOLVENTS

Information about general aspects and applications of enzymatic reactions in organic solvents with restricted amounts of water is provided in several recent reviews.[3-9] The main advantages of employing enzymes in organic solvents are as follows:

1.   Organic solvents dissolve many nonpolar substrates which are insoluble in aqueous solutions. Examples are the use of organic solvents for ester synthesis or transesterification by lipases and oxidation or reduction of steroids by dehydrogenases.
2.   Reduction of the amount of water shifts the equilibria of hydrolytic reactions in favor of synthesis of desired products. Examples are ester or peptide synthesis by lipases or proteases in nonaqueous media.

3. Change in the dissociation state of polar substrates in organic solvents often enhances enzymatic reactions. For example, equilibria of reversible hydrolytic reactions of amino acid esters or peptides shift toward synthesis by diminution of dissociation of amino acids in organic solvents.
4. There are possibilities for diminishing undesirable side reactions in organic solvents, such as nonenzymatic hydrolysis in enzymatic hydrolysis for optical resolution of racemic substrates.
5. In general, reaction products can be easily separated from organic solvents. If the solubilities of products are low in organic solvents, the products can be separated by simple filtration.
6. The use of organic solvents can eliminate the danger of bacterial contamination.

It should be noticed that the characteristic features of reactions in organic media stated before are mainly the results of specific interactions between solvents and substrate/products. One must keep in mind that organic solvents can also influence enzyme activity and specificity in a number of ways. Organic solvents may be competitive inhibitors of enzymes through specific interactions with enzymes. They may change dissociation states of polar groups on the surface of enzymes. Furthermore, since hydrophobic interactions play the dominant role in maintaining the catalytically active conformation of enzymes, organic solvents may distort or even destroy higher structures of enzymes by the unfolding of peptide chains. These would lead to changes in reaction kinetics and substrate specificities. Stereospecificities may also change as a consequence of the distortion or partial unfolding of the enzymes.

Also important are the effects of organic solvents on interactions between enzymes and substrates. For example, the primary driving force in the binding of substrates to serine proteases, such as α-chymotrypsin or subtilisins, is recognized as hydrophobic interaction between the substrates and the binding sites of the enzymes. There is ample evidence to indicate that hydrophobic interactions are significantly reduced in organic solvents, which are manifested with the decrease in relative reactivity of hydrophobic amino acids or their derivatives in transesterification or peptide formation.[9,10]

Water content is another important factor which can affect both enzyme activity and reaction equilibrium.[11-13] Typical examples are ester or peptide syntheses by serine proteases in organic solvents, which are reverse processes of hydrolytic reactions of these compounds (Figure 1). Transesterification is also catalyzed by serine or thiol proteases with acyl enzyme as an intermediate (Figure 1). Under strictly anhydrous conditions, enzymes lose catalytic activity, and, in general, activity increases with water content. However, the equilibrium yields of esters or peptides decrease by increasing water content. Therefore, the yield of the product can be either kinetic or thermodynamic

Ac–Tyr–OH + E $\rightleftharpoons$ Ac–Tyr–OH·E

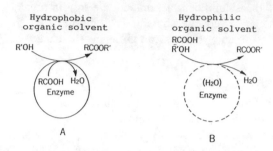

Ac–Tyr–OMe + E $\rightleftharpoons$ Ac–Tyr–OMe·E

FIGURE 1. Reaction scheme for amino acid esterification and transesterification by serine proteases.

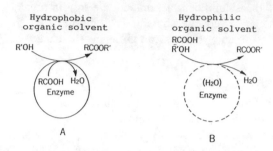

FIGURE 2. Schematic representation of (A) two-phase system and (B) cosolvent system.

depending on water content; and, for the practical applications to synthesis, optimum water content should be determined from a technical point of view.[14,15]

Several reaction systems have been employed for water-restricted enzymatic reactions: water/water-miscible (hydrophilic) organic solvent (cosolvent), water/water-immiscible (hydrophobic) organic solvent (two-phase), and nearly anhydrous organic solvent (dry) systems. Figure 2 shows a schematic representation of (A) the two-phase system and (B) the cosolvent system for esterification reactions. The characteristics of cosolvent systems may be summarized as follows:

1. Solvents with wide range of polarity are used, which allow dissolution of a variety of substrates ranging from hydrocarbons to amino acid derivatives.
2. There is no interface between water and organic solvent, and this eliminates diffusion limitations of substrates to the binding sites of enzymes.
3. Substrates and products are distributed homogeneously in solutions; this prevents inhibition by substrates or products and allows easy reaction control.
4. The analyses of reaction kinetics and specificity of enzymes are easy by virtue of no diffusion limitation and localization of substrates.
5. Nontoxic solvents can be used, such as ethanol.
6. Denaturation of enzymes may occur due to direct contact with organic solvents.

7.   Higher contents of water are required than for two-phase systems, since polar solvents are capable of solubilizing large amounts of water and tend to strip away the layers of essential water from the enzymes.

On the other hand, two-phase systems have the following characteristics.

1.   For reversible reactions, reaction rate and equilibrium are functions of nature and amount of organic solvents. By using solvents with high extractive ability for products, reaction equilibria can be shifted far to the synthesis of desired products.
2.   When substrates or products are highly soluble in water, inhibition can occur due to excess concentration of these substances around the reaction sites of enzymes.
3.   Diffusion of substrates across the water/organic interface can be rate determining.
4.   Kinetic studies and analyses of specificities are generally complicated because of inhomogeneous distribution of substrates and products in solutions.
5.   Due to the diminution of direct contact with an organic solvent, enzymes are, in general, more stable than in cosolvent systems. In some cases, however, deactivation by contact with water/organic interface occurs.
6.   Smaller amounts of water are required as compared to cosolvent systems.

The choice of solvent is crucial for successful enzyme reactions, but until now it has generally been arbitrary because of the lack of general guidelines. Most attempts to find correlations between solvent parameters, such as dielectric constant, and enzyme activity have been unsuccessful. It was found that in some cases enzyme activities are correlated with the logarithm of the partition coefficient of a solvent between octanol and water (log P). For example, Laane et al.[16] reported that catalytic activities of lipases for transesterification are high in organic solvents having high log P values. Values of log P can be easily calculated from hydrophobic fragmentary constants. However, no clear correlations between log P and catalytic activity of serine proteases for esterification and peptide formation reactions have been obtained. As stated before, since the apparent activities of enzymes are governed by many factors which originate from interactions between different reaction components, it seems difficult to correlate enzyme activity with a single solvent parameter.

In organic solvents, enzymes have been utilized in different states as summarized below:

1.   Enzymes in aqueous solutions
2.   Dispersions of hydrated enzymes

3. Dispersions of complexes of enzymes with organic or inorganic materials
4. Immobilized enzymes
5. Chemically modified enzymes which are soluble in organic solvents

In this chapter, attention will be focused on enzymes modified either chemically or physically, including enzyme complexes and immobilized enzymes in organic solvents. These techniques are highly promising for development of novel and efficient enzyme reaction systems in terms of enhancement of stability and activity of enzymes under these unusual reaction conditions. Attention will also be paid to the change in specificities of enzymes, because enzyme specificity should be sensitive to conformational changes of enzymes and changes in interactions between enzymes and substrates.

III. ENZYME MODIFICATION BY POLY(ETHYLENE GLYCOL)

Enzymes are insoluble in anhydrous organic solvents. However, it was found that when poly(ethylene glycol) (PEG) was attached to enzymes by covalent bondings, they became soluble in many organic solvents including nonpolar and highly hydrophobic solvents such as hydrocarbons and chlorinated hydrocarbons. PEG is an amphiphilic polymer, which is soluble both in water and organic solvents, and this may be responsible for the solubilization of PEG-modified enzymes in both aqueous and nonaqueous media. The experimental results on catalysis of PEG-modified enzymes in organic solvents were first reported by Inada and co-workers for lipase,[17] α-chymotrypsin,[18] catalase,[19] and peroxidase.[20] Later, PEG-modified thermolysin,[21] subtilisin,[22,23] papain,[24,25] and trypsin[26] were synthesized and used as synthetic or hydrolytic catalysts of esters, amides, or peptides. Quite interestingly, the solubilized enzymes exhibited similar or even higher catalytic activities as compared to those of native enzymes in aqueous solutions. This indicates that modified enzymes preserve their native conformations in organic solvents, and it may be assumed that solvent molecules do not penetrate into enzymes. Thus, it seems that PEG protects enzymes from direct contact with organic solvents without retarding access of substrates to the active sites of the enzymes.

Several methods have been developed for attachment of PEG to enzymes. The selection of modification method is crucial for the activity of the modified enzymes. Modifications should preserve amino acid residues which are essential for catalytic activity and also should not destroy enzyme structures significantly. Therefore, in general, modifications are carried out in aqueous solutions under mild conditions (at ambient temperatures and neutral pH). Degree of modification and molecular weight of PEG are also important factors which affect catalytic activity of modified enzymes.[27]

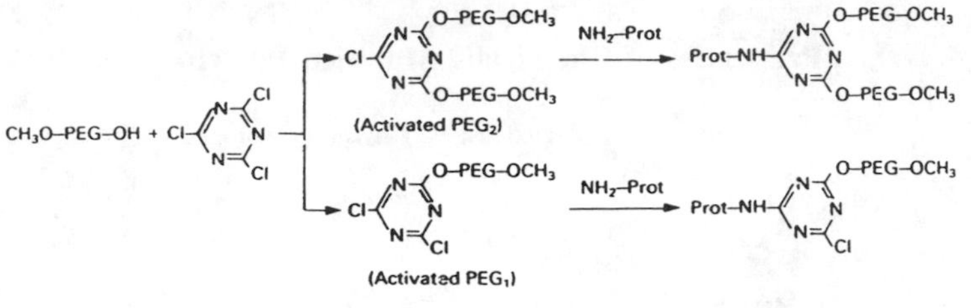

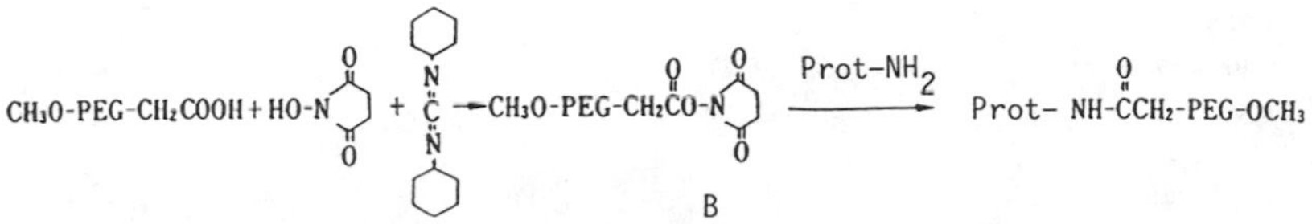

FIGURE 3. Modification of enzymes by poly(ethylene glycol) (PEG).

Amino groups of lysine residues in enzymes were chosen for covalent bonding of PEG via triazine rings (Figure 3A). PEG is reacted with cyanuric chloride (2,4,6-trichloro-*s*-triazine) to afford activated PEGs, 2-*O*-methoxy-poly(ethylene glycol)-4,6-dichloro-*s*-triazine (PEG$_1$), or 2,4-bis[*O*-methoxy-poly(ethylene glycol)]-6-chloro-*s*-triazine (PEG$_2$). Treatment of an enzyme with an activated PEG in aqueous solution gives a modified enzyme which possess PEG chains covalently attached to lysine residues via triazine rings. By using PEG$_2$, two PEG chains are introduced per triazine ring. Unreacted PEGs are removed by ultrafiltration or chromatography. The extent of modification is controlled by changing the amounts of activated PEG and is determined by quantitative analysis of residual amino groups in the enzyme by the reaction with trinitrobenzenesulfonic acid. Properties and catalysis of PEG-modified enzymes are listed in Tables 1 and 2, respectively.[27]

Alternatively, PEG can be attached to enzymes by amide bondings. PEG with a terminal carboxyl group is obtained by the reaction with succinic anhydride. The product is converted to an activated ester by reaction with *N*-hydroxysuccinimide in the presence of a carbodiimide as a dehydrating agent. The treatment of an enzyme with the activated ester affords PEG-modified enzymes (Figure 3B).

A. LIPASES

Lipases have been known as catalysts for hydrolysis of glycerides which are insoluble in water, and reactions are usually carried out in emulsions with aqueous solutions containing lipases. Since ester hydrolysis is reversible and lipases have relatively wide substrate specificities, they can be good catalysts for syntheses of glycerides and other esters from a variety of carboxylic acids

TABLE 1
Properties of Chemically Modified Proteins

	Catalase	Peroxidase	Chymotrypsinogen	Lipase
EC	1.11.1.6	1.11.1.7	(3.4.21.1)	3.1.1.3
Source	Bovine liver	Horseradish	Bovine pancreas	*Pseudomonas fluorescens*
MW	248,000	40,000	26,000	33,000
Subunit	4	1	1	1
Heme	+	+	−	−
Total number of amino groups in the molecule	112	6	15	7
Degree of modification (%)	42	60	80	50
Solvents	Benzene, toluene, chloroform, 1,1,1-Trichloroethane, trichloroethylene, perchloroethylene			
Reaction in organic solvents	Decomposition of H_2O_2	Oxidation with H_2O_2	Formation of acid-amide linkage	Ester synthesis[a] Ester exchange[b]
Activity (μmol/min/mg of protein)	70,000 in benzene at 37°C	30 in benzene at 37°C	0.8 in 1,1,1-trichloroethane at 37°C	

[a]   28.4 in 1,1,1-trichloroethane.
[b]   2.6 at 65°C.

From Matsushima, A. and Inada, Y., *Yukigoseikagaku Kyokaishi (J. Synth. Org. Chem. Jpn.)*, 46, 232, 1988. With permission.

and alcohols. The yields of esters will be thermodynamic (equilibrium) control, and therefore amounts of water should be restricted in order to obtain esters in high yields. Reactions with free lipases have been utilized for synthesis of esters in aqueous-organic mixed solutions[28,29] and transesterification in almost dry organic solvents.[30] Lipase-catalyzed synthetic and hydrolytic reactions have been used for optical resolution of carboxylic acids and alcohols.[7,31]

Inada and co-workers[17] reported on the catalysis of PEG-modified lipases in organic solvents; for example, *Pseudomonas fluorescens* lipase has seven amino groups in a molecule, about 50% of which were reacted with activated PEG$_2$. The PEG-modified lipase is soluble in many organic solvents and catalyzes hydrolysis, esterification, transesterification, and amide synthesis (Table 2). The reactions proceed in homogeneous solutions under mild conditions. As for esterification reactions, acids with larger carbon numbers are better substrates, but acids with substituents on carbons adjacent to carbonyl groups are inactive and inhibit the reactions.[32] Among the solvents used, 1,1,1-trichloroethane was the best in terms of reaction velocity.[33] Esters from terpene alcohols, such as citronellol, geraniol, farnesol, and phytol, were synthesized by the reactions with aliphatic fatty acids by the catalysis of PEG-modified lipase from *P. fragi* (Table 3).[34]

One of the advantages of using lipases is the low possibility of undesirable side reactions. For example, transesterification of retinyl acetate with palmitic acid or oleic acid in benzene afforded retinyl palmitate or oleate of low peroxide values as compared to those obtained by chemical methods.[35] It is known that C-C double bonds in fatty acids and their analogues tend to be oxidized easily. These results suggest that biocatalysis is preferable for conversion of this kind of unstable substances which include many of biologically important compounds.

PEG-modified lipases are also soluble in glycerides and other esters, and this enables reactions in the absence of conventional organic solvents.[27] For example, ester exchange reaction between amyl laurate and lauryl alcohol gave lauryl laurate and amyl alcohol by the catalysis of PEG lipase. Also ester exchange reaction between trilaurin and triolein at 58°C in the absence of solvents afforded dilauroylmonooleylglycerol and monolauroyldioleyl glycerol. As a result, the melting point was lowered from 33—36°C to 11—13°C, and this method may be applicable to the modification of fatty oils.

Lipases can be incorporated or adsorbed to PEG-modified magnetic materials, and this endows the enzymes with magnetic properties. For example, lipase from *Pseudomonas fragi* 22.39B was adsorbed to a mixture of a magnetite (Fe_3O_4) and a copolymer of PEG and maleic acid (Figure 4). The enzyme preparations form stable suspensions in benzene and catalyze esterification reactions.[36] Furthermore, the hybrid enzymes can be easily separated from reaction solutions by applying a magnetic field of 6000 Oe. The recovered hybrid enzymes can be reused without loss of activity.

B. α-CHYMOTRYPSIN, SUBTILISIN, AND TRYPSIN

α-Chymotrypsin (CT) from bovine pancreas is a serine protease which catalyzes hydrolysis of peptides mainly at the carboxyl side of aromatic amino acid residues. *N*-Protected aromatic amino acid esters are also good substrates of CT for hydrolysis. Conversely, CT catalyzes transfer of acyl groups from suitable donors to a wide array of acceptors. Depending on the type of acyl donors and acceptors, CT-catalyzed reactions include esterification, transesterification, amidation, and aminolysis. Peptides can be synthesized by using the latter two reactions. The synthetic reactions may be carried out either in aqueous solutions, monophasic media composed of hydrophilic organic solvents and minimum amounts of water, or in aqueous-organic biphasic systems using nonpolar organic solvents.

PEG was grafted on CT by using activated PEG_2 by a similar method to that for lipases.[18] Since the amino group in the terminal lysine residue is essential for catalytic activity, the activated PEG was coupled to the precursor chymotrypsinogen and then it was activated by trypsin for form PEG-modified

TABLE 2
Enzymic Reactions in Organic Solvents by Polyethylene Glycol Modified Enzymes

Enzymic reaction	Activity (μmol min^{-1} [mg protein]$^{-1}$)	Solvent	Temperature (°C)
Modified Lipase			
Ester hydrolysis: $R_1COOR_2 + H_2O \rightarrow R_1COOH + R_2OH$	1,200	Benzene	37
Indoxyl acetate + $H_2O \rightarrow$ 3 hydroxyindole + acetic acid		Benzene	37
Ester synthesis: $R_1COOH + R_2OH \rightarrow R_1COOR_2 + H_2O$			
1-Monolaurin + lauric acid $\rightarrow$ trilaurin + H_2O	2.8	Benzene	25
Pentanoic acid + amylalcohol $\rightarrow$ amyl pentanate + H_2O	7.9	Benzene	25
Lauric acid + lauryl alcohol $\rightarrow$ lauryl laurate + H_2O	28.4	TCE[a]	25
Stearic acid + cholesterol $\rightarrow$ cholestrol stearate + H_2O	0.3	Benzene	37
10-Hydroxydecanoic acid $\rightarrow$ polymer + H_2O		Benzene	37
Lauric acid + methyl alcohol $\rightarrow$ methyl laurate + H_2O	13.2	Benzene	−3
Ester exchange:			
$R_1COOR_2 + R_3OH \rightarrow R_1COOR_3 + R_2OH$			
Trilaurin + ethanol $\rightarrow$ ethyl laurate + dilaurin	0.7	Benzene	37
Amyl laurate + lauryl alcohol $\rightarrow$ lauryl laurate + amyl alcohol	2.6	Substrates	65
$R_1COOR_2 + R_3COOH \rightarrow R_3COOR_2 + R_1COOH$			
Methyl butyrate + lauric acid $\rightarrow$ methyl laurate + butyric acid	1.1	Benzene	37
Amyl laurate + linoleic acid $\rightarrow$ amyl linoleate + lauric acid	0.4	Substrates	58
$R_1COOR_2 + R_3COOR_4 \rightarrow R_1COOR_4 + R_3COOR_2$			
Amyl laurate + butyl caprylate $\rightarrow$ amyl caprylate + butyl laurate	1.0	Substrates	58
Trilaurin + triolein $\rightarrow$ dilauroyl-monooleoyl-glycerol + monolauroyl-dioleoyl-glycerol		Substrates	58

Acid-amide bond formation:
$R_1COOR_2 + R_3NH_2 \rightarrow R_1CONHR_3 + R_2OH$

Lauryl stearate + laurylamine → N-laurylstearamide + lauryl alcohol	0.07	Benzene	37

Modified Chymotrypsin

Acidamide bond formation:
$R_1COOR_2 + R_3NH_2 \rightarrow R_1CONHR_3 + R_2OH$

N-Benzoyl-L-tyrosine ethyl ester + butylamine → N-benzoyltyrosine butylamide + ethyl alcohol	0.5	Benzene	37
	0.6	TCE[a]	37
N-Benzoyl-L-tyrosine ethyl ester + phenylalanine ethyl ester → N-benzoyltyrosine-oligophenylalanine ethyl ester		Benzene	37

Modified Catalase

Decomposition of hydrogen peroxide: $H_2O_2 \rightarrow H_2O + O_2$	70,000	Benzene	30
	40,000	TCE[a]	30

Modified Peroxidase

Oxidation with hydrogen peroxide: $AH^{[b]} + H_2O_2 \rightarrow A + H_2O$	30%[c]	Benzene	37

[a] TCE, 1,1,1-trichloroethane.
[b] AH, o-phenylene diamine.
[c] 30% of the activity of native enzyme in aqueous solution.

From Matsushima, A. and Inada, Y., *Yukigoseikagaku Kyokaishi (J. Synth. Org. Chem. Jpn.)*, 46, 232, 1988. With permission.

TABLE 3
Synthesis of Ester from Terpene Alcohol and Carboxylic Acid by PEG Lipase in Benzene at 25°C

Terpene alcohol	Acetic acid	Propionic acid	n-Butyric acid	Valeric acid
		Yield (%)		
Citronellol	18	52	88	74
Geraniol	19	81	94	83
Farnesol	19	78	90	81
Phytol	29	86	95	92

Note: Yields were measured after 8-h reaction. Concentrations of terpene alcohol, carboxylic acid, and PEG-lipase were 0.1 M, 0.1 M, and 0.33 mg/ml, respectively.

From Nishio, T., Takahashi, K., Yoshimoto, T., et al., *Biotechnol. Lett.*, 9, 187, 1987. With permission.

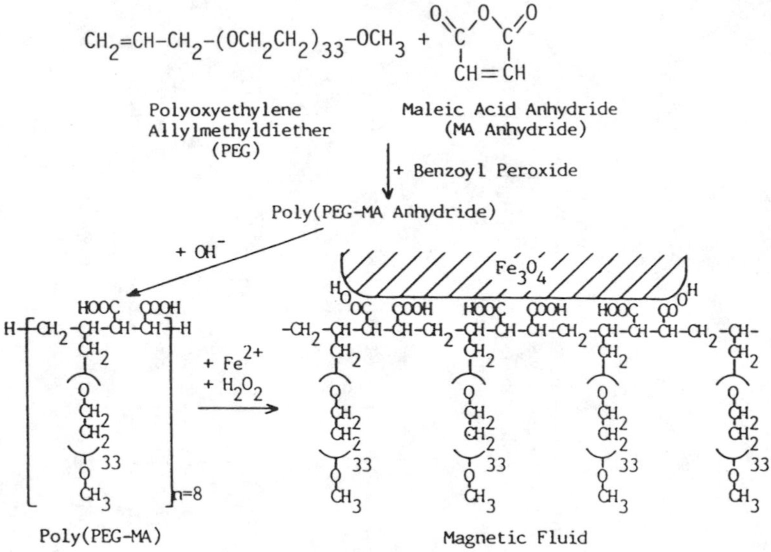

FIGURE 4.   A magnetic fluid prepared by oxidation of ferrous ions (Fe^{2+}) in the presence of an alternating copolymer of PEG and maleic acid. From Mihama, T. et al., *J. Biotechnol.*, 7, 141, 1988. With permission.

CT. Of the 15 amino groups in CT 12 were coupled to PEG via triazine rings. The modified CT is soluble in many of organic solvents including benzene and 1,1,1-trichloroethane and catalyzes amide bond formation reactions (Equations 1 and 2). The time course of the aminolysis reaction between *N*-benzoyl-L-tyrosine ethyl ester and butyl amine is shown in Figure 5.[18]

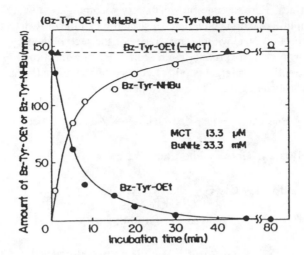

FIGURE 5. Time course of reaction of Bz-Tyr-OEt with butylamine by PEG-modified α-chymotrypsin. From Matsushima, A. et al., *FEBS Lett.*, 178, 275, 1984. With permission.

$$\text{Bz-Tyr-OEt} + \text{BuNH}_2 \longrightarrow \text{Bz-Tyr-NHBu} \tag{1}$$

$$\text{Bz-Tyr-OEt} + \text{Phe-NH}_2 \longrightarrow \text{Bz-Tyr-Phe-NH}_2 \tag{2}$$

The rate of peptide formation (Equation 2) depends on the nature of solvents, and the rate decreases in the order, toluene > benzene > 1,1,1-trichloroethane > chloroform, ethyl acetate.[22] A rate study revealed that the reaction follows Michaelis-Menten type kinetics. The rate constant of synthesis (k_{cat}) increases with increase in water content, while the Michaelis constant (K_m) does not change significantly by changing solvent and water content. Interestingly, it was found that the reaction of Bz-Lys-OMe with Phe-NH$_2$ was faster than Bz-Tyr-OEt. This means that substrate specificity is reversed as compared to that of native CT for peptide hydrolysis in aqueous solutions. The results imply that hydrophobic interaction between the enzyme and substrates is minimized in organic solvents. As will be described later, different substrate specificities were also observed for native CT in organic solvents and in aqueous solutions.

PEG was coupled to CT by amide bonds after activation of the carboxyl group of MeO-PEG-OCH$_2$COOH with DCC/N-hydroxysuccinimide.[37] The substitution degree of amino groups of lysine residues was 75%, and the remaining esterase and peptidase activities of the modified CT were 50 and 30%, respectively. The modified CT catalyzes peptide synthesis from aromatic amino acid derivatives and leucinamide in mixed solvents of *t*-amyl alcohol and benzene or trichloroethylene (50/50) in the presence of small amounts of water (e.g., 0.5%). One of the advantages of employing PEG-modified CT for peptide synthesis is diminution of hydrolysis of ester substrates and produced peptides. For example, no secondary recombination or hydrolysis occurs in coupling reaction (Equation 3), showing that the only detected product is Z-Phe-Phe-Leu-NH$_2$ without any traces of the recombined peptide Z-Phe-Leu-NH$_2$ or hydrolysis product Z-Phe-Phe-OH. The reaction was applied to fragment coupling of a larger peptide Leu-enkephalinamide (Equation 4) without any hydrolysis of the CT-sensitive peptide bonds.

$$Z\text{-Phe-Cam} + \text{Phe-Leu-NH}_2 \longrightarrow Z\text{-Phe-Phe-Leu-NH}_2 \quad (3)$$

$$Boc\text{-Tyr-Gly-Gly-Phe-Cam} + \text{Leu-NH}_2 \longrightarrow \quad (4)$$

$$Boc\text{-Tyr-Gly-Gly-Phe-Leu-NH}_2$$

$$Cam = \text{carboxamidomethyl}$$

Subtilisin (ST) from *Bacillus subtilis* was modified with PEG (MW = 5000), and its catalytic activity and specificity were investigated in benzene.[22,23] The catalytic activity of modified ST for peptide synthesis from *N*-protected amino acid esters and phenylalaninamide (Equation 5) was lower than corresponding CT preparations. As for substrate specificity, Ser and Asp derivatives gave dipeptides in higher yields than hydrophobic amino acid derivatives. This suggests again that significant change in substrate specificity occurs when the enzyme is in organic solvents.[23]

Apparent kinetic parameters, k_{cat} and K_m, were determined for PEG-ST catalyzed transesterification in benzene (Equation 6). The result shows that serine ester is more reactive than phenyl alanine ester, and this is another example which suggests the increase in enzyme activity toward hydrophilic substrates compared to hydrophobic ones in organic solvents.[23]

$$P\text{-AA-OR} + \text{Phe-NH}_2 \longrightarrow P\text{-AA-Phe-NH}_2$$

$$P = Bz, Z; R = Me, Et, Bz;$$

$$AA = Tyr, Phe, Ala, Ser, Thr, Lys, Arg, Asp \quad (5)$$

$$Z\text{-AA-OMe} + \text{EtOH} \longrightarrow Z\text{-AA-OEt}$$

$$AA = Phe, Ser \quad (6)$$

Trypsin was modified with PEG and used to catalyze digestion of bovine serum albumin or insulin in water-immiscible organic solvents.[26] Since hydrophilic protein substrates are insoluble in organic media, the reaction mixtures form multiphasic systems. It was found that the presence of an additional nucleophile is a prerequisite for catalysis of the enzyme, and therefore both water and nucleophile concentrations have some influence on trypsin activity. Phe-NH$_2$ was the most potent nucleophile for proteolysis catalyzed by PEG-modified trypsin in organic media containing 1 to 2% water (v/v). The results strongly suggest that modified trypsin catalyzes peptide bond hydrolysis at the interface of protein substrate and organic solvent. The analysis of the hydrolysis products of bovine insulin showed that enzyme-catalyzed proteolysis occurs in organic solvents with a concomitant and significant transpeptidation reaction.

C. PAPAIN

Papain is a thiol protease which has both peptidase and esterase activities. The modification of papain by activated PEG$_2$ was carried out by the method described before, and the product, in which 37% of the total amino group were coupled with PEG$_2$, was soluble in organic solvents such as benzene and chlorinated hydrocarbons.[24] The modified papain retained 72% of the enzymatic activity of native papain for peptide hydrolysis. As for amide bond formation by aminolysis reaction of an alanine derivative with fatty amines in benzene (Equation 7), the reactivity of amines increased with the length of carbon chains.

The aminolysis reaction by PEG-modified papain in organic solvents was applied to the synthesis of a derivative of a salty peptide, ornithyl-β-alanine (Equation 8).[25] The principal advantage of this reaction is that the product, insoluble in organic solvents, precipitates out of the solution making it easily separated by simple filtration. The precipitation of the product also enhances the reaction rate and shifts the equilibrium in favor of synthesis.

$$\text{Bz-Ala-OMe} + \text{R-NH}_2 \longrightarrow \text{Bz-Ala-NH-R} \tag{7}$$

$$\text{Z-Orn(Z)-OEt} + \text{β-Ala-OBz} \longrightarrow \text{Z-Orn(Z)-β-Ala-OBz} \tag{8}$$

D. THERMOLYSIN

Thermolysin has been utilized as a synthetic catalyst of small peptides by condensation reactions in organic solvents. The reactions are carried out in multiphasic systems consisting of powdery or immobilized enzymes and substrate solutions. Thermolysin-catalyzed reactions are of practical importance for the synthesis of biologically active peptides such as a precursor of a sweetener (aspartame) (Equation 9).

PEG-modified thermolysin catalyzes peptide synthesis in benzene from a variety of amino acid derivatives (Equation 10).[21] Interestingly, substrate specificity is different from that for peptide synthesis in water. The reactivities of hydrophilic acyl donors, Z-Ser or Z-Asp, are comparable with hydrophobic Z-Phe. Kinetic studies suggest that the result may be the consequence of stronger binding of hydrophilic substrates at the S_1 subsite of PEG-thermolysin than hydrophobic ones, which is in contrast to what has been observed in aqueous systems.

$$Z\text{-Asp-OH} + Phe\text{-OMe} \longrightarrow Z\text{-Asp-Phe-OMe} \tag{9}$$

$$P\text{-AA}_1 + AA_2\text{-NH}_2 \longrightarrow P\text{-AA}_1\text{-AA}_2\text{-NH}_2$$

$$P = Z, Boc; AA_1 = Phe, Leu, Met, Ala, Ser, Asp;$$

$$AA_2 = Phe, Leu, Val, Met \tag{10}$$

E. CATALASE AND PEROXIDASE

Catalase from bovine liver has a molecular weight of 248,000 and consists of four subunits with a hememolecule in each subunit. By treatment with activated PEG_2, up to 55% of amino groups in catalase were linked to PEG.[19] The catalytic activity of PEG-catalase for decomposition of hydrogen peroxide in benzene increased with degree of amino group modification up to 42%, but with higher degrees of modification, activity dropped sharply. The maximum activity was approximately 5 times greater than the activity of the modified catalase in aqueous solutions and was 1.6 times higher than the activity of nonmodified catalase in aqueous solutions. The activity is strongly dependent on the nature of organic solvents and decreases in the order, benzene > chloroform > toluene > dioxane, acetone, ethanol = 0.

Horseradish peroxidase (HRPO) is a heme protein which has a molecular weight of 40,000 and six amino groups in a molecule. A modified HRPO (modification degree 60%) is soluble in benzene and catalyzes the oxidation of *o*-phenylenediamine with hydrogen peroxide, though the catalytic activity is lower (about 21%) than that of native peroxidase in aqueous solutions.

IV. COMPLEXATION OF ENZYMES WITH POLYMERS

With few exceptions, enzymes are insoluble and inactive in dry organic solvents. There is a general agreement that enzymes require hydration for catalytic activity. In hydrophobic (water-immiscible) organic solvents, less than 1% of water is enough to fully hydrate enzymes at concentrations used in ordinary reactions. Excess water forms a separate aqueous phase which contains most of the enzymes. If an organic solvent is hydrophilic and miscible in water, the solubility of the enzyme increases with increase in water concentration. The apparent activity of an enzyme in these monophasic reaction media is strongly dependent on water content.

It may be assumed that, in these reaction mixtures, part of the enzyme plays the role of support material for immobilization of hydrated active en-

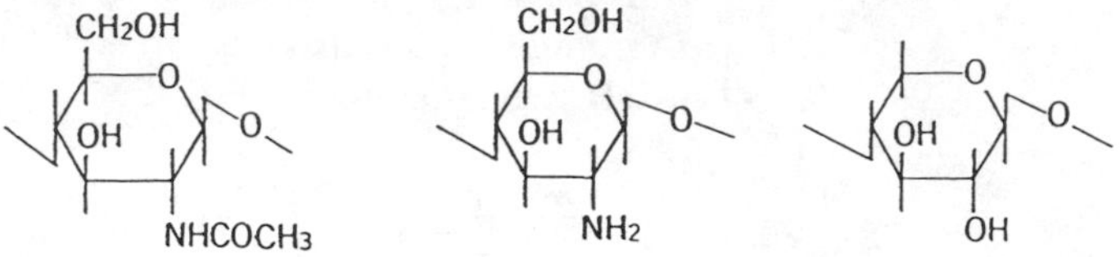

FIGURE 6.   Structure of (1) chitin, (2) chitosan, and (3) xylan.

zyme. Therefore, it may be reasonable to consider that apparent enzyme activity would increase by immobilization or complex formation with solid materials. Actually, it was found that complexation of CT with inorganic salts, such as phosphate salts, increases its catalytic activity for esterification of aromatic amino acids in organic solvents.[14,38] It was also reported that lipophilization of subtilisin or CT from phosphate buffer solutions or solutions of *N*-acetyl-L-phenylalanine afforded catalyst preparations which work even in DMF, one of the poorest solvents for enzymatic reactions.[39] This technique was utilized for preparation of dipeptides containing D-amino acid residues and sugar esters by subtilisin (Equations 11 and 12).[40,41] In organic solvents, there should be strong interactions between enzymes and hydrophilic materials, and these enzyme preparations may be regarded as active enzyme complexes.

$$\text{Ac-D-Phe-OCH}_2\text{CH}_2\text{Cl} + \text{L-Leu-NH}_2 \longrightarrow \text{Ac-D-Phe-L-Leu-NH}_2 \quad (11)$$

$$\text{Sugars} + \text{RCOOCH}_2\text{CCl}_3 \longrightarrow \text{Sugar monoesters} \quad (12)$$

Based on this consideration, the effects of addition of a variety of materials on enzyme catalysis for synthetic reactions have been studied. Of these, several polysaccharides, such as chitin, chitosan, or xylan (Figure 6), are of special interest because marked increases in enzyme activity were observed by addition of these compounds (Table 4).[9,42]

Figure 7 shows the effects of the amount of chitin or chitosan on esterification of *N*-acetyl-L-tryptophan (Ac-Trp-OH) in ethanol by the catalysis of CT.[8] The reaction rate dramatically increases by adding chitin or chitosan, but excess amounts of these materials, especially chitosan, retard the reaction. This can be attributed to the adsorption of both CT and the substrate, Ac-Trp-OH, to the additives. The rate profiles are very similar to those of bimolecular reactions in aqueous micellar systems, where substrates are adsorbed to micelles in which reactions take place.[43]

This interpretation is supported by the fact that (as shown in Figure 7) "recovery", *R*, which is defined as molar ratio of substrate plus product in solution to initial substrate, decreases with increase of the added materials, especially for chitosan. The results indicate that the substrate is more strongly bound to chitosan than the product, and this would increase the reaction rate

TABLE 4
Effects of Organic Materials on
Ac-Trp-OH Esterification[a]

Material	Ac-Trp-OEt yield (%)	
	CT	STB
—	20	0
Cellulose	20	3
Cellulose acetate	24	2
AE cellulose	18	4
DEAE cellulose	32	18
TEAE cellulose	65	31
CM cellulose	85	9
Xylan	92	14
Pullulan	25	3
Chitosan	83	34
Sodium arginate	56	15
Sephadex G25	82	6
Sephadex LH20	22	2
DEAE Sephadex	0	1
QAE Sephadex	3	3
CM Sephadex	10	2

[a]   Ac-Trp-OH 10 mM, enzyme 10 mg, organic
    material 100 mg, ethanol 20 ml, water 2.4%,
    30°C, 24 h.

From Nishio, T., Takahashi, K., Yoshimoto, T.,
et al., *Biotechnol. Lett.*, 9, 187, 1987. With permission.

and shift the equilibrium toward synthesis. CT-chitin and CT-chitosan are
stable in ethanol. Loss of activity was less than 15% after 3 weeks at 4°C.
However, thermal stability was not improved significantly by complexation.

In contrast to CT and subtilisin Carlsberg, subtilisin BPN' (STB) does
not catalyze esterification of amino acids in ethanol. In the presence of chitin
or chitosan, however, STB is activated and catalyzes the reaction.[44] The
activation of STB is most likely the consequence of conformation change
through direct interaction with the polysaccharides, since increases in water
content and amount of the enzyme are ineffective for the reaction by free
STB.

Kinetic studies revealed that reaction enhancement by complexation comes
primarily from the increase in k_{cat} rather than K_m; for example, k_{cat} values for
free CT and xylan-CT were 0.006 and 0.108 s^{-1}, while K_m values were 10.4
and 17.9 mM, respectively, for Ac-Tyr-OH esterification in ethanol.[45] In
brief, complexation described before enhances apparent catalytic activity of
enzymes in organic solvents by increasing the amount of accessible enzyme
and by keeping its active conformation. Furthermore, if the complex prepa-

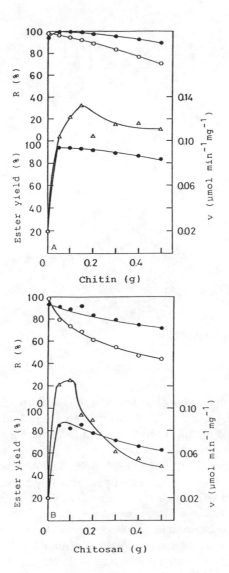

FIGURE 7. Effect of the amount of chitin (upper) or chitosan (lower) on Ac-Trp-OH esterification in ethanol by the catalysis of α-chymotrypsin. △, Reaction rate; ●, yield at 24 h; ○ and ●, R (recovery) at 0.5 and 24 h, respectively.

rations adsorb the substrates and release the products as observed for CT-chitosan, the reaction rate and equilibrium yield of the products would increase. Thus, improvement of activity and change in specificity of enzymes would be realized by complexation with suitable compounds, or, as will be described later, by immobilization to functional materials.

Recently, a new type of complex of lipase with lipids was reported as an efficient catalyst for ester synthesis.[46] By mixing aqueous solutions of a lipase and a lipid (Figure 8), lipid-coated lipase was obtained that is insoluble in

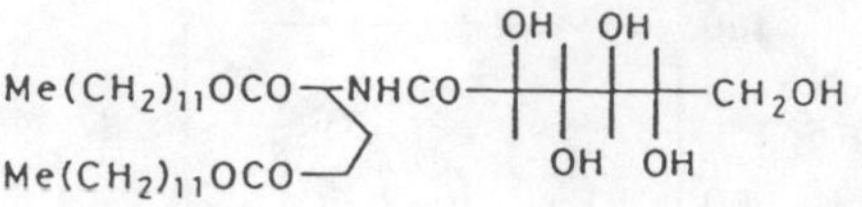

$$Me(CH_2)_{11}OCO \underset{Me(CH_2)_{11}OCO}{\overset{}{\diagdown}} NHCO \underset{OH\ OH}{\overset{OH\ OH}{\mid \mid \mid \mid}} CH_2OH$$

(1)

$$Me(CH_2)_{15} \underset{Me(CH_2)_{15}}{\overset{}{\diagdown}} \overset{+}{N} \underset{Me}{\overset{Me}{\diagup}} \quad Br^-$$

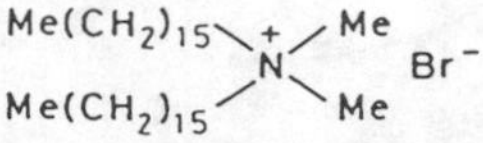

(2)

FIGURE 8.   Structure of lipids for lipase coating.

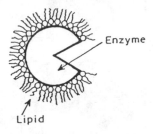

Lipid

FIGURE 9.   Schematic representation of lipid-coated lipase.

water but soluble in many organic solvents. It is assumed that lipid molecules attach to lipase through dipolar or electrostatic interactions between hydrophilic or ionic head groups of the lipids and the surface of lipase (Figure 9). Protein contents in the complexes are 10 to 29%, which means that an enzyme molecule is covered by 150 to 450 lipid molecules. Anionic and zwitterionic lipids do not form the complex with lipase probably because of weak interactions with the negatively charged surface of lipases.

The lipid-coated lipases catalyze ester synthesis from 1-monolaurin and excess amounts of lauric acid in homogeneous dry benzene at 40°C. The nonionic lipid gave the best result in terms of initial reaction rate and yield of di- and triglycerides. In the presence of molecular sieves, monolaurin was quantitatively converted to trilaurin in benzene or hexane. The technique of enzyme complexation with lipids is a promising method of enzyme modification for synthetic reactions in organic solvents.

V. IMMOBILIZATION OF ENZYMES IN ORGANIC SOLVENTS

The technique of enzyme immobilization has been extensively studied in aqueous solutions. The major profits of immobilization are easy recovery and reusage of enzymes and application of flow systems to enzyme reactions. There are several methods of immobilization: physical or chemical adsorption, entrapment, and covalent binding to support materials. Activity and stability of immobilized enzymes depend strongly on the method of immobilization and the nature of support materials. In most cases, activity decreases by immobilization, especially in covalent binding methods, due to consumption of functional groups essential for enzyme activity. Adsorption methods have, in general, the advantage of less possibility of deactivation, but detachment of enzymes from supports can be a serious drawback of this method.

The apparent activity and specificity of a biocatalyst can be modified by immobilizing it to supports which have specific interactions with substrates. Examples are entrapment of microorganisms or lipases to photocross-linking polymers with different hydrophobic-hydrophilic properties. When an enzyme is entrapped to a hydrophobic polymer matrix, hydrophobic substrates tend to enter into reactions faster than with the enzyme entrapped to hydrophilic matrices.

In organic solvents, effects of immobilization on enzyme behavior are not straightforward, and generalized interpretation of the effects is difficult. There are several principal factors which govern activity and specificity of immobilized enzymes in organic solvents:

1. Nature and strength of interactions between enzymes and support materials. Since enzymes are hydrophilic in nature, they exhibit high stability by simple adsorption or entrapment to hydrophilic supports. The molecular structures of support materials may also influence the activity of enzymes through stabilization or destabilization of active conformations of enzymes.
2. Interactions between support materials and organic solvent or water. Local concentration of organic solvent around an enzyme will be different with different nature of support materials. In addition, if support materials are hydrophilic and adsorb water, the actual concentration of water in the vicinity of the immobilized enzyme would be higher than bulk solution, and this would result in activation and stabilization of enzymes as compared to free enzymes.
3. Interactions between support materials and substrates or products. The concentration of a substrate around an immobilized enzyme may be different depending on the extent of interaction of the substrate with the support. A specific interaction between the support and substrate would enhance the reaction, or might cause enzyme inhibition. The interactions between supports and substrates or products would also affect apparent equilibrium of the reactions.

Several reaction systems have been reported in which immobilization enhances catalytic activity, but detailed studies on reaction kinetics and enzyme specificities will be required in order to elucidate specific effects of immobilization.

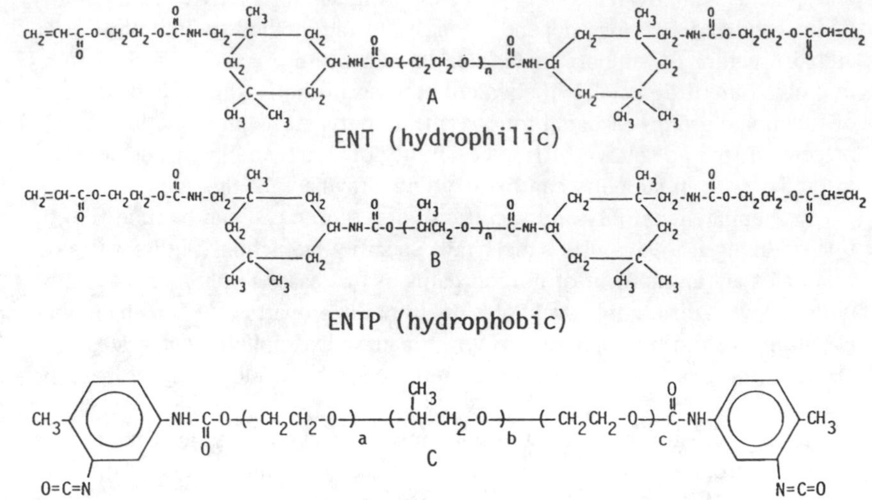

ENT (hydrophilic)

ENTP (hydrophobic)

FIGURE 10.   Structure of (A) and (B) photocross-linkable resin prepolymers and (C) poly-urethane resin prepolymers.

A. ENTRAPMENT TO SYNTHETIC POLYMERS

Photocross-linkable resin prepolymers and water-miscible urethane pre-polymers (Figure 10) have been developed for immobilization of enzymes and microorganisms. Bioconversions of various lipophilic compounds, such as steroids, in organic solvents have been successfully carried out by use of these immobilized biocatalysts.[47,48] By irradiation of UV light to the mixtures of a resin prepolymer (Figure 10A or B) and a biocatalyst solution, or by simply mixing a urethane prepolymer (Figure 10C) with an aqueous solution of a biocatalyst, three-dimensional polymer networks containing the bioca-talyst are formed. Operational stability of the biocatalyst increases signifi-cantly by entrapment to these polymer matrices probably due to multipoint interactions between entrapped biocatalysts and gel matrices. For application of such gel-entrapped biocatalysts to conversions of lipophilic compounds in organic solvents, affinity of substrates to the gels and their diffusion through gel matrices are important factors. The hydrophilic-hydrophobic nature of gels can be controlled by changing molecular structures of prepolymers or by changing mixing ratio of two prepolymers; gels from B is more hydrophobic than from A, and gels from prepolymers C with higher contents of poly(propylene oxide) moiety are more hydrophobic.

The results on model reactions of ester synthesis by gel-entrapped lipase (*Candida cylindracea* OF-360) indicate that the hydrophobic-hydrophilic na-ture of a gel has a dominant effect on reaction rate. For example, the reaction between oleic acid and heptanol is catalyzed only by the lipase entrapped to highly hydrophobic gels. A DSC study showed the presence of two different states of water molecules. It may be assumed that the behavior of water molecules is different in gels of different hydrophobic-hydrophilic natures.[48]

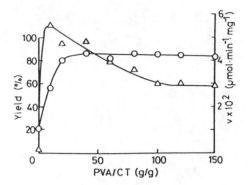

FIGURE 11.   Effect of PVA/CT ratio on Ac-Tyr-OH esterification in ethanol. △, Reaction rate; ○, ester yield after 24 h.

Thus, interactions of support materials with water, and perhaps with organic solvents, may also have significant influence on the catalysis of enzymes in organic solvents.

Proteases can be entrapped to poly(vinyl alcohol) (PVA) gel by mixing aqueous enzyme solutions with powdery PVA.[49,50] The immobilized enzymes are stable and catalyze esterification, transesterification, and peptide synthesis in hydrophilic organic solvents. With restricted amounts of water, detachment of enzymes is negligible. It is assumed that since PVA strongly adsorbs water, enzymes are in a water-rich environment; therefore, the reaction mixtures are considered to be pseudo-biphasic systems. The activity of CT for esterification of Ac-Tyr-OH in ethanol depends strongly on PVA/enzyme ratio and water content as shown in Figures 11 and 12.[49,50] One of the advantages of these reaction systems is that enzymes can be activated by hydration without shift of reaction equilibrium to hydrolysis, giving the products in high yields.

Alternatively, enzymes can be incorporated into PVA films by casting aqueous solutions of PVA and enzymes on glass plates and evaporating water under vacuum.[51] The enzyme-containing films swell in water-containing organic solvents and catalyze synthetic reactions of esters and peptides. A kinetic study on the esterification of Ac-Tyr-OH in ethanol revealed that the increase in K_m for immobilized CT as compared to free CT (25 and 10 mM, respectively) is compensated by the increase in k_{cat} (0.075 and 0.030 s^{-1} for immobilized and free CT, respectively), and second order rate constants (k_{cat}/K_m) are identical. The immobilized CT also catalyzes peptide synthesis by condensation or substitution reactions in acetonitrile or other hydrophilic organic solvents (Equation 13). The immobilized enzyme can be easily removed from the reaction solutions, and used repeatedly. After repeated reactions for 7 d, the loss of activity was negligible.

$$\text{Ac-Tyr-OH or Ac-Tyr-OEt} \; + \; \text{AA-NH}_2 \longrightarrow \text{Ac-Tyr-AA-NH}_2$$

$$\text{AA} = \text{Gly, Ala, Leu, Val, Ser, Met, Phe, Tyr} \qquad\qquad (13)$$

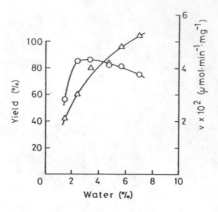

FIGURE 12.   Effect of water content on Ac-Tyr-OH esterification in ethanol. △, Reaction rate; ○, ester yield after 24 h.

Interestingly, the substrate specificities of both free and immobilized CT for peptide synthesis in acetonitrile are different from those in aqueous solutions.[51] It has been reported that in water CT exhibits strong specificity for hydrophobic amino acid derivatives as acyl acceptors. However, in acetonitrile and perhaps in other hydrophilic organic solvents, reactivities of hydrophilic amino acid amides, such as glycinamide, are higher than hydrophobic ones. This implies that, although CT is in the hydrophilic polymer matrix, the environment of the binding site is different from that in aqueous solution. It should be noted, however, that in the case of immobilized enzymes as well as chemically modified enzymes, distribution of substrates in reaction mixtures may be inhomogeneous; this should be taken into consideration in the analysis of substrate specificity.

B. IMMOBILIZATION BY ADSORPTION METHODS

A number of porous or ionic materials, many of which are commercially available, are the candidates for supports of enzyme immobilization by chemical or physical adsorption. They include inorganic materials, such as porous silica gel, porous glass, and celite, which have been most frequently used for immobilization in aqueous solutions. The selection of support materials in organic solvents has been rather arbitrary. A parameter named ''aquaphilicity'' was proposed which characterizes the hydrophilic nature of support materials and is defined as the ratio of amount of water on support to amount of water in diisopropyl ether.[52] The activity of adsorbed chymotrypsin for esterification of Ac-Phe-OH by ethanol in diisopropyl ether decreased with increased aquaphilicity of the supports. The result was interpreted in terms of competitive adsorption of water to the enzyme and the support. However, in acetonitrile, support materials with high aquaphilicity gave high enzymatic activity. It may be reasonably explained that, as mentioned before, in hydrophilic solvents, enzymes are activated and stabilized by complexation or adsorption to hydrophilic materials due to attraction of water around the enzymes.

Several problems have been recognized in two-phase reaction systems. For example, excessive partitioning of polar substrates (such as amino acid derivatives in peptide synthesis) into aqueous phase can give rise to very high aqueous phase concentrations because of the very small volume of this phase. These high concentrations may directly cause inhibition or inactivation of the enzyme.[53] Furthermore, enzymes may be subject to rapid inactivation by contact with a water-organic interface.[54]

These problems can be overcome at least partly by immobilization of enzymes to polymeric materials. A precursor of a dipeptide sweetener (aspartame) has been synthesized by the condensation reaction of Z-Asp-OH with Phe-OMe. The reaction is successfully carried out in water-saturated ethyl acetate by the catalysis of immobilized thermolysin.[55] Among the support materials investigated, Amberlite XAD-7 or -8 are the most effective in terms of the enzyme activity and stability. A stirred tank reactor[56] or a plug-flow type reactor[57] were successfully operated for over 300 and 500 h, respectively, with high yields of the product, Z-Asp-Phe-OMe.

The immobilization of proteases to commercially available porous chitosan beads in organic solvents has been studied.[58] Porous chitosan beads (Chitopearl BCW 3010, product of Fuji Spinning Co.) have an average diameter of 1 mm, a pore size of 0.1 to 0.2 μm, and a specific surface area of 120 to 150 m^2g^{-1}. They are insoluble in and highly resistant to organic solvents. Chitopearl (CP) strongly adsorbs many proteins from aqueous solutions. When CT is immobilized to CP by adsorption, the immobilized CT exhibited much higher activity for ester synthesis of amino acids in organic solvents. The kinetic parameters are summarized in Table 5. It is evident that immobilization to CP markedly increases k_{cat}, while K_m is not significantly affected by immobilization. The results may be the consequence of both adsorption of the substrates to the support and increase in enzyme activity, as described before for complexation of enzymes with chitin or chitosan. CP-immobilized enzymes can easily be separated from the reaction solutions by filtration. Stability of the immobilized enzymes is excellent in organic solvents. For example, CT immobilized on CP can be repeatedly used as a catalyst for Ac-Tyr-OH esterification in acetonitrile containing 3% water. After 3 weeks, the loss of activity was less than 40% (Figure 13). It has been argued that water-miscible organic solvents often cause reversible or irreversible inactivation of enzymes, but the difficulty can be overcome by immobilization to specific supports and by the selection of solvent and water content.

VI. CONCLUDING REMARKS

It has long been assumed that most organic solvents deactivate enzymes by unfolding the peptide chains. However, recent studies indicate that enzymes

TABLE 5
Kinetic Parameters for Ac-Tyr-OH or Ac-Trp-OH Esterification
and Ac-Tyr-OMe Transesterification in Ethanol at 30°C

Reaction	Substrate	Catalyst	H_2O (%)	K_m (mM)	k_{cat} (s^{-1})	k_{cat}/K_m ($M^{-1}S^{-1}$)
Esterification	Ac-Tyr-OH	CT	2.4	10	0.006	0.60
	Ac-Tyr-OH	CT-CP	2.8	19	0.17	8.9
	Ac-Tyr-OH	STB-CP	3.2	1.3	0.0098	7.5
	Ac-Tyr-OH	STC-CP	3.1	5.8	0.015	2.6
	Ac-Trp-OH	CT	2.4	8.3	0.0038	0.46
	Ac-Trp-OH	CT-CP	2.9	7.4	0.038	5.1
Transesterification	Ac-Tyr-OMe	CT	2.9	4.5	0.21	47
	Ac-Tyr-OMe	CT-CP	2.9	13	0.43	33

From Kise, H. and Kayakawa, A., *Enzyme Microb. Technol.*, 13, 584, 1991.

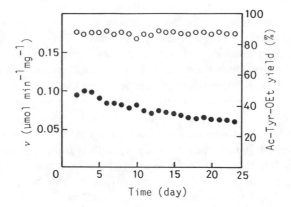

FIGURE 13.   Repeated esterification of Ac-Tyr-OH in acetonitril by Chitopearl-immobilized α-chymotrypsin. ○, Ester yield after 24 h; ●, reaction rate.

can be effective catalysts in organic solvents under selected reaction conditions. Although the effects of organic solvents on structure and catalytic behavior of enzymes have not been fully elucidated, the enzymatic reactions in organic solvents will provide synthetic chemists with an easy method for preparation of a variety of compounds with high selectivity and specificity.

In order to employ enzymes in organic solvents for practical synthetic reactions, the improvement of activity and stability of enzymes is crucial. In this chapter, as a method of activation and stabilization, the modification of enzymes with polymeric materials for use in organic solvents has been described. Especially, attention has been focused on enzyme modification by chemical coupling, complexation, and immobilization to polymers. These methods have common features in terms of protection of enzymes against organic solvents and retention of enzyme activity.

The effects of modification on enzyme activity and specificity in organic solvents are functions of a number of specific interactions between reaction components. The interaction between an enzyme and a modifier will change activity and specificity of the enzyme, and interactions between a modifier and substrates or products will change reaction kinetics, equilibrium, and apparent enzyme specificity. Furthermore, partition of organic solvents and water between the bulk solution and modified enzyme will also affect activity and stability of enzymes.

The methods of grafting poly(ethylene glycol) (PEG) to enzymes have been established and utilized for modification of lipases, proteases, and redox enzymes. Because of the amphiphilic nature of PEG, PEG-modified enzymes are soluble in organic solvents and catalyze many synthetic reactions, including ester and peptide synthesis, under water-restricted reaction conditions. Modification by PEG is a unique method in that it provides the opportunity to investigate reaction kinetics without uncertainty arising from insoluble enzymes. The change in substrate specificity by PEG modification has been reported for chymotrypsin. This may be explained either by change in enzyme structure or by change in partition of substrates to enzymes. The underlying principles governing the substrate specificity of modified enzymes are somewhat similar to those for solvent effects on enzyme specificity.

The complexation of enzymes with solid materials is an effective method for activation and stabilization of enzymes in organic solvents. It has been reported that complexation with low molecular weight salts, such as phosphate or amino acid salts, markedly activates enzymes. Although there is no direct evidence, it is assumed that enzymes maintain optimum structures by forming complexes of low mobility. The results of synthetic reactions by proteases in the presence of polysaccharides suggest that enzyme activity is strongly dependent on the structure of support materials. This indicates that enzyme structure and therefore its activity can be modified by complexation with polymeric materials.

Effects of immobilization on enzyme catalysis in organic solvents have not attracted much attention. However, as mentioned before, enzyme catalysis would be profoundly affected by immobilization. In some cases, enzymes are activated by immobilization. Excellent long-term stability has been obtained by immobilization to polymeric materials. The effect of immobilization has also some features in common with solvent effect; that is, the effects on activity and specificity should originate from diverse interactions between reaction components.

REFERENCES

1. **Jones, J. B.,** Enzymes in organic synthesis, *Tetrahedron,* 42, 3351, 1986.
2. **Ward, O. P. and Young, C. S.,** Reductive biotransformations of organic compounds by cells or enzymes of yeast, *Enzyme Microb. Technol.,* 12, 482, 1990.
3. **Klibanov, A. M.,** Enzymes that work in organic solvents, *Chemtech,* 16, 354, 1986.
4. **Kasche, V.,** Mechanism and yields in enzyme catalyzed equilibrium and kinetically controlled synthesis of β-lactam antibiotics, peptides and other condensation products, *Enzyme Microb. Technol.,* 8, 4, 1986.
5. **Khmelnitsky, Yu. L., Levashov, A. V., Klyachko, N. L., and Martinek, K.,** Engineering biocatalytic systems in organic media with low water content, *Enzyme Microb. Technol.,* 10, 710, 1988.
6. **Zaks, A. and Russell, A. J.,** Enzymes in organic solvents: properties and applications, *J. Biotechnol.,* 8, 259, 1988.
7. **Dordick, J. S.,** Enzymatic catalysis in monophasic organic solvents, *Enzyme Microb. Technol.,* 11, 194, 1989.
8. **Chen, C.-S. and Sih, C. J.,** General aspects and optimization of enantioselective biocatalysis in organic solvents: the use of lipases, *Angew. Chem., Int. Ed. Engl.,* 28, 695, 1989.
9. **Kise, H., Hayakawa, A., and Noritomi, H.,** Protease-catalyzed synthetic reactions and immobilization-activation of the enzymes in hydrophilic organic solvents, *J. Biotechnol.,* 14, 239, 1990.
10. **Kise, H. and Fujimoto, K.,** Enzymatic reactions in aqueous-organic media. VIII. Medium effect and nucleophile specificity in subtilisin-catalyzed peptide synthesis in organic solvents, *Biotechnol. Lett.,* 10, 883, 1988.
11. **Martinek, K., Semenov, A. N., and Berezin, I. V.,** Enzymatic synthesis in biphasic aqueous-organic systems. I. Chemical equilibrium shift, *Biochim. Biophys. Acta,* 658, 76, 1981.
12. **Hahn-Hagerdal, B.,** Water activity: a possible external regulator in biotechnical processes, *Enzyme Microb. Technol.,* 8, 322, 1986.
13. **Eggers, D. K., Blanch, H. W., and Prausnitz, J. M.,** Extractive catalysis: solvent effects on equilibria of enzymatic reactions in two-phase systems, *Enzyme Microb. Technol.,* 11, 84, 1989.
14. **Kise, H., Shirato, H., and Noritomi, H.,** Enzymatic reactions in aqueous-organic media. II. Effects of reaction conditions and selectivity in the esterification of aromatic acids by α-chymotrypsin in alcohols, *Bull. Chem. Soc. Jpn.,* 60, 3613, 1987.
15. **Reslow, M., Adlercreutz, P., and Mattiasson, B.,** The influence of water on protease-catalyzed peptide synthesis in acetonitrile/water mixture, *Eur. J. Biochem.,* 177, 313, 1988.
16. **Laane, C., Boeren, S., Vos, K., and Veeger, C.,** Rules for optimization of biocatalysis in organic solvents, *Biotechnol. Bioeng.,* 30, 81, 1987.
17. **Inada, Y., Nishimura, H., Takahashi, K., Toshimoto, T., Saha, A. R., and Saito, Y.,** Ester synthesis catalyzed by polyethylene glycol-modified lipase in benzene, *Biochem. Biophys. Res. Commun.,* 122, 845, 1984.
18. **Matsushima, A., Okada, M., and Inada, Y.,** Chymotrypsin modified with polyethylene glycol catalyzes peptide synthesis reaction in benzene, *FEBS Lett.,* 178, 275, 1984.
19. **Takahashi, K., Ajima, A., Yoshimoto, T., and Inada, Y.,** Polyethylene glycol-modified catalase exhibits unexpectedly high activity in benzene, *Biochem. Biophys. Res. Commun.,* 125, 761, 1984.
20. **Takahashi, K., Nishimura, H., Yoshimoto, T., Saito, Y., and Inada, Y.,** A chemical modification to make horseradish peroxidase soluble and active in benzene, *Biochem. Biophys. Res. Commun.,* 121, 261, 1984.

21. **Ferjancic, A., Puigserver, A., and Gaertner, H.,** Unusual specificity of polyethylene glycol-modified thermolysin in peptide synthesis catalyzed in organic solvents, *Biotechnol. Lett.,* 10, 101, 1988.

22. **Gaertner, H. and Puigserver, A.,** Kinetics and specificity of serine proteases in peptide synthesis catalyzed in organic solvents, *Eur. J. Biochem.,* 181, 207, 1989.

23. **Ferjancic, A., Puigserver, A., and Gaertner, H.,** Subtilisin-catalyzed peptide synthesis and transesterification in organic solvents, *Appl. Microbiol. Biotechnol.,* 32, 651, 1990.

24. **Lee, H., Takahashi, K., Kodera, Y., Ohwada, K., Tsuzuki, T., Matsushima, A., and Inada, Y.,** Polyethylene glycol-modified papain catalyzes peptide bond formation in benzene, *Biotechnol. Lett.,* 11, 403, 1989.

25. **Ohwada, K., Aoki, T., Toyota, A., Takahashi, K., and Inada, Y.,** Synthesis of N^{α},N^{γ}-dicarbobenzoxy-L-ornityl-β-alanine benzyl ester, a derivative of salty peptide, in 1,1,1-trichloroethane with polyethylene glycol-modified papain, *Biotechnol. Lett.,* 11, 499, 1989.

26. **Gaertner, H. F. and Puigserver, A. J.,** Protein-protein interaction in low water organic media, *Biotechnol. Bioeng.,* 36, 601, 1990.

27. **Matsushima, A. and Inada, Y.,** Synthetic macromolecule-modified enzymes, *Yukigoseikagaku Kyokaishi (J. Synth. Org. Chem. Jpn.),* 46, 232, 1988.

28. **Tsujisaka, Y., Okumura, S., and Iwai, M.,** Glyceride synthesis by four kinds of microbial lipase, *Biochim. Biophys. Acta,* 489, 415, 1977.

29. **Okumura, S., Iwai, M., and Tsujisaka, Y.,** Synthesis of various kinds of esters by four microbial lipases, *Biochim. Biophys. Acta,* 575, 156, 1979.

30. **Zaks, A. and Klibanov, A. M.,** Enzymatic catalysis in organic media at 100°C, *Science,* 224, 1249, 1984.

31. **Klibanov, A. M.,** Asymmetric transformations catalyzed by enzymes in organic solvents, *Acc. Chem. Res.,* 23, 114, 1990.

32. **Nishio, T., Takahashi, K., Tsuzuki, T., Yoshimoto, T., Kodera, Y., Matsushima, A., Saito, Y., and Inada, Y.,** Ester synthesis in benzene by polyethylene glycol-modified lipase from *Pseudomonas fragi* 22.39B, *J. Biotechnol.,* 8, 39, 1988.

33. **Takahashi, K., Ajima, A., Yoshimoto, T., Okada, M., Matsushima, A., Tamaura, Y., and Inada, Y.,** Chemical reactions by polyethylene glycol modified enzymes in chlorinated hydrocarbons, *J. Org. Chem.,* 50, 3414, 1985.

34. **Nishio, T., Takahashi, K., Yoshimoto, T., Kodera, Y., Saito, Y., and Inada, Y.,** Terpene alcohol ester synthesis by polyethylene glycol-modified lipase in benzene, *Biotechnol. Lett.,* 9, 187, 1987.

35. **Ajima, A., Takahashi, K., Matsushima, A., Saito, Y., and Inada, Y.,** Retinyl esters synthesis in polyethylene glycol-modified lipase in benzene, *Biotechnol. Lett.,* 8, 547, 1986.

36. **Mihama, T., Yoshimoto, T., Ohwada, K., Takahashi, K., Akimoto, S., Saito, Y., and Inada, Y.,** Magnetic lipase adsorbed to magnetic fluid, *J. Biotechnol.,* 7, 141, 1988.

37. **Babonneau, M.-T., Jacquier, R., Lazaro, R., and Viallefont, P.,** Enzymatic peptide syntheses in organic solvent mediated by modified α-chymotrypsin, *Tetrahedron Lett.,* 30, 2787, 1989.

38. **Kise, H. and Shirato, H.,** Enzymatic reactions in aqueous-organic media. V. Medium effect on the esterification of aromatic amino acids by α-chymotrypsin, *Enzyme Microb. Technol.,* 10, 582, 1988.

39. **Zaks, A. and Klibanov, A. M.,** Enzymatic catalysis in nonaqueous solvents, *J. Biol. Chem.,* 263, 3194, 1988.

40. **Margolin, A. L., Tai, D.-F., and Klibanov, A. M.,** Incorporation of D-amino acids into peptides via enzymatic condensation in organic solvents, *J. Am. Chem. Soc.,* 109, 7885, 1987.

41. **Riva, S., Chopineau, J., Kieboom, A. P. G., Klibanov, A. M.,** Protease-catalyzed regioselective esterification of sugars and related compounds in anhydrous dimethylformamide, *J. Am. Chem. Soc.*, 110, 584, 1988.

42. **Kise, H., Hayakawa, A., and Noritomi, H.,** Enzymatic reactions in aqueous-organic media. IV. Chitin-alpha-chymotrypsin complex as a catalyst for amino acid esterification and peptide synthesis in organic solvents, *Biotechnol. Lett.*, 9, 543, 1987.

43. **Fendler, J. H. and Fendler, E. J.,** *Catalysis in Micellar and Macromolecular Systems,* Academic Press, New York, 1975, chap. 5.

44. **Kise, H.,** Difference in catalytic activities of subtilisin Carlsberg and subtilisin BPN' and immobilization-activation for ester synthesis and transesterification in ethanol, *Bioorg. Chem.*, 18, 107, 1990.

45. **Kise, H. and Hayakawa, A.,** unpublished data, 1990.

46. **Okahata, Y. and Ijiro, K.,** A lipid-coated lipase as a new catalyst for triglyceride synthesis in organic solvents, *J. Chem. Soc., Chem. Commun.*, 1392, 1988.

47. **Fukui, S. and Tanaka, A.,** Application of biocatalysts immobilized by prepolymer methods, in *Advances in Biochemical Engineering/Biotechnology,* Vol. 29, Fiechter, A., Ed., Springer-Verlag, Berlin, 1984, 1.

48. **Fukui, S., Tanaka, A., and Iida, T.,** Immobilization of biocatalysts for bioprocesses in organic solvent media, in *Biocatalysis in Organic Media,* Laane, C., Tramper, J., and Lilly, M. D., Eds., Elsevier, Amsterdam, 1987, 21.

49. **Noritomi, H. and Kise, H.,** Enzymatic reactions in aqueous-organic media. III. Peptide synthesis by α-chymotrypsin immobilized with polyvinyl alcohol in ethanol, *Biotechnol. Lett.*, 9, 383, 1987.

50. **Noritomi, H., Watanabe, A., and Kise, H.,** Enzymatic reactions in aqueous-organic media. VII. Peptide and ester synthesis in organic solvents by α-chymotrypsin immobilized through non-covalent binding to poly(vinyl alcohol), *Polym. J.*, 21, 147, 1989.

51. **Watanabe, A., Noritomi, H., Nagashima, T., and Kise, H.,** Enzymatic reactions in aqueous-organic media. X. Amino acid esterification and peptide synthesis by α-chymotrypsin entrapped to PVA films, *Koubunshi Ronbunshyu (Jpn. J. Polym. Sci.),* 48, 247, 1991.

52. **Reslow, M., Adlercreutz, P., and Mattiasson, B.,** On the importance of the support material for bioorganic synthesis. Influence of water partition between solvent, enzyme, and solid support in water-poor reaction media, *Eur. J. Biochem.*, 172, 573, 1988.

53. **Cassells, J. M. and Halling, P. J.,** Protease-catalyzed peptide synthesis in low-water organic two-phase systems and problems affecting it, *Biotechnol. Bioeng.*, 33, 1489, 1989.

54. **Cassells, J. M. and Halling, P. J.,** Protease-catalyzed peptide synthesis in aqueous-organic two-phase systems: reactant precipitation and interfacial inactivation, *Enzyme Microb. Technol.*, 12, 755, 1990.

55. **Oyama, K., Nishimura, S., Nonaka, Y., Kihara, K., and Hashimoto, T.,** Synthesis of an aspartame precursor by immobilized thermolysin in an organic solvent, *J. Org. Chem.*, 46, 5241, 1981.

56. **Nakanishi, K., Kamikubo, T., and Matsuno, R.,** Continuous synthesis of N-(benzyloxycarbonyl)-L-aspartyl-L-phenylalanine methyl ester with immobilized thermolysin in an organic solvent, *Bio/Technol.*, 3, 459, 1985.

57. **Nakanshishi, K., Takeuchi, A., and Matsuno, R.,** Long-term continuous synthesis of aspartame precursor in a column reactor with an immobilized thermolysin, *Appl. Microbiol. Biotechnol.*, 32, 633, 1990.

58. **Kise, H. and Hayakawa, A.,** Immobilization of proteases to porous chitosan beads and their catalysis for ester and peptide synthesis in organic solvents, *Enzyme Microb. Technol.*, 13, 584, 1991.

4.2 CHEMICAL AND GENETIC MUTATION OF ENZYMES FOR CREATING NOVEL CATALYSTS

S. Kimura

TABLE OF CONTENTS

I. CREATION OF NEW PROTEINS

The study of structure-function relationships in proteins has attracted much attention from a large number of researchers in various fields. The major goal in protein engineering is to clarify the molecular mechanism of protein function in constituting each amino acid. However, industrial demands for new proteins, which exhibit improved properties or different functions from naturally occurring protein, have been increasing rapidly in recent years. New proteins will be very useful in application to various fields such as biosensors, bioelectronics, bioreactors, biomaterials, and medicine etc. The methods for modifying or creating proteins can be summarized as shown in Figure 1. They can be classified roughly into two groups, using gene technique and chemical synthesis. Both techniques have disadvantages as well as advantages for preparing a protein at the present stage.

Gene technique as well as semisynthesis usually handles naturally occurring proteins. On the other hand, artificial proteins with completely new primary sequences have been successfully constructed by chemical synthesis. For example, in template-assembled synthetic proteins, peptide segments with a secondary structure were connected to side chains of a template molecule, yielding a branched polypeptide, which is in contrast to the linear polypeptides of naturally occurring proteins. Meanwhile the catalytic antibody has stemmed from a different idea, because the molecular design of the protein is unnecessary, but the design of a hapten molecule is needed. Studies on incorporation of unnatural amino acids into a protein are also important for developing a new functional protein, because the method enlarges the possible protein function upon incorporation of a new functional group.

However, it should be emphasized that the naturally occurring proteins have been produced by nature after long evolutionary processes, and the structure is so sophisticated that we still have many things to learn from the proteins. In the following sections, each subject in Figure 1 is explained in detail.

II. CHANGE IN pK_a BY MODIFICATION OF PROTEIN SURFACE

Since many enzyme reactions proceed via charged transition states, stabilization of charges in the protein is essential for realizing efficient catalytic activities. Therefore, a point mutation of the charged amino acid residue involved in the active site of a protein will generally affect the enzymatic activity of the protein considerably. From the beginning of protein research, when a point mutation technique was not available, the significance of electrostatic effects on the protein activities was recognized. For example, trypsin was subjected to acetylation with N-acetylimidazole which led to modification of both accessible amino and tyrosyl hydroxyl groups.[1] The resulting N,O-

Methods

1. Chemical Modification
     Modification of Protein Surface
     Incorporation of Prosthetic Group
2. Semisynthesis
3. Gene Technique
     Deletion Mutagenesis
     Point Mutation
     Cassette Mutagenesis
     Construction of Chimeric Protein

Other Approaches

4. Artificial Protein
     Chemical Synthesis
     Gene Technique
5. Catalytic Antibody
6. Incorporation of Unnatural Amino Acid
     Semisynthesis
     Gene Technique

FIGURE 1.   Methods and various approaches for creation of a new protein.

acetyltrypsin showed increased catalytic activity compared to the native trypsin with respect to the hydrolysis of specific ester and amide substrates in the
optimal pH range. The pH-independent constants corresponding to K_m (dissociation constant) and k_{cat} (rate constant for hydrolysis) for trypsin were 3.4
mM and 0.7 s^{-1}, while those for acetyltrypsin were 2.1 mM and 1.2 s^{-1}.
The apparent pK values for the free and modified enzymes have been estimated
from plots of log (k_{cat}/K_m) vs. pH. The pK values for trypsin were 7.0 and
10.1, while those for acetyltrypsin were 7.2 and 10.5. The results indicate
that the acetylation of exposed tyrosyl residues increased the magnitude of
the acylation rate constant in the trypsin-catalyzed hydrolysis of specific amide
substrates. It can be reasonably speculated that acetylation of the exposed
amino and tyrosyl residues should affect the surface charges and the hydration
state of the protein surface, resulting in changes of the effects of surface
charges on pK_a values of groups in active sites.

Succinylated and ethylenediamine-amidated chymotrypsins were also prepared, showing that the charged groups of the protein exert a strong influence
on the dissociation of ionizable groups of the active site.[2] Upon succinylation,
k_{cat} for specific ester substrates increased about 50 to 60%, and the apparent
pK_a of the ionizing group (probably histidine 57) shifted from 7.0 in the native
enzyme to 8.0 in the succinylated enzyme. On the other hand, upon amidation,
shift in the pK_a of the histidine from 7.0 to 6.1 and from 7.8 to 7.1 was
observed.

Effects of charges on the protein surface are transmitted to the charged
group at the active site through the interior of the protein. However, evaluation

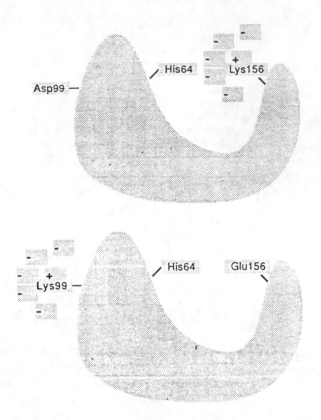

FIGURE 2.   Schematic representation of location of His 64, Asp 99, and Lys 156 in subtilisin. (From Russell, A. J. and Fersht, A. R., *Nature (London)*, 328, 496, 1987. With permission).

of the effects is considerably difficult. The dielectric constant of the protein is still a controversial subject. Russell and Fersht[3] have proposed highly effective dielectric constants by studying the relation of the surface charge of subtilisn with the pK_a of its active-site histidine (His 64). Subtilisin is a serine protease, and residue His 64 acts as a general base in catalysis with a pK_a of 7. Site-directed mutagenesis was utilized to produce mutant enzymes (Figure 2). In one of them, a negatively charged Glu 156, which is 1.4 to 1.5 nm away from the imidazole of His 64, was replaced by Ser 156. The environment between the two is predominantly aqueous. In the other, Asp 99, which is separated from His 64 by 1.2 to 1.3 nm of an environment that is predominantly protein, was replaced with Ser 99. Changing either Asp 99 or Glu 156 to serine lowered the pK_a by about 0.4 units at an ionic strength of 0.001 M, which is effectively zero. Changing both simultaneously to give the double mutant with change of two charge units lowered the pK_a by 0.65 units. Thus, the changes in coulombic interactions were cumulative. The effective dielectric constant (D) between the group concerned and His 64 was calculated from the changes in pK_a. The value was high and independent of whether there was protein or water between the two groups: Asp 99-His 64, D = 50; Glu 156-His 64, D = 40. The results support the idea that most of the dielectric effect comes from water around the protein and not from the interior of the protein.[4]

Changes in pH-activity profiles of a protein can be designed according to the following rules: (1) making the surface more negatively charged raises the pK_a values of acidic groups in proteins because they all lose a proton on ionization, (2) changes will be maximized at low ionic strengths, (3) significant changes will be manifested at ionic strengths as high as $0.1\,M$ if multiply-charged counterions are avoided, and (4) mutations should be designed so that they do not concentrate counterions in the active site cavity.

It should be noted when a protein is modified by the chemical method using electrophilic reagents, a few ten amino acid residues usually located at the protein surface will be electrically altered at the same time. Such non-specific modification was not effective to change a pK_a value compared with the point mutation. The importance of the distribution of charged amino acid residues on the protein surface in the catalytic activity has been pointed out by Dao-Pin et. al.,[5] who have analyzed the relation between the catalytic activity of lysozyme and the asymmetric distribution of charged residues on the protein surface. From the very earliest works, electrostatic interactions of specific active site residues with the substrate were considered to be an important feature of the catalytic mechanism. However, they have pointed out that the charge distribution of the enzyme as a whole plays an important role in the enzymatic mechanism. For example, the electrostatic potential for human lyzozyme was calculated by using the Klapper algorithm. Contoured images of the electrostatic potential of the human lysozyme showed the obvious gradient between the positive and negative equipotential surfaces that coincides exactly with the active site cleft (Figure 3). The proximity of the positive and negative equipotential surfaces indicates that there is an electrostatic potential difference of 6 kcal mol^{-1} across a distance of 4 Å in the active site cleft. This corresponds to an electric field of 6×10^6 V cm^{-1} in the appropriate direction to promote movement of positive charge from the catalytic glutamate toward the substrate.

III. SUBSTRATE SPECIFICITY OF ENZYME

Chemical binding forces between enzyme and substrate include electrostatic interactions, steric, and hydrophobic effects. Among them, electrostatic interactions are believed to be a general binding force; however, evaluation of the free energy is very difficult due to lack of a knowledge about the dielectric constant throughout a protein molecule. Wells et al.[6] have evaluated electrostatic interactions between charged peptide substrates and subtilisin, in which two different sites were modified by using a cassette mutagenesis method.

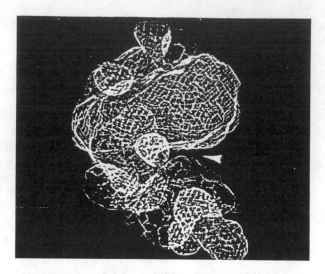

FIGURE 3.   The electrostatic potential for human lysozyme. The active site cleft is indicated by the arrowhead. The cleft is located between contours at 5 kT (upper face) and -5 kT (bottom face), producing a large potential gradient across the cleft. (From Dao-Pin, S., Liao, D.-I., and Remington, S. J., *Proc. Natl. Acad. Sci. U.S.A.*, 86, 5361, 1989. With permission.)

X-ray crystal structures of subtilisin containing bound transition state analogues show two possible modes of substrate binding depending on the P1 substrate side chain (the scissile peptide bond is between P1 and P1'). For a phenylalanine P1 substrate, the side chain extends inward the α-carbon of glycine 166. On the other hand, for a lysine P1 substrate, the side chain extends across the hydrophobic P1 binding cleft to form an ion pair with glutamate 156 at the entrance to the cleft. Because side chains from either residue 156 or 166 have the potential to form an ion pair with a complementary charge at the P1 position of the substrate, single- and double-mutant proteins were prepared to evaluate the effects on substrate specificity. Mutations at positions 156 and 166 produced changes in k_{cat}/K_m toward glutamate, glutamine, methionine, and lysine P1 substrates of up to 4000, 60, 200, and 80 times, respectively. The changes in k_{cat}/K_m were dominated by changes in the $1/K_m$ term. Changes in substrate preference are best attributed to electrostatic effects. As shown in Figure 4, when the P1 binding cleft becomes more positively charged, the average catalytic efficiency increases much more for the glutamate P1 substrate than for its neutral and isosteric P1 homologue, glutamine. In contrast, as the P1 site becomes more positively charged, the catalytic efficiency toward the lysine P1 substrate decreases and diverges from its neutral and steric homologue, methionine. From kinetic analysis of 16 mutants of subtilisin and the wild type, the average free energies for enzyme-substrate ion-pair interactions at the two different sites were calculated to be -1.8 and -2.3 kcal/mol, from which the effective dielectric constants ob-

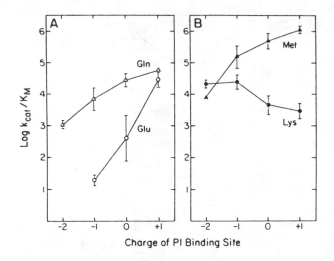

FIGURE 4. Effect of charge in the P1 binding site due to substitutions at positions 15 and 166 on log(k_{cat}/K_m) for the P1 substrate: (A) Glutamine ($\triangle$) and glutamate ($\bigcirc$); (B) methionine ($\blacktriangle$) and lysine ($\bullet$). (From Wells, J. A., Powers, D. B., Bott, R. R., Graycar, T. P., and Estell, D. A., *Proc. Natl. Acad. Sci. U.S.A.*, 84, 1219, 1987. With permission.)

tained were 61 and 48. The obtained free energies were lower than the values estimated for more buried ion pairs in chymotrypsin and phenylalanine tRNA synthetase.

In order to evaluate steric and hydrophobic effects on the binding specificity of subtilisin, a conserved glycine (Gly 166), located at the bottom of the substrate binding cleft, was replaced by 12 nonionic amino acids by the cassette mutagenesis method.[7] As the volume of the side chain at position 166 increased, the substrate preference shifted from large to small P1 side chains. Enlarging the side chain at position 166 caused k_{cat}/K_m to decrease in proportion to the size of the P1 substrate side chain (Figure 5). Catalytic efficiency for the Ala substrate reached a maximum for Ile 166, and for the Met substrate it reached a maximum between Val 166 and Leu 166. In summary, the optimal substitution at position 166 decreased in volume with increasing volume of the P1 substrate. The combined volumes for these optimal pairs may approximate the volume for productive binding in the P1 cleft. The average combined side-chain volume was calculated to be 160 Å³. Exceeding the optimal binding volume of the cleft, by enlarging either the substrate side chain or the side chain at position 166, evoked precipitous drops in catalytic efficiency as a result of steric hindrance. It may appear surprising that Gly 166 is conserved in five known extracellular *Bacillus subtilisins*, and yet Gly 166 can be freely substituted to produce highly functional mutant proteases with generally more narrowed substrate specificities.

It has been shown that substrate specificity can be altered by the modification of residues in direct contact with a bound substrate.[8] For example,

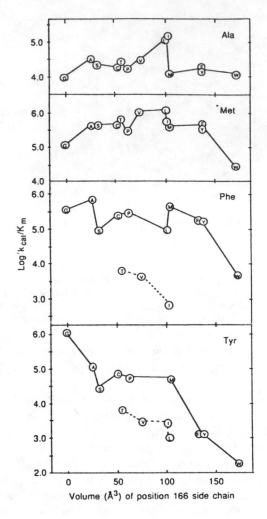

FIGURE 5.   Effect of position 166 side chain molecular volume on the $\log(k_{cat}/K_m)$ for the P1 substrates: Ala (top), Met, Phe, and Tyr (bottom). (From Estell, D. A., Graycar, T. P., Miller, J. V., Powers, D. B., Burnier, J. P., Ng, P. G., and Wells, J. A., *Science,* 233, 660, 1986. With permission.)

the *B. amyloliquefaciens* and the *B. licheniformis* subtilisins differ by more than 60 times in catalytic efficiency toward substrates containing a glutamate residue in the P1 position. The wild-type *B. amyloliquefaciens* subtilisin (Glu 156/Gly 169/Tyr 217) was subjected to site-directed mutagenesis to give a triple mutant (Ser 156/Ala 169/Leu 217) which possess the same amino acid residues of *B. licheniformis* subtilisin at the three sites. Residues at 156 and 217 are in contact distance with the substrate and residue at 169 is within 7 Å. It was shown that the substrate specificity of the mutant resembled to that

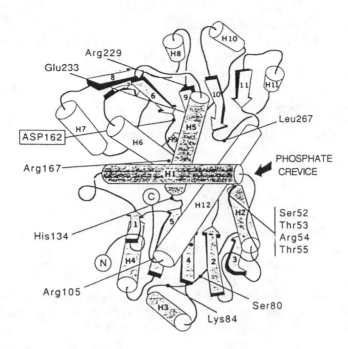

FIGURE 6. Schematic representation of the structure of single catalytic polypeptide of ATCase. The chimeric ATCase was formed from the fusion of the polar domain (shaded structures) of OTCase and the equatorial domain of ATCase (clear structures). The fusion-junction corresponds to Asp 162 of the ATCase sequence. The active site of the enzyme lies within the phosphate crevice and is formed by the asymmetrical insertion of a polar domain from one catalytic subunit into the phosphate crevice of a second subunit. (From Houghton, J. E., O'Donovan, G. A., and Wild, J. R., *Nature (London)*, 338, 172, 1989. With permission.)

of the *B. licheniformis* wild-type subtilisin. This result indicates that large effects occur on substrate specificity by modification of substrate contact residues.

IV. CHIMERIC ENZYME

Houghton et al.[9] have succeeded in switching the substrate specificity of an enzyme producing a chimeric enzyme. The polar domains of the two transcarbamoylases, aspartate transcarbamoylase (ATCase) and ornithine trans-carbamoylase (OTCase), bind the common substrate carbamoyl phosphate and share extensive amino acid sequence homology. The equatorial domains of the two enzymes differ in their substrate specificity (ATCase binds aspartate, OTCase binds ornithine). When the DNA encoding the polar domain of OTCase was fused to DNA encoding the equatorial domain of ATCase *in vitro,* the chimeric enzyme showed ATCase but not OTCase activity (Figure 6). The result shows that by exchanging protein domains between two functionally divergent enzymes it is possible to switch substrate specificity.

On the other hand, ligand binding specificity of receptors also has been changed by formation of chimeric receptors.[10] The adrenergic receptors which mediate the physiological effects of catecholamines belong to the family of plasma membrane receptors that are coupled to guanine nucleotide regulatory proteins. All of the subtypes of adrenergic receptors are activated by epinephrine. The β_2-adrenergic receptors couple to G_s (the stimulatory G protein for adenyl cyclase) while the α_2-adrenergic receptors couple to G_i (the inhibitory G protein for adenyl cyclase), so as to have the opposite effects on the adenyl cyclase system. To determine which structural domains of these two receptors confer specificity for agonist and antagonist binding as well as G protein coupling, ten chimeric receptor genes were constructed from the human β_2-adrenergic receptor and human platelet α_2-adrenergic receptor genes (Figure 7). Of the chimeric receptors, chimera receptor 8 was found to activate adenyl cyclase although it contained the shortest stretches of the β_2-adrenergic receptor. The agonist order of potency for both cyclase activation and ligand binding for the chimera receptor 8 resembled the α_2-adrenergic receptor. Therefore, portions of the fifth and sixth hydrophobic domains are considered to be required for determining the specificity of β_2-adrenergic receptor coupling to G_s. Chimeric receptors were shown to be very useful for elucidating the structural basis of receptor function, which is in contrast to more standard mutagenesis approaches where the end point is the loss of function resulting from amino acid deletions or substitutions.

The chimeric receptor 8 stimulated adenyl cyclase activity upon binding of the α_2-specific agonist p-aminoclonidine, which inhibits cyclase activity in the parent α_2-adrenergic receptor. Thus, α_2 and β_2-adrenergic receptors should use similar mechanisms for turning on the receptor upon ligand binding.

Similarly, insulin and epidermal growth factor (EGF) receptors were found to employ closely related or identical mechanisms for signal transduction across the plasma membrane. Both receptors appear to share a common evolutionary origin, as suggested by structural similarity of cysteine-rich regions in their extracellular domains and a highly conserved tyrosin-specific protein kinase domain. Riedel et al.[11] have designed a chimeric receptor molecule comprising the extracellular portion of the insulin receptor joined to the transmembrane and intracellular domains of the EGF receptor (Figure 8). The EGF receptor kinase domain of the chimeric protein, expressed transiently in simian cells, was shown to be activated by insulin binding. The ability of the chimeric receptors to transduce a signal suggests that the structural differences between EGF and insulin receptors are variations on a common mechanism for signal transfer. With respect to a signal transduction across the plasma membrane, it has been postulated that a ligand-induced conformational change in the receptor extracellular domain triggers receptor dimerization resulting in activation of intracellular enzymatic activities. An alternative mechanism of transmembrane signal transfer, in which domains

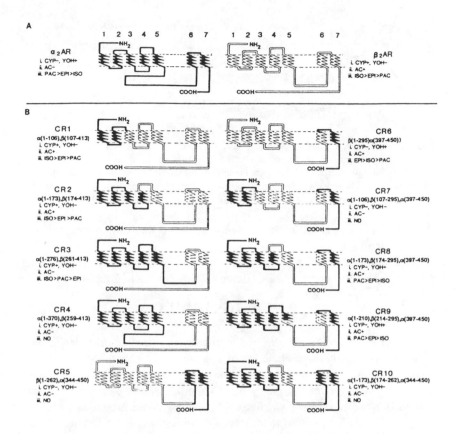

FIGURE 7.   (A) Diagram of the wild-type, α_2-adrenergic receptor (α_2-AR) and the wild type, β_2 receptor (β_2-AR). The hydrophobic domains are shown as forming α-helices that span the plasma membrane. (B) Chimeric receptors made from combinations of the wild-type α_2-AR and β_2-AR. The α_2-AR sequence is indicated by a solid line and β_2-AR sequence is indicated by an open line. The α_2-AR (α) and β_2-AR (β) amino acid sequences from NH$_2$- to COOH-terminus are indicated in parentheses beside each chimeric receptor. (From Kobilka, B. K., Kobilka, T. S., Daniel, K., Regan, J. W., Caron, M. G., and Lefkowitz, R. J., *Science*, 240, 1310, 1988. With permission.)

of the receptor communicate through a transmembrane α-helix, could exist. According to this view, alterations in the transmembrane region of the receptor should affect the ability of the extracellular domain to transmit the appropriate conformational change to the catalytic domain on the cytoplasmic side of the plasma membrane. To explore the role of the transmembrane region in the process of receptor activation, Kashles et al.[12] have generated EGF-receptor mutants with altered transmembrane regions by utilizing *in vitro* site-directed mutagenesis. It was shown that EGF was able to stimulate the kinase activity and receptor dimerization of wild-type and four different EGF-receptor mutants with altered transmembrane regions. The results are compatible with an intermolecular allosteric oligomerization model for receptor activation rather

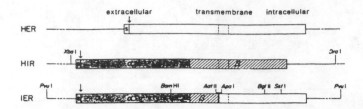

FIGURE 8.   Schematic comparison of normal and chimeric receptor constructs. Coding regions of human EGF receptor (HER), human insulin receptor (HIR) and insulin-EGF receptor chimeric (IER). Shaded regions, HIR subunit sequences; striped regions, HIR sequences; unshaded regions, HER sequences. The sequences are aligned at their transmembrane domains. (From Riedel, H., Dull, T. J., Schlessinger, J., and Ullrich, A., *Nature (London)*, 324, 68, 1986. With permission.)

than with a model based on an intramolecular mechanism for receptor activation (Figure 9).

The chimera receptor composed of insulin and insulin-like growth factor receptors was also prepared to evaluate the insulin binding region.[13]

V. THERMOSTABILITY

Enzyme inactivation at elevated temperatures often accompanies irreversible loss of activity due to deleterious changes in the enzyme structure. It is suggested that inactivating reactions may include deamidation of aspargine residues, hydrolysis of the polypeptide chain at aspartic acid residues, and elimination of cystine residues. Site-directed mutagenesis is possible to replace the "weak links" in the protein molecules with other, more thermoresistant amino acid residues. In order to improve the thermostability, point mutations have been engineered in the dimeric enzyme yeast triosephosphate isomerase.[14] The refined crystal structure of triosephosphate isomerase reveals that 3 of 12 aspargine residues (Asn 14, 65, and 78) for each subunit are present at the interfacial Van der Waals surface. Deamidation of such residues results in the formation of charged aspartic acid residues, thereby promoting dissociation to monomers, which have been shown to be catalytically inactive. Cumulative replacement of asparagine residues at the subunit interface by residues resistant to heat-induced deterioration and approximating the geometry of asparagine (Asn 14 → Thr 14 and Asn 78 → Ile 78) nearly doubled the half-life of the enzyme at 100°C, pH 6. In contrast, replacement of interfacial Asn 78 by an aspartic acid residue increased the rate constant of irreversible thermal inactivation.

The stabilization of a protein was also improved by introduction of a disulfide bond into T4 lysozyme, a disulfide-free enzyme.[15] T4 lysozyme is a muramidase which has a biological function of breaking down the cell wall of an infected bacterium to facilitate the lytic release of replicated phage. The wild-type T4 lysozyme has a cysteine at position 97. X-ray analysis of the T4 lysozyme crystal structure revealed that the residue at position 3 is located

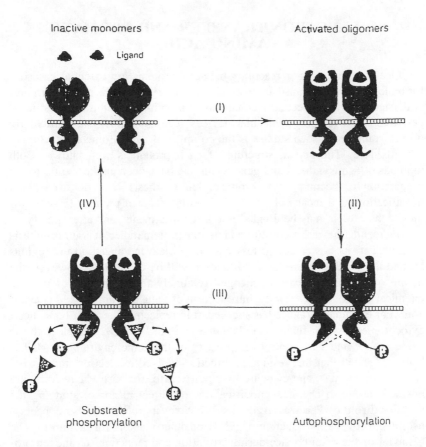

FIGURE 9. Molecular steps in the activation of the EGF-receptor kinase. (I) The binding of EGF stabilizes an oligomeric state with enhanced protein tyrosine kinase activity. (II) The activated receptor oligomers undergo autophosphorylation, which is mediated by an intermolecular cross-phosphorylation mechanism. (III) The intrinsic autophosphorylation sites and the exogenous substrates compete for the substrate binding region of the protein tyrosine kinase. Autophosphorylation renders the EGF-receptor kinase more accessible toward exogenous substrates. (IV) A tyrosine-specific phosphatase removes the phosphate groups from the EGF-receptor. (From Schlessinger, J., *TIBS*, 13, 443, 1988. With permission.)

favorably for formation of a disulfide bond with the cysteine at position 97. The mutant T4 lysozyme (Ile 3 → Cys 3) was prepared by site-directed mutagenesis. Oxidized T4 lysozyme (Ile 3 → Cys 3) exhibited specific activity identical to that of the wild-type enzyme when measured at 20°C in a cell-clearing assay. At the elevated temperature, 67°C, the decay in activity of the wild-type enzyme showed a half-life of 11 min. On the other hand, the disulfide-linked mutant appeared to decay more slowly, with an initial half-life of 28 min. More dramatically, the activity of the disulfide form did not fall below 50% of the starting activity. In contrast, the wild-type enzyme exhibited only about 0.2% of its starting activity after 180 min.

VI. ELECTRON TRANSFER AND AROMATIC
AMINO ACID

One-electron transfer reactions between proteins are essential cascades for biological energy transduction and intermediary metabolism. Furthermore, understanding the molecular mechanism of protein electron transfer is indispensable for developing molecular devices such as biotips. One particularly useful example for such studies is the complex of cytochrome c-cytochrome c peroxidase. The crystal structure of each protein is well known. Both methods of semisynthesis and gene technique have been applied so far to the preparation for the mutant cytochrome c. Semisynthesis is superior to chemical modification as a means of providing a point-mutated protein. In semisynthesis, the protein may be divided into a few fragments, and after modifying the fragments, they are reconstituted into one protein molecule. Semisynthesis is limited in its power due to less choice of cleavage agents to the regions around the residue where the modification will be conducted. However, the nature of the residue can be changed relatively freely because of using a chemical method, and even the introduction of unnatural amino acid is available. For example, Wallace has succeeded in replacing Phe 82 of horse heart cytochrome c with *o*-fluorophenylalanine (F-Phe).[16] Phe 82 lies close to the heme group, and may control the polarity of the heme environment. It has been suggested that this residue is directly involved in electron transfer between cytochrome c and cytochrome c peroxidase. Because Phe residues are not susceptible to chemical modification, no direct means of studying the functional role of Phe was available. Whatever its role is, it also appears a useful site for the introduction of NMR or other markers for monitoring the behavior of the hemi crevice during oxidation and reduction. Cyanogen bromide was used to cleave cytochrome c at Met 80. The obtained fragment starting from Ile 81 was subjected to the Edman reaction for two cycles, resulted in removement of Ile 81 and Phe 82. To the N terminal of the fragments were coupled F-Phe and Ile successively; then the fragments were reconstituted into the mutant cytochrome c. The general scheme of semisynthetic preparation for cytochrome c is shown in Figure 10. Unfortunately, the biological activity of the product was not assayed, probably due to low yield of the mutant protein.

Theoretical and experimental investigations indicate that protein can influence the redox potential of a prosthetic group or stabilize separated charge after electron transfer. Important factors that affect the reduction potential of a protein include (1) the nature of ligands at the redox center, (2) conformational changes associated with reduction, and (3) electrostatic interactions of the redox center with charged groups both on the surface and in the interior of the protein. In order to assess the influences of the local electrostatic field on the redox properties of the protein, Val 68 of wild-type human myoglobin has been replaced by Glu, Asp, and Asn.[17] Residue Val 68 is situated below

I.   Side-chain protection
       Acetimidation of ε-amino groups.
II. Cleavage
       120-fold excess of CNBr in 0.1 *N* HCl
III. Separation of fragments
       Sephadex G50 in 7% formic acid
IV. Functional group protection of fragments
       Boc group on α-amino of fragment B (residues 66-80)
       Methylation on side-chain carboxyl groups of B
V.   Fragment modification
       Stepwise degradation on fragment B.
       Stepwise degradation and condensation on fragment C (residues 81-104)
VI. Fragment condensation
       Active ester couplings of modified fragments B and C.
VII. Deprotection
       Removal of methyl and Boc groups from semisynthetic BC.
VIII. Resynthesis of cytochrome c
       Coupling of natural fragment A (residues 1-65) and semisynthetic BC.
IX. Final deprotection
       Removal of acetimidino groups, if desired.

FIGURE 10.   A semisynthetic scheme for cytochrome c. (From Wallace, C. J. A., in *Semisynthetic Peptides and Proteins,* Offord, R. E. and Bello, C. D., Eds., Academic Press, London, 1978, 101. With permission.)

heme ring I on the distal side of the hemi pocket and within Van der Waals contact of the heme. Replacement of Val 68 by the potentially charged residues Glu and Asp decreased $E^{0'}$ by about 200 mV, whereas replacement by the polar but uncharged residue Asn lowers $E^{0'}$ by 82.7 mV. At pH 7.0, reduction of Fe(III) to Fe(II) in the Glu and Asp mutants is accompanied by uptake of a proton by the protein. These studies demonstrate that myoglobin can tolerate substitution of a buried hydrophobic group by potentially charged and polar residues and that such amino acid replacements can lead to substantial changes in the redox thermodynamics of the protein.

Pielak et al.[18] have replaced Phe 87 of yeast iso-1-cytochrome c (corresponding to Phe 82 in horse heart cytochrome c) either with a Ser, a Tyr, or a Gly by using site-directed mutagenesis. The mutated genes have been introduced into a yeast strain lacking both isozymes of cytochrome c. The purified mutant proteins were similar to the wild type with respect to their visible spectra. The oxidation of reduced cytochrome c with 0.6 n*M* cytochrome c peroxidase was measured. At a cytochrome c concentration of 5 μ*M*, the Ser, Tyr, and Gly 87 variants were approximately 70, 30, and 20%, respectively, as active as the wild-type protein. The reduction potentials of the Ser 87 and Gly 87 proteins were found to be equal and 50 mV lower than the Tyr 87 and the wild-type protein, indicating that the function of Phe 87 is to control the polarity of the heme environment by excluding water. It follows that replacement of Phe 87 with Gly or Ser allows greater solvent access to the heme. Thus, Phe 87 is not essential for cytochrome c to transfer electrons, but is involved in determining the reduction potential.

However, steady state kinetic studies were imperfect windows on the electron-transfer step because they inevitably convolve bimolecular processes and the intracomplex transfer event. Liang et al.[19] have studied the same mutant cytochrome c proteins as described in the previous paragraph on the kinetics of long-range electron transfer within their complexes with zinc-substituted cytochrome c peroxidase. They demonstrated that substitution of Phe 87 by serine and tyrosine had no effect on the photostimulated forward reaction; glycine substitution appeared to perturb the complex and reduced k_t. However, the thermal return process was more than 104 times faster for the wild-type protein and the Tyr 87 variant than for the serine or glycine variants.

The three-dimensional atomic structures of yeast iso-1-cytochrome c and the Ser 87 mutant protein have been elucidated.[20] Replacement of Phe 87 with a serine residue resulted in conformational changes both near and remote from the mutation site. Placement of a serine side chain at position 87 also led to the formation of a large solvent channel which substantially increased the solvent accessibility of the heme group. The results indicate that Phe 87 in the wild-type protein has at least three roles: (1) limiting solvent accessibility to the heme, thereby regulating the reduction potential of the protein di-electrically; (2) facilitating the rate of electron transfer by providing the optimal medium along the transfer route; and (3) forming contact face interactions with redox partners to assist in the formation of the productive electron-transfer complex. It should be emphasized that the mutation of a single amino acid can induce conformational changes in regions remote from the mutation site. Thus, proteins are composed of specifically designed integral units, in which the placement and role of any given individual amino acid can be dependent on the conformation of other amino acids even if the latter residues are at some distance in the tertiary structure.

With regard to (3) just described, the importance of side chains of amino acid residues present at the interface between cytochrome c and cytochrome c peroxidase was studied.[21] Hazzard et al.[21] investigated the effect of single amino acid replacements within cytochrome c on the reactivity of this protein with cytochrome c peroxidase in order to obtain information on residues involved in the complex formation of cytochrome c peroxidase and in the electron-transfer pathway.

In the hypothetical model, Lys 13 of cytochrome c forms a salt bridge with Asp 37 of cytochrome c peroxidase. As a result of this ionic interaction, His 181 of cytochrome c peroxidase, which in the crystal structure of free cytochrome c peroxidase is hydrogen bonded to Asp 37, increases freedom of motion and plays a key role as an electron-transfer conduit into the peroxidase. On the basis of this model, it would be reasonable to presume that converting this lysine to isoleucine would have a deleterious effect on the electron-transfer rate constant. To the contrary, the site-specific mutant of cytochrome c (Arg 13 → Ile) showed the increased electron-transfer rate

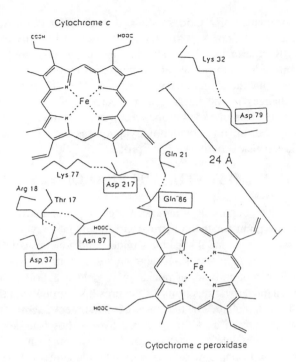

Cytochrome c

Cytochrome c peroxidase

FIGURE 11. Model of the cytochrome c/cytochrome c peroxidase complex focusing on the heme-heme interface. Cytochrome c peroxidase amino acids are within boxes. The dotted lines represent hydrogen bonds between the two proteins. (From Poulos, T. L. and Finzel, B., *Pept. Protein Rev.*, 4, 115, 1984. With permission.)

constant at low ionic strength, suggesting that when the Lys 13-Asp 37 electrostatic interaction is lost, the complex can more easily adopt a structure in which the heme-to-heme distance and/or the orientation is improved in terms of the kinetics of electron transfer. The results clearly demonstrate that the electrostatically stabilized 1:1 complex is not optimized for electron transfer and that by neutralization of key positively charged residues or by an increase in the ionic strength thereby masking the ionic interactions, the two proteins can orient themselves to allow the formation of a more efficient electron-transfer complex.

Das et al.[22] conducted mutational studies of amino acids thought to be involved in the binding and recognition of cytochrome c with its physiological partners such as cytochrome b_2, cytochrome c peroxidase, and cytochrome oxidase.[22] One key residue in this regard is Lys 32, which is invariant in cytochrome c from 88 of 92 species sequenced. Lys 32 was generally believed to form part of an ionic "binding domain" by which cytochrome c binds to other proteins. Lys 32 of iso-1-cytochrome c was replaced by Leu 32, Gln 32, Trp 32, and Tyr 32 (Figure 11).[23] For the Leu 32- and Gln 32-altered proteins, steady state kinetic studies with cytochrome c peroxidase, cytochrome b_2, and cytochrome oxidase showed that neither of the steady state kinetic parameters, K_m nor V_{max} (reactivity), were substantially modified by mutation. NMR experiments demonstrated that the Gln 32-altered protein can still bind strongly to a physiological partner, cytochrome c peroxidase. The findings suggest that the evolutionary invariance of Lys 32 reflects only small quantitative changes in the binding and reactivity of cytochrome c. This demonstration suggests that evolutionarily conserved residues may be altered by genetic engineering for biotechnology purposes without necessarily diminishing the intrinsic reactivity of the protein.

VII. SEMISYNTHETIC FLAVOENZYME

Tertiary structures present in naturally occurring proteins will provide an excellent starting point for the construction of artificial enzymes, since they bind a wide range of potential substrates under mild aqueous conditions. For example, covalent attachment of a foreign catalytic group to a specific amino acid residue in a protein-binding pocket will yield a hybrid molecule that combines the characteristic reactivity of the prosthetic group with the binding specificity of the template. On the basis of the strategy, semisynthetic flavoenzymes were prepared by using glyceraldehyde-3-phosphate dehydrogenase (GAPDH) as the template.[24] GAPDH catalyzes the reversible oxidation of glyceraldehyde-3-phosphate with concomitant reduction of NAD^+. Holo-glyceraldehyde-3-phosphate dehydrogenase containing three to four molecules of bound NAD^+ was isolated from *Bacillus stearothermophilus* cell paste. Since it is known that the bound cofactor inhibits alkylation of the active-site sulfhydryl group, NAD^+ was removed from the holoenzyme by chromatography. Bacterial apo-GAPDH reacted readily with 7-(α-bromoacetyl)-10-methylisoalloxazine. As judged by amino acid analysis and dithionite titration of the flavin chromophore, 0.80 isoalloxazine moiety was incorporated into each subunit of the bacterial enzyme. Under aerobic conditions at 25°C oxidation of the enzyme is characterized by apparent negative cooperativity; however, at low concentrations of NADH an acceleration rate of about 6000-fold is observed over the analogous model reaction catalyzed by 7-acetyl-10-methylisoalloxazine.

Kaiser has proposed five criteria for designing a semisynthetic enzyme prepared from an appropriate enzyme: (1) the enzyme to be used should be readily available in highly purified form; (2) the X-ray structure of the enzyme should be known; (3) the enzyme should have a suitably reactive amino acid functional group at or near the active site; (4) the covalent modification of the enzyme should result in a significant change in the activity characteristic of the native enzyme; and (5) the attachment of the coenzyme analogue should not cause the entry of substrates to the binding site to be blocked.[25] Papain also has been converted into a highly effective oxidoreductase by covalent modification of the sulfhydryl group of the active site cysteine residue with flavins such as 8-bromoacetyl-10-methylisoalloxazine.[26-28]

VIII. CATALYTIC ANTIBODY

Antibodies possess a highly specific recognition site as enzymes, however, without ability to catalyze chemical changes in a selectively bound molecule. Nevertheless, when the combining site of the immunoglobulin possesses a nucleophilic group nearby, the antibody will enhance hydrolysis of the bound molecule. Kohen et al.[29] have prepared the antibody that binds 5α-dihydrotestosterone and testosterone with similar affinity (antidihydrotestosterone IgG). Testosterone-1α-carboxyethyl thioether was conjugated through an ester bond with the fluorescent compound umbelliferone and used as a substrate for the IgG. Testosterone-1α-carboxyethyl thioether umbelliferone conjugate was devoid of fluorescence, but yielded a fluorescent product upon incubation with the IgG, indicating hydrolyis of the substrate. The hydrolytic reaction was inhibited by the homolgous hapten but not by unrelated steroids. Moreoever, the antidihydrotestosterone IgG failed to promote the hydrolysis of testosterone-7-umbelliferone, indicating that the antibody recognized the steroid specifically. The results indicate that the antibody can have enzyme-like properties, but the turnover was low due to slow dissociation of the reaction product from the antibody.

The low turnover was improved by using a haptenic group resembling the presumed transition state of a given reaction. Pollack et al.[30] have produced an antibody, which binds the transition state analogue *p*-nitrophenylphosphorylcholine with high affinity. The antibody catalyzed the hydrolysis of the corresponding carbonate. The antibody-catalyzed hydrolysis of carbonate was inhibited by the addition of the transition state analogue, *p*-nitrophenylphosphorylcholine. The rate of reaction was first order in hydroxide ion concentration between pH 6.0 and 8.0, indicating that the role of the antibody combining site is to stabilize the transition state formed by attack of an external nucleophile on the IgG-complexed carbonate. The acceleration of hydrolysis by the antibody combining site was evaluated to be 770.

Specific monoclonal antibodies were prepared by using monoaryl phosphonate esters as haptens, which were designated as analogues of the transition state in the hydrolysis of carboxylic esters.[31] It was shown that antibodies reacted with cognate aryl carboxylic esters to release a fluorescent alcohol. The haptenic phosphonate was a competitive inhibitor of the reaction. Chemical modification of side-chain groups in the protein showed a partial reduction in activity upon acylation of lysine or nitration of tyrosine and a dramatic quenching upon modification of histidine. The results suggest that the imidazole group is participating in the transacylation reaction. This might involve the direct formation of a stable acylated histidine as the product or perhaps its transient involvement in a two-step mechanism with acyl transfer to tyrosine, for example. Alternatively, the imidazole may act as a general base catalyst in the acylation of tyrosine directly (Figure 12).

There have been many other reports on catalytic antibodies.[32-48] Interestingly, a semisynthetic antibody was also reported.[49,50] Pollack et al. pre-

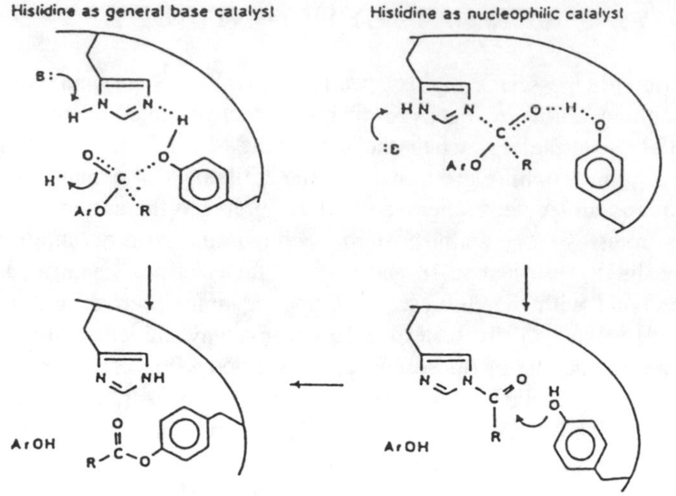

Histidine as general base catalyst   Histidine as nucleophilic catalyst

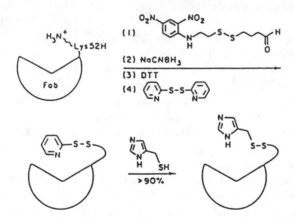

FIGURE 12.   Possible role of histidine and tyrosine in the antibody-induced hydrolysis of an ester. (From Tramontano, A., Janda, K. D., and Lerner, R. A., *Proc. Natl. Acad. Sci. U.S.A.*, 83, 6736, 1986. With permission.)

FIGURE 13.   Modification of the Fab fragment of MOPC 315 to introduce an imidazole at the binding site. (From Pollack, S. J. and Schultz, P. G., *J. Am. Chem. Soc.*, 111, 1929, 1989. With permission.)

pared a semisynthetic antibody that incorporated a catalytic group in the antibody-combining site. The antibody MOPC 315 bound substituted 2,4-dinitrophenyl ligands. Lys 52, which is proximal to the combining site of MOPC 315, has been selectively modified with a unique thiol through the use of cleavable affinity labels (Figure 13). In order to derivatize this modified antibody with imidazole, the thiopyridyl disulfide adduct of the antibody was treated with 4-mercaptomethylimidazole. The second-order rate constant for the antibody-catalyzed reaction, k_{cat}/K_m, was compared to the rate constant for the reaction catalyzed by 4-methylimidazole at pH 7.0. The ratio $[k_{cat}/K_m]/k_{4\text{-methylimidazole}}$ gave an acceleration factor of 1.1×10^3 for hydrolysis of coumarin ester by the antibody. Multiple (>10) turnovers were observed without any loss of catalytic activity.

The method of site-specific mutation has been also applied to introduce a catalytic side chain into an antibody to enhance the catalytic efficiency.[51] Antibody S107 bound phosphorycholine mono- and diesters. The antibody was shown to catalyze the hydrolysis of 4-nitrophenyl *N*-trimethylammonioethyl carbonate. The gene encoding the V^H binding domain of S107 was inserted into plasmid pUC-fl, and *in vitro* mutagenesis was performed. Mutation at Tyr 33 by histidine had an enhanced effect on catalytic activity. That is, the histidine mutant had an eightfold higher k_{cat} for the substrate relative to wild-type antibody.

IX. SYNTHETIC PROTEINS

Since the protein structure is constituted with several elements of characteristic secondary structures, it is a rational approach to the creation of a new protein to assemble elemental secondary structures in a desired spatial arrangement. To overcome this difficulty in constructing a desirable tertiary structure, a novel strategy of using a template molecule has been proposed, in which elemental secondary structures are connected to a template molecule to induce a specific three-dimensional arrangement.[52]

Sasaki and Kaiser[53] have synthesized a designed heme protein (helichrome), in which four amphiphilic α-helices were connected to a porphyrin ring. The helichrome was designed to create a hydrophobic pocket for substrate binding above the porphyrin ring after the expected folding of the peptide chains (Figure 14). The aniline hydroxylase activity of the Fe(III) complex of the helichrome was measured, and k_{cat} and K_m were calculated to be 0.02 min^{-1} and 5.0 m*M*, respectively. On the other hand, Fe(III) coproporphyrin showed negligible aniline hydroxylase activity under the same conditions, demonstrating a significant contribution of the peptide segments to catalysis, most probably by providing a binding pocket for the substrate.

Hahn et al.[54] have reported the design and synthesis of a 73-residue peptide "chymohelizyme" that showed characteristics resembling those of chymotrypsin. Computer modeling was used to design a bundle of four short parallel amphiphilic helical peptides bearing the serine protease catalytic site residues serine, histidine, and aspartic acid at the amino end of the bundle in the same spatial arrangement as in chymotrypsin (Figure 15). The four chains were linked covalently at their carboxyl ends. The peptide has affinity for chymotrypsin ester substrates similar to that of chymotrypsin and hydrolyzes them at rates 0.01 that of chymotrypsin; total turnovers >100 have been observed.

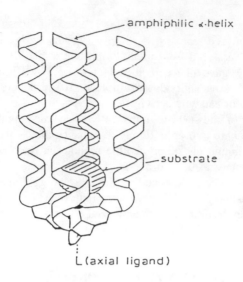

FIGURE 14.   Proposed structure of helichrome after folding of the peptide chains. (From Sasaki, T. and Kaiser, E. T., *J. Am. Chem. Soc.*, 111, 380, 1989. With permission.)

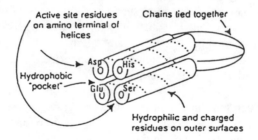

FIGURE 15.   Schematic presentation of the structure of chymohelizyme. (From Hahn, K. W., Klis, W. A., and Stewart, J. M., *Science,* 248, 1544, 1990. With permission.)

The model of allosteric proteins was synthesized by combining cyclic octapeptide and four amphiphilic α-helices.[55] Cyclo(Lys(Z)-Sar)$_4$ forms a complex selectively with Ca^{2+}. Carbobenzoxy groups of cyclo(Lys(Z)-Sar)$_4$ were deblocked, and to four side chains of the cyclic peptide were introduced four Boc-(Lys-Aib-Leu-Aib)$_4$-OH (Aib represents 2-aminoisobutyric acid). The peptide took a partially helical structure in 95% methanol. The helical content increased in the presence of Ca^{2+} due to interaction of the helical macrodipole moment with Ca^{2+}, which was captured by the cyclic moiety and located at the C terminal of helices. The fluorescence intensity of 8-anilinonaphthanlene-1-sulfonate increased with the addition of the peptide in aqueous solution. In the presence of Ca^{2+}, the fluorescence intensity increased further, indicating that formation of the hydrophobic pocket was regulated by Ca^{2+} (Figure 16).

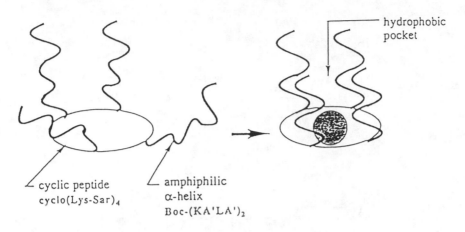

cyclic peptide
cyclo(Lys-Sar)₄

amphiphilic
α-helix
Boc-(KA'LA')₂

hydrophobic
pocket

FIGURE 16.   Schematic presentation of the amphiphilic helix/cyclic peptide conjugate as a model of allosteric proteins.

X. INCORPORATION OF UNNATURAL AMINO ACIDS INTO PROTEINS

Although the technique of oligonucleotide-directed mutagenesis is a powerful tool for the production of mutant proteins, the substitution was limited to amino acid side chains from the pool of 20 common naturally occurring amino acids. Thus, the incorporation of unnatural amino acids into proteins by site-specific mutagenesis has been studied intensively because it provides a valuable new methodology for the generation of novel proteins that possess unique structural and functional features.

Historically, transfer RNA (tRNA) has been misacylated with an unnatural amino acid, which was used for protein biosynthesis. However, the unnatural amino acid was uniformly incorporated into every site of the same codon. Furthermore, the resulting mutants contain both natural and unnatural amino acids at each site of incorporation. On the other hand, Noren et al.[56] have succeeded in the site-specific incorporation of unnatural amino acids into protein. The outline of the mutagenesis method is shown in Figure 17: (1) replacement of the codon encoding the amino acid of interest with the "blank" nonsense codon, TAG, by oligonucleotide-directed mutagenesis; (2) chemical aminoacylation of a suppressor tRNA directed against this codon with the desired unnatural amino acid; and (3) addition of the aminoacylated tRNA to an *in vitro* protein synthesizing system. With regard to preparation of misacylated tRNA, Heckler et al.[57] have reported the following method. The dinucleotide pCpA was chemically acylated and enzymatically ligating to the 3′ terminus of a truncated tRNA with T4 RNA ligase. A major drawback of this method is that the α-amino protecting group was not removed. On the other hand, Noren et al. have used a minimal protection scheme shown in Figure 18, in which only the exocyclic amine of cytidine was protected by

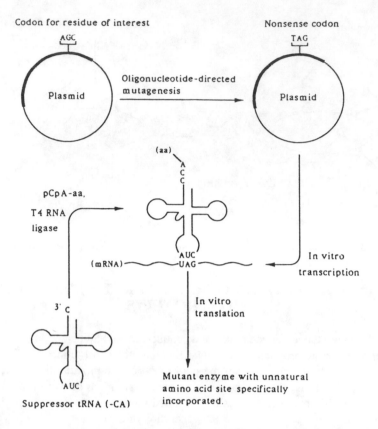

Codon for residue of interest   Nonsense codon

FIGURE 17.   Scheme of site-specific incorporation of unnatural amino acids into proteins. (From Noren, C. J., Anthony-Cahill, S. J., Griffith, M. C., and Schultz, P. G., *Science,* 244, 182, 1989. With permission.)

o-nitrophenyl sulfenyl chloride (NPS-Cl). The α-amino group of the amino acid was also protected with the NPS group. The NPS-protecting groups were removed in high yield from cytidine and the amino acid with aqueous thiosulfate. Fully deprotected pCpA-Phe was ligated directly to tRNA$^{Phe}_{CUA}$(-CA) with T4 RNA ligase. This acylation strategy was used to acylate tRNA$_{CUA}$ (-CA) with phenylalanine, *p*-fluorophenylalanine, *p*-nitrophenylalanine, and homophenylalanine, etc. These aminoacyl tRNA$_{CUAs}$ were added to in vitro reactions charged with the plasmid to produce wild-type and mutant β-lactamases. A similar method was also reported by Bain et al.[58] A drawback of this method is that the level of suppressor-mediated biosynthesis of β-lactamase was 15 to 20% compared to the level of β-lactamase synthesized from the wild-type plasmid. It is postulated by Yokoyama et al. to use minor tRNA instead of suppressor tRNA to improve the yield.

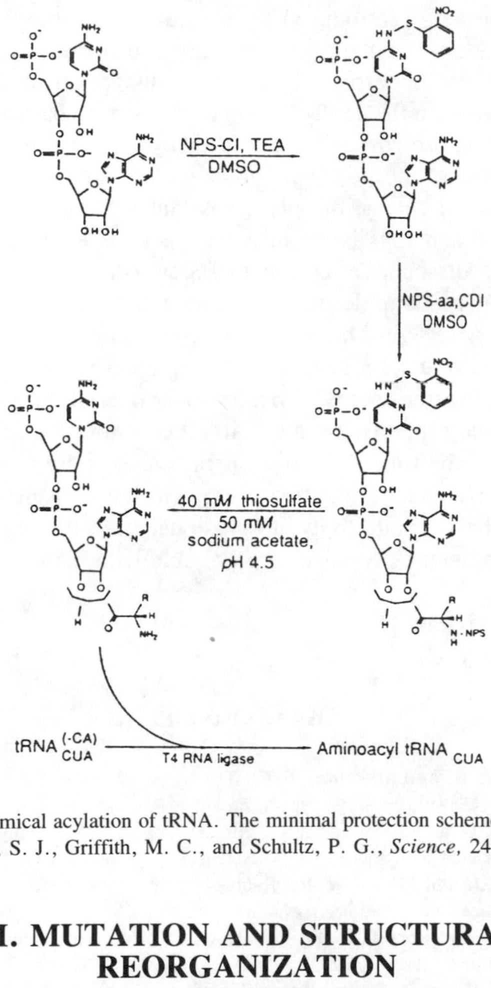

FIGURE 18. Chemical acylation of tRNA. The minimal protection scheme. (From Noren, C. J., Anthony-Cahill, S. J., Griffith, M. C., and Schultz, P. G., *Science,* 244, 182, 1989. With permission.)

XI. MUTATION AND STRUCTURAL REORGANIZATION

So far, the major goal in protein engineering using site-directed mutagenesis etc. has been to study relationships between protein structure and function. Mutation of a protein usually causes the following changes in the protein: (1) loss of interactions from the original side chain, (2) gain of modified interactions from the new side chain, and (3) structural reorganization, which may be local or propagated elsewhere in the protein. Interpretation of the functional significance of changes in structure is greatly simplified if the mutation is carefully designed to minimize the latter two effects. Ward et al.[59] have proposed the following guidelines to avoid the difficulties stemming from the latter two effects:[59] (1) the amino acid residue is changed to one that is smaller than that in the wild type, and (2) a single functional group is deleted without introducing a new one. Larger side chains or new functional groups would be better introduced to surface loops rather than within areas of defined secondary structure or hydrogen bond networks. Upon mutation, one must pay attention especially to avoid disruption of pairing opposite charges and changes in the conformation of the peptide backbone. Otherwise, the alteration in the protein structure induced by the mutation will be too large to evaluate the role of the side chain in the protein function. One should keep in mind that a small change of only a few tenths of an angstrom may break a hydrogen bond and thus be readily detectable as a change in the protein function such as dissociation constant of a substrate, etc. However, we do not have to get nervous in designing a mutant protein by incorporation of an unnatural amino acid. Wells has pointed out that removal of a single molecular contact by a point mutation causes relatively small reductions (typically 0.5 to 5 kcal/mol) in the free energy compared to the overall free energy associated with the functional property (usually 5 to 20 kcal/mol).[60] Furthermore, in the majority of cases, the sum of the free energy changes derived from the single mutations is nearly equal to the free energy change measured in the multiple mutant. On the basis of additivity of mutational effects in proteins, molecular designing of a mutant enzyme will be possible with high confidence.

REFERENCES

1. **Spomer, W. E. and Wootton, J. F.,** The hydrolysis of α-N-benzoyl-l-argininamide catalyzed by trypsin and acetyltrypsin, *Biochim. Biophys. Acta,* 235, 164, 1971.
2. **Valenzuela, P. and Bender, M. L.,** Kinetic properties of succinylated and ethylene-diamine-amidated δ-chymotrypsins, *Biochim. Biophys. Acta,* 250, 538, 1971.
3. **Russell, A. J. and Fersht, A. R.,** Rational modification of enzyme catalysis by engineering surface charge, *Nature (London),* 328, 496, 1987.
4. **Warshel, A. and Russell, S. T.,** Calculations of electrostatic interactions in biological systems and in solutions, *Q. Rev. Biophys.,* 17, 284, 1984.
5. **Dao-Pin, S., Liao, D.-I., and Remington, S. J.,** Electrostatic fields in the active sites of lysozymes, *Proc. Natl. Acad. Sci. U.S.A.,* 86, 5361, 1989.
6. **Wells, J. A., Powers, D. B., Bott, R. R., Graycar, T. P., and Estell, D. A.,** Designing substrate specificity by protein engineering of electrostatic interactions, *Proc. Natl. Acad. Sci. U.S.A.,* 84, 1219, 1987.
7. **Estell, D. A., Graycar, T. P., Miller, J. V., Powers, D. B., Burnier, J. P., Ng, P. G., and Wells, J. A.,** Probing steric and hydrophobic effects on enzyme-substrate interactions by protein engineering, *Science,* 233, 660, 1986.
8. **Wells, J. A., Cunningham, B. C., Graycar, T. P., and Estell, D. A.,** Recruitment of substrate-specificity properties from one enzyme into a related one by protein engineering, *Proc. Natl. Acad. Sci. U.S.A.,* 84, 5167, 1987.
9. **Houghton, J. E., O'Donovan, G. A., and Wild, J. R.,** Reconstruction of an enzyme by domain substitution effectively switches substrate specificity, *Nature (London),* 338, 172, 1989.
10. **Kobilka, B. K., Kobilka, T. S., Daniel, K., Regan, J. W., Caron, M. G., and Lefkowitz, R. J.,** Chimeric α_2-, β_2-adrenergic receptors: delineation of domains involved in effector coupling and ligand binding specificity, *Science,* 240, 1310, 1988.

11. **Riedel, H., Dull, T. J., Schlessinger, J., and Ullrich, A.**, A chimaeric receptor allows insulin to stimulate tyrosine kinase activity of epidermal growth factor receptor, *Nature (London)*, 324, 68, 1986.

12. **Kashles, O., Szapary, D., Bellot, F., Ullrich, A., Schlessinger, J., and Schimdt, A.**, Ligand-induced stimulation of epidermal growth factor receptor mutants with altered transmembrane regions, *Proc. Natl. Acad. Sci. U.S.A.*, 85, 9567, 1988.

13. **Andersen, A. S., Kjeldsen, T., Wiberg, F. C., Christensen, M., Rasmussen, J. S., Norris, K., Moller, K. B., and Moller, N. P. H.**, Changing the insulin receptor to possess insulin-like growth factor I ligand specificity, *Biochemistry*, 29, 7363, 1990.

14. **Ahern, T. J., Casal, J. I., Petsko, G. A., and Klibanov, A. M.**, Control of oligomeric enzyme thermostability by protein engineering, *Proc. Natl. Acad. Sci. U.S.A.*, 84, 675, 1987.

15. **Perry, L. J. and Wetzel, R.**, Disulfide bond engineered into T4 lysozyme: stabilization of the protein toward thermal inactivation, *Science*, 226, 555, 1984.

16. **Wallace, C. J. A.**, Semisynthetic analogues of cytochrome c: modifications to fragment (66-104), in *Semisynthetic Peptides and Proteins*, Offord, R. E. and Bello, C. D., Eds., Academic Press, London, 1978, 101.

17. **Varadarajan, R., Zewert, T. E., Gray, H. B., and Boxer, S. G.**, Effects of buried ionizable amino acids on the reduction potential of recombinant myoglobin, *Science*, 243, 69, 1989.

18. **Pielak, G., Mauk, G., and Smith, M.**, Site-directed mutagenesis of cytochrome c shows than an invariant Phe is not essential for function, *Nature (London)*, 313, 152, 1985.

19. **Liang, N., Pielak, G. J., Mauk, G., Smith, M., and Hoffman, B. M.**, Yeast cytochrome c with phenylalanine or tyrosine at position 87 transfers electrons to (zinc cytochrome c peroxidase)$^+$ at a rate ten thousand times that of the serine-87 or glycine-87 variants, *Proc. Natl. Acad. Sci. U.S.A.*, 84, 1249, 1987.

20. **Louie, G. V., Pielak, G. J., Smith, M., and Brayer, G. D.**, Role of phenylalanine-82 in yeast iso-1-cytochrome c and remote conformational changes induced by a serine residue at this position, *Biochemistry*, 27, 7870, 1988.

21. **Hazzard, J. T., McLendon, G., Cusanovich, M. A., Das, G., Sherman, F., and Tollin, G.**, Effects of amino acid replacements in yeast iso-1 cytochrome c on heme accessibility and intracomplex electron transfer in complexes with cytochrome c peroxidase, *Biochemistry*, 27, 4445, 1988.

22. **Das, G., Hickey, D. R., Principio, L., Conklin, K. T., Short, J., Miller, J. R., McLendon, G., and Sherman, F.**, Replacements of lysine 32 in yeast cytochrome c, *J. Biol. Chem.*, 263, 18290, 1988.

23. **Poulos, T. L. and Finzel, B. C.**, Heme enzymes structure and function, *Pep. Protein Rev.*, 4, 115, 1984.

24. **Hilvert, D., Hatanaka, Y., and Kaiser, E. T.**, A highly active thermophilic semisynthetic flavoenzyme, *J. Am. Chem. Soc.*, 110, 682, 1988.

25. **Kaiser, E. T. and Lawrence, D. S.**, Chemical mutation of enzyme active sites, *Science*, 226, 505, 1984.

26. **Slama, J. T., Radziejewski, C., Oruganti, S. R., and Kaiser, E. T.**, Semisynthetic enzymes: characterization of isomeric flavopapains with widely different catalytic efficiencies, *J. Am. Chem. Soc.*, 106, 6778, 1984.

27. **Radziejewski, C., Ballou, D. P., and Kaiser, E. T.**, Catalysis of *N*-alkyl-1,4-dihydronicotinamide oxidation by a flavopapain: rapid reaction in all catalytic steps, *J. Am. Chem. Soc.*, 107, 3352, 1985.

28. **Kaiser, E. T.**, Catalytic activity of enzymes altered at their active sites, *Angew. Chem., Int. Ed. Engl.*, 27, 913, 1988.

29. **Kohen, F., Kim, J.-B., Barnard, G., and Lindner, H. R.**, Antibody-enhanced hydrolysis of steroid esters, *Biochim. Biophys. Acta*, 629, 328, 1980.

30. **Pollack, S. J., Jacobs, J. W., and Schultz, P. G.,** Selective chemical catalysis by an antibody, *Science,* 234, 1570, 1986.
31. **Tramontano, A., Janda, K. D., and Lerner, R. A.,** Chemical reactivity at an antibody binding site elicited by mechanistic design of a synthetic antigen, *Proc. Natl. Acad. Sci. U.S.A.,* 83, 6736, 1986.
32. **Tramontano, A., Janda, K. D., and Lerner, R. A.,** Catalytic antibodies, *Science,* 234, 1566, 1986.
33. **Tramontano, A., Ammann, A. A., and Lerner, R. A.,** Antibody catalysis approaching the activity of enzymes, *J. Am. Chem. Soc.,* 110, 2282, 1988.
34. **Napper, A. D., Benkovic, S. J., Tramontano, A., and Lerner, R. A.,** A sterospecific cyclization catalyzed by an antibody, *Science,* 237, 1041, 1987.
35. **Janda, K. D., Lerner, R. A., and Tramontano, A.,** Antibody catalysis of bimolecular amide formation, *J. Am. Chem. Soc.,* 110, 4835, 1988.
36. **Cochran, A. G., Sugaawara, R., and Schultz, P. G.,** Photosensitized cleavage of a thymine dimer by an antibody, *J. Am. Chem. Soc.,* 110, 7888, 1988.
37. **Schultz, P. G.,** The interplay between chemistry and biology in the design of enzymatic catalysts, *Science,* 240, 426, 1988.
38. **Iverson, B. L. and Lerner, R. A.,** Sequence-specific peptide cleavage catalyzed by an antibody, *Science,* 243, 1184, 1989.
39. **Janijic, N. and Tramontano, A.,** Antibody-catalyzed redox reaction, *J. Am. Chem. Soc.,* 111, 9109, 1989.
40. **Shokat, K. M., Leumann, C. J., Sugasawara, R., and Schultz, P. G.,** An antibody-mediated redox reaction, *Angew. Chem., Int. Ed. Engl.,* 27, 1172, 1988.
41. **Shokat, K. M., Leumann, C. J., Sugasawara, R., and Schultz, P. G.,** A new strategy for the generation of catalytic antibodies, *Nature (London),* 338, 269, 1989.
42. **Janda, K. D., Schloeder, D., Benkovic, S. J., and Lerner, R. A.,** Induction of an antibody that catalyzes the hydrolysis of an amide bond, *Science,* 241, 1188, 1988.
43. **Benkovic, S. J., Napper, A. D., and Lerner, R. A.,** Catalysis of a stereospecific bimolecular amide synthesis by an antibody, *Proc. Natl. Acad. Sci. U.S.A.,* 85, 5355, 1988.
44. **Hilvert, D., Carpenter, S. H., Nared, N. K., and Auditor, M.-T., M.,** Catalysis of concerted reactions by antibodies: the Claisen rearrangement, *Proc. Natl. Acad. Sci. U.S.A.,* 85, 4953, 1988.
45. **Jackson, D. Y., Jacobs, J. W., Sugasawara, R., Reich, S. H., Bartlett, P. A., and Schultz, P. G.,** An antibody-catalyzed Claisen rearrangement, *J. Am. Chem. Soc.,* 110, 4841, 1988.
46. **Janda, K. D., Benkovic, S. J., and Lerner, R. A.,** Catalytic antibodies with lipase activity and R or S substrate selectivity, *Science,* 244, 437, 1989.
47. **Hilvert, D., Hill, K. W., Nared, K. D., and Auditor, M.-T. M.,** Antibody catalysis of a Deils-Alder reaction, *J. Am. Chem. Soc.,* 111, 9261, 1989.
48. **Blackburn, G. M., Kang, A. S., Kingsbury, G. A., and Burton, D. R.,** Catalytic antibodies, *Biochem. J.,* 262, 381, 1989.
49. **Pollack, S. J. and Shultz, P. G.,** A semisynthetic catalytic antibody, *J. Am. Chem. Soc.,* 111, 1929, 1989.
50. **Pollack, S. J., Nakayama, G. R., and Schultz, P. G.,** Introduction of nucleophiles and spectroscopic probes into antibody combining sites, *Science,* 242, 1038, 1988.
51. **Jackson, D. Y., Prudent, J. R., Baldwin, E. P., and Schultz, P. G.,** A mutagenesis study of a catalytic antibody, *Proc. Natl. Acad. Sci. U.S.A.,* 88, 58, 1991.
52. **Mutter, M. and Vuilleumier, S.,** Chemical approach to protein design — Template-assembled synthetic proteins, *Angew. Chem., Int. Ed. Engl.,* 28, 535, 1989.
53. **Sasaki, T. and Kaiser, E. T.,** Helichrome: synthesis and enzymatic activity of a designed heme protein, *J. Am. Chem. Soc.,* 111, 380, 1989.
54. **Hahn, K. W., Klis, W. A., and Stewart, J. M.,** Design and synthesis of a peptide having chymotrypsin-like esterase activity, *Science,* 248, 1544, 1990.

55. **Kimura, S. and Imanishi, Y.,** *Peptide Chemistry 1990,* in press.
56. **Noren, C. J., Anthony-Cahill, S. J., Griffith, M. C., and Schultz, G. P.,** A general method for site-specific incorporation of unnatural amino acids into proteins, *Science,* 244, 182, 1989.
57. **Heckler, T. G., Chang, L.-H., Zama, Y., Naka, T., Chorghade, M. S., and Hecht, S. M.,** T4 RNA ligase mediated preparation of a novel "chemically misacylated" tRNA^{Phe}s, *Biochemistry,* 23, 1468, 1984.
58. **Bain, J. D., Diala, E. S., Glabe, C. G., Dix, T. A., and Chamberlin, A. R.,** Biosynthetic site-specific incorporation of a non-natural amino acid into a polypeptide, *J. Am. Chem. Soc.,* 111, 8013, 1989.
59. **Ward, W. H. J., Timms, D., and Fersht, A. R.,** Protein engineering and the study of structure-function relationships in receptors, *TiPS,* 11, 280, 1990.
60. **Wells, J. A.,** Additivity of mutational effects in proteins, *Biochemistry,* 29, 8509, 1990.

Biocomposite Materials for Cell Culture

5.1 CELL ADHESION FACTOR IMMOBILIZED MATERIALS

Yoshihiro Ito

TABLE OF CONTENTS

I. INTRODUCTION

Cell adhesion is a ubiquitous process that influences many aspects of cell behavior. For example, proliferation, migration, secretion, and differentiation of cells are triggered by adhesion to a matrix. The effects of the surface properties of polymer film on cell adhesion have been extensively investigated over many years.

These investigations can be classified into three categories from the perspective of material science, i.e., (1) physicochemical, (2) morphological, and (3) biochemical approaches.

II. PHYSICOCHEMICAL AND MORPHOLOGICAL APPROACHES

A. PHYSICOCHEMICAL APPROACHES

The physicochemical approach is based on the control of surface free energy, wettability, or surface electric charges. Cell adhesion onto polystyrene is enhanced by hydrophilization by glow discharge or sulfuric acid treatment.[1-3] Curtis et al.[3-5] emphasized the importance of hydroxy groups on the polystyrene surface. Wildevuur's group[6,7] reported that human endothelial cells adhere to more hydrophilic surfaces as shown in Figure 1. Cell adhesion onto hydrogels is difficult as shown in Figure 2.[8,9] Therefore, van Wachem et al.,[10-14] Kang et al.[15] and Ikada and Tamada,[16] indicated that excessively hydrophilic or hydrophobic surfaces suppress cell adhesion or spreading as shown in Figure 3. These opposite conclusions have been the subject of controversy.

Absolom et al.[17-22] have demonstrated that the extent of cell-surface interactions is determined to a significant degree by the wettability of the substrate material and the surface tension of the liquid suspension medium. When the surface tension (γ_{LV}) of the medium is larger than that of the surface tension (γ_{CV}) of the adhering cells, then the extent of cell adhesion (i.e., number of cells per unit surface area) decreases with increasing substrate surface tension (γ_{SV}). Under the experimental conditions where $\gamma_{LV} < \gamma_{CV}$, the reverse pattern is observed, i.e., cell adhesion increases with increasing γ_{SV}.[17-22] The nature of cells[23] and the conditions of the adhesion, such as concentrations of serum and ions,[24] should be taken into consideration. Figure 4 shows the different behavior of two cell types. On the other hand, Horbett et al. demonstrated that cell adhesion and spreading differ as shown in Figures 2 and 3d.

Furthermore, a most significant parameter is the rapid adsorption of serum proteins.[13,14,19,25-27] Protein adsorption has two consequences: (1) it gives rise to markedly altered surface properties of the polymer material; and (2) it may also result in an altered conformation of the protein molecules depending on the nature of both the substrate material and the adsorbed protein. Horbett

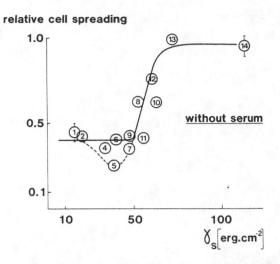

FIGURE 1.   Relative cell spreading without serum as a function of the substratum surface-free energy. The dotted line represents the possible bioadhesive minimum. 1, Fluoroethylenepropylene copolymer; 2, fluoroethylenepropylene copolymer; 3, polytetrafluoroethylene; 4, silicone rubber; 5, paraffin; 6, polyethylene; 7, polycarbonate; 8, polyetherurethane; 9, polyesterurethane; 10, polysulfone; 11, polyvinylidenefluoride; 12, poly(methyl methacrylate); 13, tissue culture polystyrene; 14, glass. (From Schakenraad, J. M., Busscher, H. J., Wildevuur, C. R. H., and Arends, J., *J. Biomed. Mater. Res.*, 20, 773, 1986. With permission.)

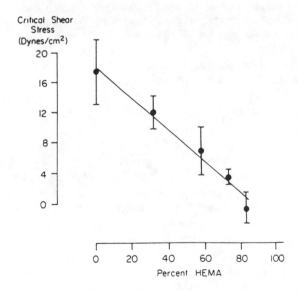

FIGURE 2.   Relationship between surface-free energy and cell adhesion. Critical shear stress for 3T3 cell detachment from HEMA-EMA copolymers. (From Horbett, T. A., Waldeburger, J. T., Ratner, B. D., and Hoffman, A. S., *J. Biomed. Mater. Res.*, 22, 383, 1988. With permission.)

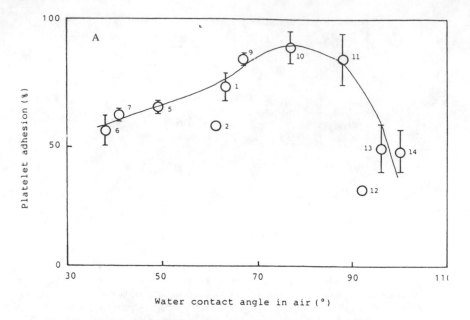

FIGURE 3.   Relationship between surface wettability and cell adhesion. (A) Human platelet adhesion to polypeptide derivatives as a function of water contact angle. 1, poly(γ-benzyl L-glutamate) (PBLB); 2, poly(ε-N-benzyloxycarbonyl L-lysine) (PBCL); 3, hydrolyzed PBLG; 4, PBCL treated with HBr; 5-7, aminolyzed PBLG; 8, 9, PBLG-PBCL copolymers; 10, 11, PBLG containing trimethylsilyl side groups; 12, PBLG containing polydimethylsiloxane; 13, 14, PBLG containing fluoroalkyl side groups; 14, glass, (From Kang, I.-K., Ito, Y., Sisido, M., and Imanishi, Y., *J. Biomed. Mater. Res.*, 23, 223, 1989. With permission.)

and Schway[26] described that the spreading of cells in serum was well correlated with the amount of fibronectin adsorption to the substrates. Attachment was not correlated with fibronectin adsorption, especially on glass preadsorbed with diluted serum. They considered that for cells that have a receptor in the medium, the amount of adsorption of this protein to the substrate appeared to be a critical factor controlling cell interactions with the substrate.

Charge effect is also important for cell attachment.[28] Generally speaking, the cell surface is negatively charged. Therefore, polylysine and other cationic polymers have been used as an accelerator of cell adhesion, which is based on electrostatic interactions.[29-31] Polyethyleneimine was also utilized. Recently, polyallylamine was discovered to be a very efficient cell immobilization agent.[31] The results of an attachment assay using baby hamster kidney cells (BHK-21, ATCC CCL 10); human histocytic lymphoma cells (U-937, ATCC CRL 1593); and lymphocytic cells (P3X63-Ag8.653, ATCC CRL 1580) are shown in Table 1. This result indicates that polyallylamine and poly L-lysine are efficient mediators of cell attachment and that polyallylamine can be used effectively at much lower concentrations that either CELL TAKR adhesive or poly L-lysine.

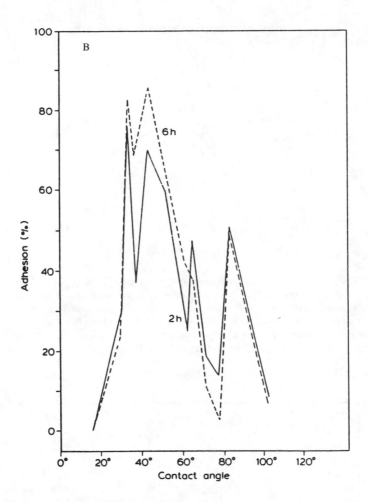

FIGURE 3(B).   Adhesion of human endothelial cells onto various wettable surfaces. (From van Wachem, P. B., Beugeling, T., Feijen, J., Bantjes, A., Detmers, J. P., and van Aken, W. G., *Biomaterials,* 6, 403, 1985. With permission.)

On the other hand, Nagai and Machida[32] and others[33-36] indicated that controlled systems containing poly(acrylic acid) could be considered as mucoadhesives, probably due to the interactions of their carboxylatic groups with functional groups of the mucus. Vogler and Bussian[37] and McAuslan and Johnson[27] reported that cell adhesion appeared to correlate with the introduction of surface COOH. McAuslan et al.[38] reported that endothelial cells adhered to a Nafion membrane, which contains perfluorosulfonic acid, less than to tissue culture polystyrene dishes, but more than to a Teflon membrane. Aubert et al.[39] cultured insulin-secreting cells on $PSSO_3Na$ microbeads. The cell growth rate was either identical to that of cells grown on plastic wells or slower, depending on the initial cell concentration. However, no spreading

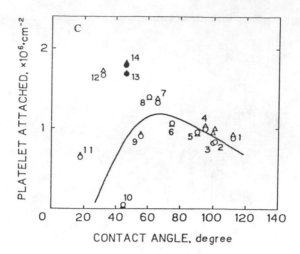

FIGURE 3(C). Adhesion of human platelets onto materials. (◯) Measured by LDH activity measurement, (△) measured by a radioactive method. 1, Polytetrafluoroethylene; 2, silicone rubber; 3, tetrafluoroethylene-hexafluoroethylene copolymer; 4, polyethylene; 5, polypropylene; 6, polystyrene; 7, polyterephtalate; 8, nylon; 9, VAECO; 10, poly(vinyl alcohol); 11, cellulose; 12, glass; 13, collagen-coated high density polyethylene; 14, fibronectin-coated high density polyethylene. (From Tamada, Y., Kyoto University, Ph.D. thesis, 1988. With permission.)

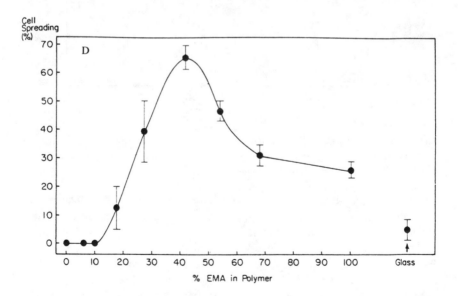

FIGURE 3(D). Relative cell spreading on HEMA-EMA copolymers. (From Horbett, T. A., Waldeburger, J. T., Ratner, B. D., and Hoffman, A. S., *J. Biomed. Mater. Res.*, 22, 383, 1988. With permission.)

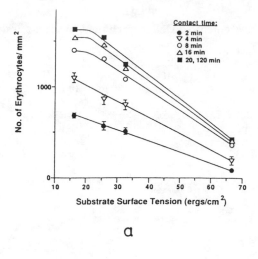

a

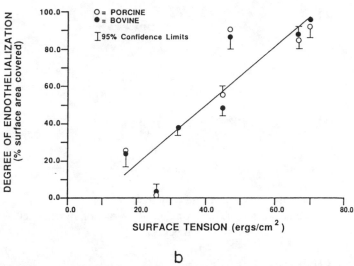

b

FIGURE 4. Relationship between surface-free energy and cell adhesion. (a) Fixed human erythrocyte adhesion as a function of substrate surface tension for various contact times as indicated. (From Absolom, D. R., Thomson, C., Hawthorn, L. A., Zingg, W., and Neumann, A. W., *J. Biomed. Mater. Res.*, 22, 215, 1988, Figure 7. With permission.) (b) Degree of endothelial cell adhesion as a function of substrate surface tension, γ_{SV}. Substrate surface tension is determined from contact angle measurements via an equation of state approach. Degree of cell adhesion is defined as the extent of cell coverage expressed as a percentage of total available substrate surface area. Substrates are preconditioned by incubation in nutrient media containing 10% fetal calf serum. (From Absolom, D. R., Hawthorn, L. A., and Chang, G., *J. Biomed. Mater. Res.*, 22, 271, 1988. With permission.)

TABLE 1
Enhancement of Cell Attachment by Bioadhesives

	Cell type				
	BHK-21	P3X		U-937	
Adhesive	15 μg	25 μg	10 μg	15 μg	25 μg
CELL-TAK® adhesive	83	99	0	15	89
Poly-L-lysine	97	93	91	97	94
Polyallylamine	98	92	97	98	98
Plastic	19	17	20	13	14

Note: Values represent percent cell attachment.

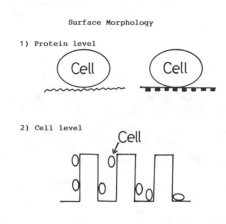

FIGURE 5.   Morphological approaches to control cell adhesion.

was observed when cells were cultured on the microbeads. Furthermore, their study provided an example of cell-biomaterial interaction in which cell growth was possible, but with altered cell function.

B. MORPHOLOGICAL APPROACHES

The morphological approach is based on the generation of phase structure, the size of which is measured either at the protein level or the cellular level as shown in Figure 5. The protein level approach was invoked by consideration of the relatively high nonthrombogenicity of polyetherurethanes. It has been suggested that the formation of a domain structure in the range of some tens to some hundreds of angstroms, is related to high blood compatibility.[40] This view has been extended to adhesion of platelets and lymphocytes onto block copolymer films having a microphase-separated structure on the order of some thousands of angstroms.[41,42] However, formation of a microphase-separated structure on the film surface of block copolymers[43] as well as polyetherurethanes[44] has recently become suspect. These situations already have been described in Chapter 2.1.

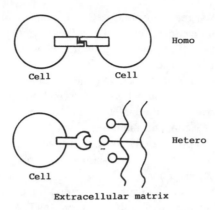

FIGURE 6.   Communications between cell-cell, and cell-extracellular matrices.

At another level of morphology, the roughness of the surface on the cellular level has been of recent interest.[45-58] Cells recognize and respond to the curvature of rough surfaces. These situations are similar to the seeding of endothelial cells or the promotion of endothelialization mentioned in Chapter 2.1.

III. BIOCHEMICAL APPROACHES

In the third approach, biological components such as proteins or polysaccharide derivatives containing cell-adhesive activity have been utilized to enhance cell adhesion. Generally, it has been suggested that cell-cell communication is based on homogeneous polysaccharide-polysaccharide interaction, and that cell-extracellular matrix communication is based on heterogenous protein-protein interactions as shown in Figure 6.

A. BIOLOGICAL MACROMOLECULE IMMOBILIZATION
1. Protein Immobilization
a. Collagen

Over the years, collagen has been used in such diverse medical applications as hemostasis, nerve regeneration, injectable collagen for tissue augmentation, burn and wound dressing, hernia repair, urinary tract surgery, drug delivery system, ocular surfaces, vaginal contraceptive barrier, bioprosthetic heart valves, and vascular grafts as reviewed by several researchers.[59-68] The majority of uses require that the collagen be cross-linked, rendering it temporarily nondegradable. Cross-linking increases its strength and stiffness, and decreases its degradation rate and antigenicity.

Among the cross-linking agents, glutaraldehyde (GTA) has been most popular, and the nature of the cross-linking of collagen with GTA has been discussed in detail by several researchers.[69-76] GTA autopolymerizes and subsequently hydrolyzes, releasing free, toxic glutaraldehyde into the local environment. Huang-Lee et al.[77] reported that local cytotoxicity of GTA cross-linked bioprostheses may be due to unstable GTA polymers that persist in the interstices of cross-linked tissues.

Nimni et al.[78] attempted to overcome these problems by enhancing cross-linking through (1) bridging of activated carboxyl groups with diamines and (2) using GTA to cross-link the ϵ-NH_2 groups in collagen and the unreacted amines introduced by aliphatic diamines. This cross-linking reduced tissue degradation and nearly eliminated humoral antibody induction. Covalent binding of diphosphonates, specifically 3-amino-1-hydroxypropane-1,1-diphosphonic acid (3-APD), and chondroitin sulfate to collagen or to the cross-link-enhanced collagen network reduces its potential for calcification. The optimum quantity and conditions for the addition of GTA for cross-linking of collagen gels which can be used as biomaterials has been determined.[79,80]

This phenomenon induced by GTA has spurred the search for other, less toxic, cross-linking agents, including dehydration,[81,82] chromium, irradiation by ultraviolet light,[83] and carbodiimide cross-linking.[84] Warcquier-Clerout et al.[85] reported the *in vitro* stability of molecules of collagen cross-linking as a function of time. They indicated improved stability of the carbodiimide-cross-linked proteins compared to those with GTA. The acyl azide method has been applied by some researchers.[86] Carboxyl groups of collagen were transformed in acyl azide via a three-step reaction and then were allowed to react with the amino group of an adjacent collagen chain. Petite et al.[87] found the greatest increase in the thermal stability of collagen. Collagen, chemically modified by Kasai et al.,[88] influenced cell attachment and growth.

The nature of the cross-linker and cross-linking density are important in drug delivery devices. Gilbert and Kim[89] examined the effect of the quaternary structure on the release rates of insulin from monolithic devices. Rosenblatt et al.[90] also investigated the effect of collagen fiber size on the release rate of proteins. The mesh size of the collagen matrix was computed from the collagen fiber size distribution, and geometric arguments were based on the collagen fiber as a rigid cylinder. Regions of the release curve which agreed with predictions based on a diffusive mechanism were identified. They considered that fibrillar collagen matrices were capable of moderating the release rates of very large proteins only (such as fibrinogen) and a significant non-fibrillar content was necessary to modulate the diffusivity of smaller proteins such as chymotrypsinogen.

b. Hybridization of Collagen with Other Components

Collagen-chondroitin sulfate was prepared by Umemura et al.[91] Chondroitin-4-sulfate (CH-4-S) and chondroitin-6-sulfate (CH-6-S), predominant components of the intima among glycosaminoglycans, have shown little inhibitory effect on collagen-induced platelet aggregation.[92-95] In the work of Klein et al.,[96] chondroitin sulfates did not reduce the ability of collagen to aggregate human platelets, but this may have been due to insufficient time for collagen-chondroitin sulfate interaction. Umemura et al.[91] reported that the collagen-CH-6-S complex had a greater effect than the CH-6-S/collagen layer on platelet activation. Collagen glycosaminoglycan composites have been used in the design of an artificial skin for the treatment of burns.[97-100]

The presence of hyaluronic acid and/or fibronectin in collagen sponges was shown to enhance the repair of dermal wounds, as well as promoting the extent of cellular infiltration and colonization by fibroblasts *in vitro,* as measured by cell replication and increased collagen biosynthesis.[101] Srivastava et al.[102-104] prepared hyaluronic acid-collagen composites. They showed that 2.5% hyaluronic acid was the optimum concentration for cell proliferation. At low concentrations of hyaluroic acid, the collagen fibers were not fully coated, and areas of both hyaluroic acid and collagen were available for cell adhesion, either directly or through any mediating protein. It was considered that the reduced cell proliferation at higher concentrations of hyaluronic acid might be attributed to the glycosaminoglycan acting as a physical barrier between the cells and the collagen. Similar phenomena were observed by another group.

Factor XIII collagen was synthesized by Blanchy et al.[107] Factor XIII was grafted onto collagen membranes by the acyl-azide procedure. Factor XIII or fibrin-stabilizing factor is the last plasma zymogen of the coagulation cascade to become activated during blood clot formation. This proenzyme is converted by thrombin into an active enzyme, factor XIIIa, also termed transamidase, transglutaminase, or fibrinoligase. This activated enzyme stabilizes the fibrin clot in the presence of calcium ions by inducing the formation of covalent bonds between neighboring fibrin molecules. Besides its major role in the coagulation process, factor XIII also promotes both wound and fracture healing. The activity of factor XIII grafted onto collagen membranes was almost constant over a period of 8 months. Blanchy et al.[107] reported that the potential of such a material for biomedical use was presently under investigation. Collagen-heparin conjugates are reviewed in Chapter 2.1.

Collagen has been hybridized with synthetic polymers by many researchers.[108-112] Amudeswari et al.[108] prepared a hydrogel by copolymerization of 2-hydroxyethylmethacrylate (HEMA), methylmethacrylate (MMA), and glycidylmethacrylate (GDMA) onto soluble collagen using cross-linking agents. Jeyanthi and Rao[109] synthesized collagen-polyHEMA hydrogen matrices. In controlled drug delivery, especially in cancer therapy, a prolonged and site-specific effect is very important. Collagen hybridization is considered to be useful for this system. Three potent cancer chemotherapeutic agents, namely, 5-fluorouracil, bleomycin, and mitomycin, were entrapped in the matrices. They reported that zero-order release was achieved.

Gilbert and Lyman[111] fabricated composites composed of collagen coated on urethane and Silastic Rubber R films to give improved tear resistance. Tran and Walt[112] developed new methods for covalently binding collagen to

PTFE grafts modified initially by plasma deposition to form an alkylamine surface. Collagen was attached to the graft via a glutaraldehyde cross-linker. Specific staining procedures for collagen revealed that the adventitial or luminal surfaces of the graft could be modified selectively and only at the surface.

Collagen-mineral gel implants have been synthesized as biomaterials.[113] Lizarbe et al.[114-116] described that sepiolite, a clay (hydrated magnesium silicate), forms stable complexes with collagen, and they investigated the interaction of the complexes with human fibroblasts.

Apatite-collagen was synthesized by Okazaki et al.[117] A carbon-containing hydroxyapatite with a chemical composition and crystalline structure similar to that of bone was synthesized. The apatite mixed with collagen solution, from which antigenicity had been removed by enzymatic treatment, was formed into apatite-collagen pellets. After insolubilization by UV irradiation, the composites showed remarkably reduced disintegration and good biocompatibility when implanted into the rat abdomen.

c. Gelatin

Gelatin has also been used by many researchers. Gelatin is water-soluble and is a good substrate for fibroblasts or epithelial cells. Therefore, Yoshikawa et al. tried to use gelatin cross-linked by glycerol polyglycidylether as artificial skin.[118] Gourevitch et al.[119] reported that cultured human adult endothelial cells adhere most effectively to prosthetic surfaces precoated with gelatin, and remain attached following exposure to shear forces.

d. Fibronectin

Fibronectin (FN) is an extracellular matrix glycoprotein that mediates the adhesion of many mesenchymal cells to their collagen environment.[120-122] This occurs by the binding of FN to a cell surface glycoprotein receptor complex, called the integrins,[123-125] as well as to cell surface heparan sulfate proteoglycans[126,127] in order to facilitate the complete physiological response from some cells. Studies of the molecular mechanisms by which FNs bind to artificial matrices and whether the composition of biomaterials can modulate the biological activities of FNs have been limited. Grinnell and colleagues described different adhesion-promoting activities of fibroblasts upon plasma FN binding to tissue culture or bacteriological polystyrene surfaces. The stoichiometry and stability of FN binding to some artificial surfaces have also been examined.[128-134]

Recently, it has been reported that the binding of plasma FN to derivatized glass surfaces alters FN conformation so that the adhesion response of fibroblasts and neuroblastoma cells are modulated in distinctive ways.[135] These altered responses are cell type specific, i.e., fibroblast changes on FN-coated surfaces differed from those of neuroblastoma cells, demonstrating differences in the permutation and complementation of different cell surface receptors in mediating these effects by mesenchymally derived fibroblasts or nervous system-derived neuroblastoma cells.[135] Human neuroblastoma neurites are particularly useful in such studies and easy to enumerate because of their long linear surface extensions. They require a complex array of signals from the FNs in order to achieve growth cone migration on the substratum.[126,127,136-144] The ability to deposit cellular fibronectin onto a polymer surface is necessary for the spreading and proliferation of human endothelial cells, as indicated by van Wachem.[145]

Chinn et al.[146] reported that plasma treatment of poly(tetrafluoroethylene) and polystyrene enhanced FN adsorption, which led to cell adhesion on their surfaces. Preadsorbed FN resulted in a greater number of bonds between the material surface and the cells, which in turn promoted cell spreading and increased the adhesive strength of the cell.[147] Klein-Soyer et al.[30] compared two culture surfaces, FN-coated tissue culture grade polystyrene and a surface-modified polystyrene called Primaria (Falcon). Adhesion and growth of endothelial cells at low and clonal density were identical on both substrates, and the biological properties were preserved. Regeneration of injured endothelium was less simple to study on Primaria polystyrene because the extracellular matrix was damaged during the lesion process. Nevertheless, Primaria polystyrene can easily be substituted for FN coating in grow experiments, especially at a very low seeding density.

e. Vitronectin

Vitronectin (VN) is of interest because of its roles in cell adhesion and spreading, in coagulation, in complement activation, and in three important aspects of artificial surface induced thrombosis. VN accounts for most of the cell adhesion and spreading activity of serum.[148] VN participates in the coagulation pathway by binding to heparin,[149,150] and by binding to antithrombin III in the thrombin-antithrombin III complex in a manner which slows down the inactivation of thrombin.[151] Preadsorption of vitronectin on polymeric arteriovenous shunts stimulated thrombosis in a canine *ex vivo* experimental model as measured by increased platelet and fibrinogen deposition compared to uncoated shunts.[152] In the complement pathway, VN has a regulatory role in the membrane attack complex. VN binds to the newly exposed hydrophobic region of the C5b-7 complex,[153] which maintains its solubility thus preventing adsorption to and lysis of cell membranes. In addition to protecting bystander cells against lysis, VN inhibits C9 polymerization.[153]

Pitt et al.[154] investigated the adsorption of vitronectin onto polystyrene and oxidized polystryrene. They did not find significantly different adsorption behavior. The secondary structure of vitronectin contains considerable β- and random structures. Upon adsorption, vitronectin loses its β-sheet structure.

Serotonin-immobilized material was found to enhance vitronectin adhesion by Hannan and McAuslan.[155]

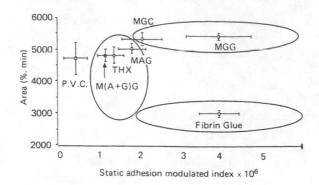

FIGURE 7.   Cell adhesion (static adhesion modulated index) vs. cell spreading (area) on various materials. (From Duval, J. L., Letort, M., and Sigot-Luizard, M. F., *Biomaterials*, 9, 155, 1988. With permission.)

f. Fibrin

Fibrin has been used as a natural "glue". Duval et al.[156] compared membranes of fibrin glue with those of gelatin and albumin cross-linked by glutaldehyde (MGG and MAG), gelatin cross-linked by carbodiimide (MGC), mixed gelatin and albumin cross-linked by glutaraldehyde (M[A + G]G), Thermanox plastic coverslips and poly(vinyl chloride) (PVC). The fibrin glue enhanced cell growth and adhesion as shown in Figure 7.[156]

Rabaud et al.[157-162] reported a new biomaterial resulting from the basic reaction of fibrin monomers on elastin. Thrombin releases two molecules of fibrinopeptide A (FPA) and two molecules of fibrinopeptide B (FPB) from the Aα- and Bβ-chains of fibrinogen, and FPA is released at a much faster rate than PFB. When fibrinogen was treated by reptilase, only FPA was released. The fibrin monomers behaved differently when generated either by thrombin or reptilase. Using thrombin, adduct synthesis was more rapid. However, whatever the enzyme, monomers bound to elastin were the same. Since reptilase released only FPA, it can be suggested that the reaction with elastin must occur from the A site of the Aα-part of the fibrinogen molecule. The rapid reaction with thrombin could be explained as follows; if FPB is released after FPA, it would produce an inducing effect on the Aα-site of fibrinogen. When fibronectin was added to fibrinogen, it markedly improved the quality of the material, as did the addition of a small amount of collagen, yielding a crude matrix. This has been used in digestive surgery to repair experimental fistulae in rat cecum and in vascular surgery to repair experimental arteriotomies in rabbit abdominal aortas. The material induced restoration of the tissue to its former condition and disappeared after 3 months, leaving no trace.

g. Lectin

A growing body of data supports the hypothesis that cell-surface glyco-conjugates can act as specific cell-to-cell and cell-surface recognition molecules.[163] Lectins have been widely used for characterization of the cell surface and subcellular distribution of glycoconjugate moieties at the tissue level and in cultured cells.[164]

Parhizgar et al.[165] investigated several kinds of lectins as binding materials for endothelial cell adhesion onto polystyrene surfaces. The lectins were immobilized by carbodiimide. Their results showed that α-L-fucose binding lectin was a specific and fast-attaching substrate for human umbilical vein endothelial cells.

2. Polysaccharide-Based Material

It seems that proteoglycans and hyaluronic acid play a crucial role during morphogenesis. As mentioned before, polysaccharide components are useful for hybridization with other cell adhesive proteins.

Chitin, a polysaccharide composed of GlcNAc, has been proposed as an absorbable suture material which is widely acceptable because the tissue reaction is not specific and the healing which ensues provides evidence for satisfactory biocompatibility; there are generally no adverse effects.[166-168]

Chitosan, the partially deacetylated form of chitin, has been used to inhibit firboplasia in wound healing, and to promote tissue growth and differentiation in tissue culture.[169,170] Chitosan provides a nonprotein matrix for three-dimensional tissue growth.

B. ACTIVE SITE IMMOBILIZATION
1. Peptide Immobilization

In 1984, Pierschbacher and Ruoslahti[171] found that the Arg-Gly-Asp-Ser (RGDS) sequence constitutes the active site of fibronectin. It was then clarified that this sequence is generally contained in cell adhesive proteins, such as vitronectin,[172] laminin,[173] von Willebrand factor,[174] type I collagen,[175] fibrinogen,[176] thrombospondin,[177] osteopontin,[178] and tenascin[179] as shown in Figure 8.

This finding provoked the design and synthesis of new cell-adhesive materials composed of this peptide RGD. The ideas are summarized in Figure 9.

Imanishi et al.[180] immobilized the peptide RGDS onto silicone films, and remarkably enhanced cell adhesion. They introduced amino groups on the surface by immersing silicone rubber film into aminoalkylated polydimethylsiloxane. The RGDS peptide was then coupled using water-soluble carbodiimide. The number of adherent cells increased with the increasing peptide concentration linked to the surface of the silicone film.

Matsuda et al.[181] immobilized the peptide onto poly(vinyl alcohol) (PVA) film. Activation of surface hydroxyl groups by carbonyl diimidazole successfully incorporated GRGDSP (Gly-Arg-Gly-Asp-Ser-Pro) onto PVA films. The resultant film surface was bioactive, molecularly recognizing the adhesive receptor of bovine endothelial cells.

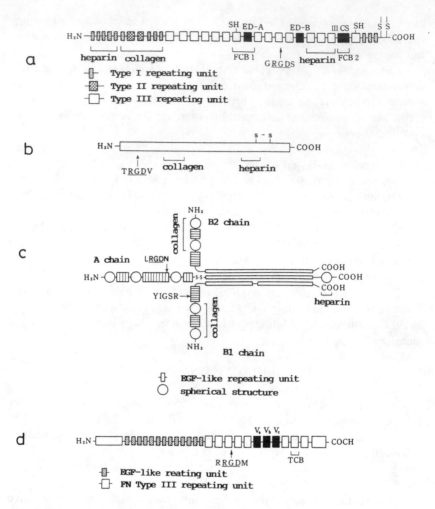

FIGURE 8. Adhesive proteins, collagen, fibronectin, fibrinogen, vitronectin, thrombospondin, tenacine.

Massia and Hubbell[182] immobilized the synthetic peptides Gly-Arg-Gly-Asp-Tyr and Gly-Tyr-Ile-Gly-Ser-Arg-Tyr, which contain Arg-Gly-Asp (RGD) and Tyr-Ile-Gly-Ser-Arg (YIGSR), the ligands for two important classes of cell adhesion receptors. They were covalently coupled to a nonadhesive modified glass surface by the N terminal Gly. Glycerolpropylsilane-bonded glass (glycophase glass) was used as the substrate. The glass contains a covalently bound organic layer that imbibes water and reduces protein adsorption similarly to hydrogens without the problems of swelling and bulk permeation of aqueous solutes. Since glycophase glass adsorbs proteins poorly, it alone is not suitable for supporting cell adhesion, even with serum in the medium.

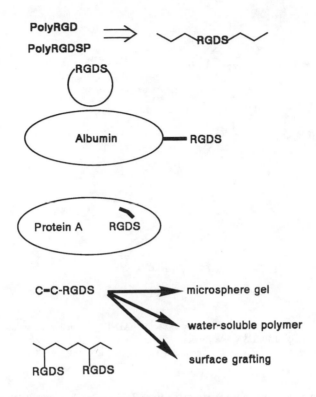

FIGURE 9. Materials using RGDS peptide sequences.

Human foreskin fibroblasts spread on both immobilized peptide substrates, but at much slower rates on grafted YIGSR glass surfaces than on the RGD-containing substrates. Cells formed focal contacts on the RGD-derivatized substrates in the presence or absence of serum. Focal contacts formed on the YIGSR-grafted surfaces only when serum was present in the medium and then morphologies differed from those observed on the RGD-containing substrates. Serum influenced the organization of microfilaments and the extent of spreading of adherent cells, although adsorption of adhesion proteins was minimal on all substrates.

On the other hand, the peptide RGDS was inserted into nonadhesive protein, albumin, or protein A by chemical modification[183] of gene technology,[184] respectively. Danilov and Juliano[183] prepared protein-peptide conjugates composed of bovine serum albumin (BSA) derivatized with short peptides containing the Arg-Gly-Ser (RGD) sequence derived from the adhesion site of fibronectin. The RGD-BSA conjugates were used to coat tissue culture plastic surfaces which then served as substrata in cell adhesion experiments. Their results indicated that the efficiency of adhesion to RGD-BSA-coated surfaces was highly dependent on the valency of the (RGD)n-BSA conjugates. For example, on surfaces with approximately equal amounts of RGD ligand,

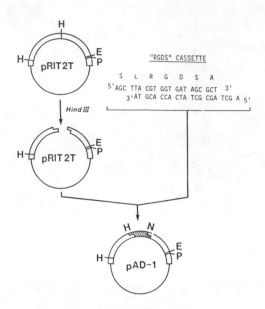

FIGURE 10. Construction of the plasmid vector pAD-1 by insertion of an oligonucleotide cassette encoding SLRGDSA. Restriction sites indicated: H, *Hind*III; P, *Pst*I; E, *Eco*RI; N, *Nhe*I. (From Maeda, T., Oyama, R., Ichihara-Tanaka, K., Kimizuka, F., Kato, I., Titani, K., and Sekiguchi, K., *J. Biol. Chem.*, 264, 15165, 1989. With permission.)

CHO cells adhered virtually 100% to the (RGD)n-BSA (n = 20.8) conjugate and not at all to the (RGD)n-BSA (n = 3.8) conjugate. Adhesion on (RGD)n-BSA-coated substrate and on fibronectin- or vitronectin-coated substrate was also examined in terms of the relationship between cell adhesion and the intermolecular distances of adsorbed proteins. It was observed that for substrate coated with relatively compact, symmetric molecules, such as RGD-BSA or vitronectin, adhesion dropped off sharply as intermolecular distances increased; by contrast, for fibronectin, a large asymmetric molecule, adhesion declined more gradually as intermolecular distances increased. In addition, competition and blocking experiments with antibodies and with soluble competing proteins suggested that the vitronectin receptor rather than the fibronectin receptor mediated adhesion to RGD-BSA.

Maeda et al.[184,185] introduced the RGDS tetrapeptide into a truncated form of protein A, a staphylococcal immunoglobulin-binding protein, by inserting an oligonucleotide cassette encoding the tetrapeptide into the coding region of the protein A expression vector pRIT2T as shown in Figure 10. The mutagenized protein was capable not only of binding to immunoglobulin G but also of mediating cell attachment and spreading onto an inserted substrate. Cell adhesion mediated by the mutagenized protein was inhibitable by a synthetic peptide Gly-Arg-Gly-Asp-Ser but not by a related peptide Gly-Arg-Gly-Glu-Ser, confirming that the inserted RGDS tetrapeptide served as a

1. **Co-immobilization with other active-site of adhesion proteins**

2. **Secondary structure of the RGDS**

   a. **Polymerization**

   b. **Cyclization**

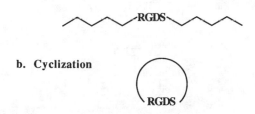

FIGURE 11.   Strategies to enhance the activity of the RGDS unit.

recognition signal for cell adhesion. This unique feature of these cell adhesive proteins offers an excellent model for "active site transplantation" by insertional mutation. The technology is essentially applicable to any protein and allows the design of a wide variety of artificial cell proteins possibly without destroying their intrinsic biological activities.

In order to enhance the unit activity of the peptide RGDS, there are two strategies (Figure 11). Cyclization was performed by Pierschbacher and Ruoslahti.[186] However, the activity was not enhanced. On the other hand, Ozeki et al.[187] synthesized a different cyclo(RGDS)$_2$. They reported that the adhesion activity of the latter was 100 times more than that of GRGDSP. They considered that this enhancement was caused by the rigid structure of the cyclic peptide.

RGD was first polymerized by Saiki et al.[188,189] They showed that the polypeptide dramatically inhibited the aggregation of platelets induced by adenosine diphosphate malignant melanoma cells, or experimental lung metastases. Figure 12 shows the effect on cell adhesion. The RGD polypeptide can inhibit cell adhesion onto fibronectin-coated surfaces effectively. Suzuki et al.[190] synthesized a polypeptide containing Arg-Gly-Asp-Ser-Pro (RGDSP). Figure 13 shows the circular dichroic spectra of RGDSP and polyRGDSP. RGDSP conformed to a β-bend structure, whereas polyRGDSP had no specific conformation since the spectra were not affected by temperature. In this case, the adhesion activity of RGDSP was little enhanced by the polymerization, demonstrating that the RGDS peptide itself conforms to a specific secondary structure which is moderately effective in cell attachment but is broken down by polymerization.

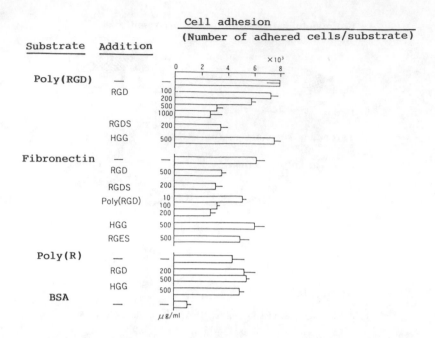

FIGURE 12.   Inhibition of cell adhesion onto substrate-coated materials by adding peptides and polypeptides.

Some researchers[191-193] synthesized the polymerizable (acryloyled) peptide to construct polymers containing the RGD peptide in the side chains. Brandley and Schnaar[191] polymerized the peptide with a cross-linking monomer into a microgel. Matsuda et al.[192] polymerized the peptide as a solution-castable polymer. However, these approaches were unsuccessful. Therefore, they introduced the peptide into a polymer by a reaction of vinyl copolymers which had an activated ester group as a side chain with GRGDSP. The surfaces of solvent-cast films of this polymer were analyzed by X-ray photoelectron spectroscopy by which surface chemical composition was quantitatively measured at the outermost layer of several tenths of angstroms, which determines biological responses including cell adhesion and growth.

Ito and Imanishi[193] immobilized the acryloyl derivative of the RGDS peptide by graft-polymerization onto a polymer membrane which was glow discharged in advance as shown in Figure 14. This method was the same as that described for synthesizing anticoagulant materials using polymerizable anticoagulant (Chapter 2.1).

RGDS-immobilized materials were compared with FN-immobilized materials in reference to cell-adhesion activity and stability by Ito et al.[194] Figure 15 shows the advantages of RGDS-immobilized materials. They immobilized the peptide onto poly(acrylic acid)-grafted polystyrene film and the activity of the biocomposite film was compared with fibronectin-immobilized film.

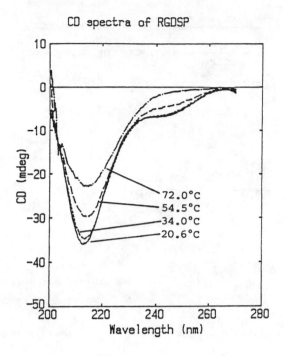

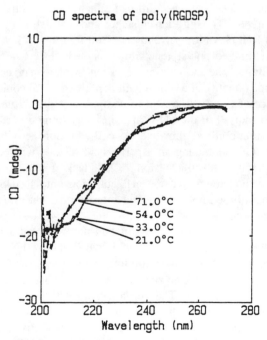

FIGURE 13. Circular dichroism spectra of RGDSP and polyRGDSP.

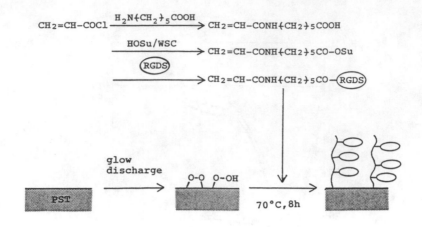

HOSu:N-Hydroxysuccinimide

WSC :Water-Soluble Carbodiimide

FIGURE 14.   Method of graft-copolymerization of acryloyled RGDS onto materials.

RGDS-immobilized polystyrene film had enhanced cell-adhesion activity as shown in Figure 16. Though Pierschbacher and Ruoslahti have reported that the activity of RGDS was only 1/200- to 1/300-fold that of FN,[171] the activity of RGDS-immobilized polystyrene film was very close to FN-immobilized polystyrene film in the current investigation, which may be explained as follows. A large quantity of a short-chain peptide (RGDS) could be integrated into the surface of polystyrene film to compensate for low unit activity. On the other hand, not all of the FN is appropriately immobilized.

RGD-OH-immobilized film had no cell-adhesion activity. It has been reported that RGD containing an aspartic acid residue with a free carboxyl terminal is not active in cell adhesion.[186] This lack of RGD-OH-immobilized film activity demonstrates that the cell adhesion onto RGDS- and FN-immobilized film is based not on physical interactions, but on specific ligand/receptor interactions.

Figure 17 summarizes the effects of thermal treatment and pH variation on the cell-adhesion properties of immobilized RGDS or FN. The cell-adhesion activity of RGDS- and FN-immobilized polystyrene films decreased with increasing temperature, and the activity decreased more sharply in FN- than in RGDS-immobilized film. The cell-adhesion activity of RGDS- or FN-immobilized film was markedly reduced at high and low pH regions. The longer the incubation time, the more extensive the activity decrease. The activity was more reduced on FN- than in RGDS-immobilized film.

Figure 18 shows the circular dichroism spectra of RGDS and FN in PBS before and after heat treatment. Though a considerable spectral change was

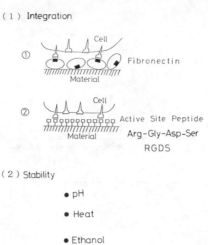

FIGURE 15. Advantages of RGDS immobilization. (1) RGDS can be integrated on the surface. (2) RGDS is more stable than protein.

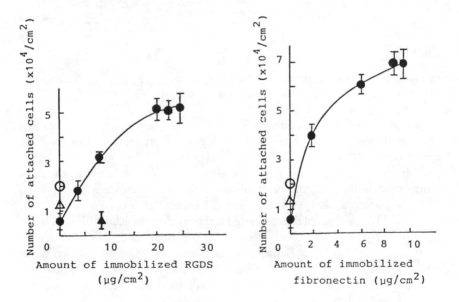

FIGURE 16. Cell adhesion onto fibronectin-immobilized and RGDS-immobilized surfaces. (○), Bare polystyrene surface; (△), poly(acrylic acid)-grafted polystyrene surface; (▲), RGD-immobilized surface.

Stability of RGDS- or Fibronectin-Immobilized Polymeric Film

Treatment	Residual Activity(%)*	
	Immobilized RGDS	Immobilized Fibronectin
Ethanol(70%) (20 s)	100 ± 2	90 ± 3
37°C	78 ± 4	41 ± 4
47°C(1 h)	56 ± 6	36 ± 2
57°C	58 ± 6	38 ± 2
pH 3.0	63 ± 8	39 ± 2
pH 7.4(7 d)	76 ± 6	66 ± 9
pH 11.5	66 ± 2	25 ± 4

*The activity of the film before treatments was taken as 100%

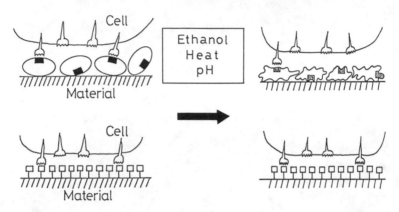

FIGURE 17.   Stability of RGDS- or fibronectin-immobilized polymer film. Immobilized fibronectin is easily denatured by ethanol, temperature, and pH.

observed in FN, the spectrum of RGDS did not change. This indicates that the sharp reduction in cell-adhesion activity of FN-immobilized film resulting from thermal treatment is ascribed to a heat-induced conformational change of the FN molecule. The minor reduction of cell-adhesion activity of RGDS-immobilized film was considered to be due to surface reorganization induced by environmental changes as reported in Chapter 2.1.

RGDS-immobilized film enhanced cell growth more than FN-immobilized film in the absence of serum. It has been reported that RGDS is somewhat different from FN in regulating cell functions.[195,196] Cells developed novel adhesion structures on peptide-coated substrate. Interference reflection microscopy showed the predominance of small round dark gray/black patches of adherent membrane (''spots'') with relatively few focal adhesions, which occurred only at the outermost cell margins in contrast to their distribution in cells spread on fibronectin. The spots were resistant to detergent extraction and stained less intensely or not at all for vinculin. Electron microscopy of vertical thin sections showed that the ventral surface of the cell was characterized by ''point-contacts'', corresponding in size to the spot structures seen by interference reflection microscopy, and which were only occasionally

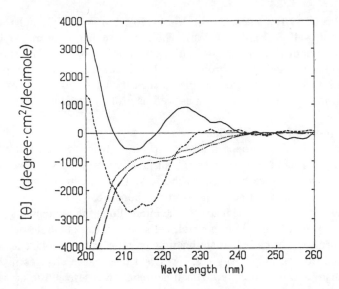

FIGURE 18. Circular dichroism of RGDS and fibronectin before and after thermal treatment. RGDS before (– · – ·) and after (· · · · ·) thermal treatment; fibronectin before (– – –) and after (——) thermal treatment.

TABLE 2
Classification of Bioadhesive Polypeptide Sequences

Type	Sequence	Example
Random	TYUOKMBDCXZWSAAUYTOPLHYESVBNEW	Barnacle
Collagen	——XY———XY————-XY——-XY—-	Clam worms
Almost regular	-ABCDEFGHI-ABCDEFGHI-ABCDEFGHI-	Mussell
Completely regular	-ABCDEFGHIJKLMN(RGD)OPQRSTUVWXY	Fibronectin

associated with microfilaments. Cells also required a higher substratum loading of peptide than fibronectin to promote spreading and did so less rapidly and to a lesser extent, developing very few and extremely fine actin cables.

Another type of polypeptide adhesive identified by Waite and Tanzer[197,198] consisted of bioadhesive polyphenolic proteins, secreted by marine mussels, which live below the water line and routinely cope with the forces of surf and tides. Formulations containing such extracted and purified bioadhesive polyphenolic proteins are currently on the market (CELL-TAK R adhesive, Biopolymers, Inc., Farmington, CT[199]). This protein has a high molecular weight and contains polar groups (-OH, -NH$_2$). Yamamoto et al.[200] categorized adhesion proteins as shown in Table 2.

Elastin-like polypeptides were synthesized by Urry's group.[201] Elastin is a major component of the connective tissue of elastic fiber. Primary repeating peptide sequences of tropoelastin were identified and consist of a tetramer (L

Val-L-Pro-Gly-Gly), a pentamer (L-Val-L-Pro-Gly-L-Val-Gly), and a hexamer (L-Ala-L-Pro-Gly-L-Val-Gly-L-Val). The repeat sequences were synthesized in their monomeric, oligomeric, and polymeric forms. *In vitro* studies on human gingival and skin fibroblasts in tissue culture showed this polypentapeptide to be quite biocompatible as measured by cell viability and cell function in terms of reproduction as well as protein and collagen synthesis.

2. Saccharide Immobilization

Schnaar et al.[202] proposed a more specific model for cell-cell adhesion in which binding occurs between cell surface glycosyl transferases and their respective complex carbohydrate acceptors, or between the sugars on the surfaces of neighboring cells via hydrogen bonding.

That carbohydrates mediate specific intercellular interactions is supported by three observations: (1) in general, cells have a carbohydrate-rich surface or glycocalyx; (2) carbohydrate-binding proteins such as plant lectins are capable of aggregating many cell types and may cause changes in cellular metabolism and morphology; (3) cell aggregation is affected in a variety of cell types (slime molds, sponges, etc.), by carbohydrate-containing or binding proteins derived from those systems or by carbohydrates themselves.

In preliminary studies to obtain direct evidence for interaction between cell surface and carbohydrates on apposing surfaces, Chipowsky et al. attached thioglycosides, containing an amino group in the aglycon, to Sephadex G-25 beads activated by cyanogen bromide, and specifically bound SV40 virus-transformed BALB/c 3T3 tissue-cultured fibroblasts to beads derivatized with a β-galactoside ligand. Although this was the first evidence that cells could respond to carbohydrates immobilized on surfaces, the use of cyanogen bromide activation led to unavoidable incorporation of a cationic charge into the product, resulting in nonspecific binding of the polyanionic cells. In addition, the lack of stability between ligand and matrix at physiological pH, the difficulty of quantitatively measuring the amount of cell binding, and possible damage to the cells on collision with the beads were featured.

Schnaar[202-204] synthesized polyacrylamide gels containing the desired ligands linked to the gel via amide bonds. Two different methods were used as shown in Figure 19. For type I gels, an active ester of acrylamide (e.g., *N*-succinimidyl acrylate) was co-polymerized with acrylamide and *N,N'*-methylenebis-acrylamide. Type II gels were synthesized by a different method in which the amino groups of the ligands were first *N*-acryloylated with acryloyl chloride, and the resulting *N*-substituted acrylamides were co-polymerized with acrylamide and *bis*-acrylamide. Schnaar reported that at 37°C, 70 to 100% of chicken hepatocyte cells adhered within 60 min to gels derivatized with *N*-acetylglucosamine, i.e., gels derivatized with 6-aminohexyl 2-acetamido-2-deoxy-β-D-glucopyranoside (or the corresponding thioglycoside). By contrast, less than 5% of the cells adhered to polyacrylamide or to gels derivatized with 6-aminohexanol or the 6-aminohexyl glycosides of β-D-

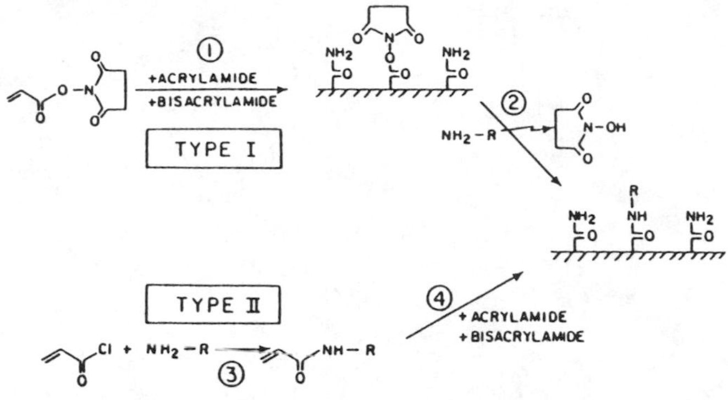

FIGURE 19. Synthesis of type I- and type II-derivatized gels. Type I, *N*-succinimidyl acrylate was co-polymerized with acrylamide and bis acrylamide (reaction 1), resulting in an activated gel. The gel was treated with ligands (R) containing amino groups (reaction 2), and a stable amide linkage between ligand and matrix was formed with concomitant release of *N*-hydroxysuccinimide. Type II, ligands containing amino groups were treated with acryloyl chloride (reaction 3) to yield the *N*-acryloylated derivatives. The derivatives were then co-polymerized with acrylamide and bisacrylamide (reaction 4), resulting in the incorporation of the ligand (R) into the gel matrix.

glucose, β-D-galactose, α-D-mannose, β-D-maltose, β-D-melibiose, β-D-cellobiose, and (α or β)-D-lactose.

It was demonstrated that: (1) cell-gel adhesion was both cell and sugar specific; (2) adhesion was remarkably dependent on the sugar concentration on the synthetic surface, exhibiting a threshold-binding phenomenon; (3) the binding was absolutely dependent on the presence of calcium ions; and (4) initial adhesion could be blocked or reversed by adding the appropriate free sugar. Thus, "cell surface analogues" consisting of simple carbohydrate ligands immobilized on synthetic surfaces were detected by intact cells.

On the other hand, Kobayashi et al.[205-207] also synthesized oligosaccharide-carrying polymers to cultivate rat hepatocytes. The chemical structures are shown in Figure 20. They reported that hepatocytes adhered specifically to dishes in which the surface was coated with lactose-carrying polystyrene (PVLA). PVLA-coated dishes markedly resembled the clearance of asialoglycoproteins into hepatocytes. Experiments on adhesion inhibition and temperature effect suggested that the interactions between polymers and hepatocytes decreased in the order of PVLA (lactose) > PVMeA (melibiose) > PVCA (cellobiose) > PVMA (maltose).

Cell-specific carbohydrate receptors have been used in experimental site-specific drug delivery for some time.[208,209] Therefore, polyacryl starch and polyacryl mannan microcapsules were synthesized and the interactions with various cells were investigated.

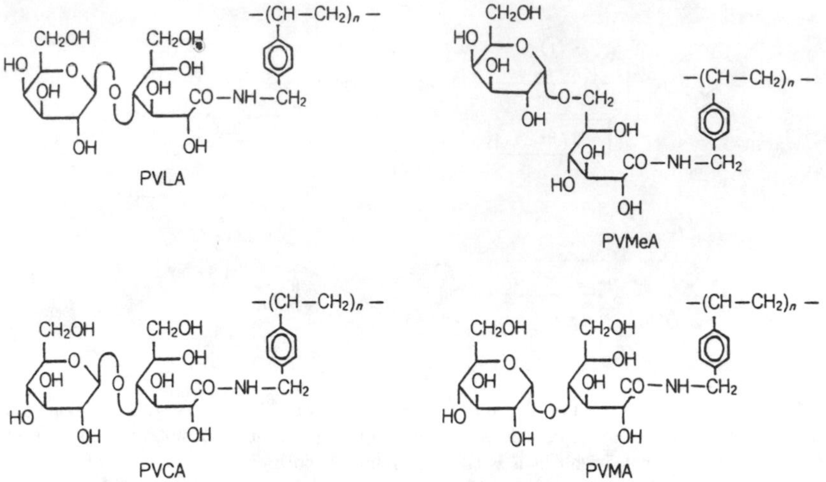

FIGURE 20.   Chemical structure of a polymer containing carbohydrate.

IV. OTHER BIOADHESIVES

A. SYNTHETIC GLUES FOR BIOMEDICAL USE

An ideal tissue adhesive should: (1) be easy to apply; (2) wet the tissue surface; (3) polymerize rapidly even in the presence of moisture; (4) adhere tenaciously to tissues; (5) maintain adhesiveness upon aging *in vivo;* (6) possess adequate shelf life; and (7) show good biocompatibility including minimum histotoxicity, and cytotoxicity. Many studies have investigated potential bonding agents such as epoxy resins, polyurethanes,[210] poly(methyl methacrylate),[211] methacrylate oligomers with pendant isocyante groups,[212,213] dental composite resins,[214] and α-cyanoacrylates.[215,216] The state of art of soft tissue adhesives has been summarized by Manly[217] and Peppas and Buri.[218]

B. ORTHOPEDIC IMPLANTS

In the development of orthopedic implants, a direct bond with bone tissue is necessary.[219] Certain bioactive ceramics such as Bioglass,[220,221] hydroxyapatite,[222-224] tricalciumphosphate,[224] and Ceravital[225] formed a strong direct bond with bone tissue.

Therefore, there were several attempts to strenghten the bone-implant interface of alumina ceramic prostheses.[228,229] Griss et al.[228] reported the firm attachment of a Bioglass-coated alumina total hip prosthesis. On the other hand, Higashi et al. evaluated the effect of granule coating (with a granular alumina ceramic, a granular hydroxyapatite and polished granular hydroxyapatite).[229,230] They found that the granular alumina ceramic coating created a bioinert, porous surface, which was most effective due to a strong mechanical bond between the bone and implant. Bioactive glass was also coated onto metallic prostheses.[231,232] A mixture of apatite-wollastonite containing glass ceramic (A-W GC, 42-60 mesh in granule size) with fibrin was synthesized as a bone defect filler.[233] Fibrin was considered to have good adhesiveness and plasticity as well as hemostatic and tissue healing activities.

Hakkinen et al.[234] investigated cell attachment and spreading on bioactive glass and cell culture polystyrene dishes. They observed that the cell attachment to and spreading on the bioglass were slower and that cell morphology was more elongated compared with the control plastic dishes.

V. FUTURE OUTLOOK

To enhance cell adhesion on the surface, immobilization of the cell adhesion factor is a very attractive method. Recent developments in biological science have revealed various new cell-adhesion proteins and their functions. It is expected that adhesion factor-immobilized materials will control cell growth, differentiation, and secretion, as well as cell adhesion.

REFERENCES

1. **Grinnell, F. and Feld, M. K.**, Adsorption characteristics of plasma fibronectin in relationship to biological activity, *J. Biomed. Mater. Res.*, 15, 363, 1981.
2. **Curtis, A. S. G., Forrester, J. V., and Clark, P.**, Substrate hydroxylation and cell adhesion, *J. Cell Sci.*, 86, 9, 1984.
3. **Pratt, K. J., Williams, S. K., and Jarrell, B. E.**, Enhanced adherence of human adult endothelial cells to plasma discharge modified polyethylene terephthalate, *J. Biomed. Mater. Res.*, 23, 1131, 1989.
4. **Curtis, A. S. G., Forrester, J. V., McInnes, C., and Lawrie, F.**, Adhesion of cells to polystyrene surfaces, *J. Cell Biol.*, 97, 1500, 1983.
5. **Curtis, A.**, Cell activation and adhesion, *J. Cell Sci.*, 87, 609, 1987.
6. **Schakenraad, J. M., Busscher, H. J., Wildevuur, C. R. H., and Arends, J.**, The influence of substratum surface free energy on growth and spreading of human fibroblasts in the presence and absence of serum proteins, *J. Biomed. Mater. Res.*, 20, 773, 1986.
7. **Wildevuur, C. R. H., van der Lei, B., and Schakenraad, J. M.**, Basic aspects of the regeneration of small-calibre neoarteries in biodegradable vascular grafts in rats, *Biomaterials*, 8, 418, 1987.
8. **Fasol, R., Zilla, P., Groscurth, P., Wolner, E., and Moser, R.**, Experimental *in vitro* cultivation of human endothelial cells on artificial surfaces, *Trans. Am. Soc. Artif. Intern. Organs*, 31, 276, 1985.
9. **Horbett, T. A., Waldburger, J. T., Ratner, B. D., and Hoffman, A. S.**, Cell adhesion to a series of hydrophilic-hydrophobic copolymers studied with a spinning disc apparatus, *J. Biomed. Mater. Res.*, 22, 383, 1988.
10. **van Wachem, P. B., Beugeling, T., Feijen, J., Bantjes, A., Detmers, J. P., and van Aken, W. G.**, Interaction of cultured endothelial cells with polymeric surfaces of different wettabilities, *Biomaterials*, 6, 403, 1986.

11. **van Wachem, P. B., Hogt, A. H., Beugeling, T., Feijen, J., Bantjes, A., Detmers, J. P., and van Aken, W. G.,** Adhesion of cultured human endothelial cells onto methacrylate polymers with varying surface wettability and charge, *Biomaterials,* 8, 323, 1987.

12. **van Wachem, P. B., Schakenraad, J. M., Feijen, J., Beugeling, T., van Aken, W. G., Blaauw, E. H., Nieuwenhuis, P., and Molenaar, I.,** Adhesion and spreading of cultures endothelial cells on modified and unmodified poly(ethylene terephthalate): a morphological study, *Biomaterials,* 10, 532, 1989.

13. **van Wachem, P. B., Vreriks, C. M., Beugeling, T., Feijen, J., Bantjes, A., Detmers, J. P., and van Aken, W. G.,** The influence of protein adsorption on interactions of cultured human endothelial cells with polymers, *J. Biomed. Mater. Res.,* 21, 701, 1987.

14. **van Wachem, P. B., Mallens, B. W. L., Dekker, A., Beugeling, T., Feijen, J., Bantjes, A., Detmers, J. P., and van Aken, W. G.,** Adsorption of fibronectin derived from serum and from human endothelial cells onto tissue culture polystyrene, *J. Biomed. Mater. Res.,* 21, 1317, 1987.

15. **Kang, I.-K., Ito, Y., Sisido, M., and Imanishi, Y.,** Attachment and growth of fibroblast cells on polypeptide derivatives, *J. Biomed. Mater. Res.,* 23, 223, 1989.

16. **Ikada, Y. and Tamada, Y.,** Cell attachment to various polymer surfaces, in *Polymers in Medicine 2,* Chiellini, E., Giusti, P., Migliaresi, C., and L. Nicolais, Eds., Plenum Press, 1986, 101.

17. **Neumann, A. W., Absolom, D. R., Francis, D. W., van Oss, C. J., and Zingg, W.,** Thermodynamic aspects of cellular engulfment and adhesion, in *Biocompatible Polymers, Metals and Composites,* Szycher, M., Ed., Technomic Publ, Lancaster, PA, 1983, 89.

18. **Absolom, D. R., Foo, M. H., Zingg, W., and Neumann, A. W.,** Thermodyanmic assessment of platelet adhesion to polyacrylamide gels, in *Polymers as Biomaterials,* Shalaby, S. W., Hoffman, A. S., Ratner, B. D., and Horbett, T. A., Eds., Plenum Press, New York, 1984, 149.

19. **Absolom, D. R., Zingg, W., van Oss, C. J., and Neumann, A. W.,** Measurement of surface tensions of cells and biopolymers with application to the evaluation of biocompatibility, in *Comprehensive Biotechnology,* Vol. 3, Cooney, C. C. and Humpherey, A. E, Eds., Pergamon Press, New York, 1985, 433.

20. **Absolom, D. R., Hawthorn, L. A., and Chang, G.,** Endothelialization of polymer surface, *J. Biomed. Mater. Res.,* 22, 271, 1988.

21. **Absolom, D. R., Thomson, C., Hawthorn, L. A., Zingg, W., and Neumann, A. W.,** Kinetics of cell adhesion to polymer surfaces, *J. Biomed. Mater. Res.,* 22, 215, 1988.

22. **Chang, G., Absolom, D. R., Strong, A. B., Stubley, G. D., and Zingg, W.,** Physical and hydrodynamic factors affecting erythrocyte adhesion to polymer surfaces, *J. Biomed. Mater. Res.,* 22, 13, 1988.

23. **Jansen, J. A., van der Waerden, J. P. C. M., and de Groot, K.,** Effect of surface treatments on attachment and growth of epithelial cells, *Biomaterials,* 10, 604, 1989.

24. **Facchini, P. J., Neumann, A. W., and DiCosmo, F.,** Adhesion of suspension-cultured *Catharanthus roseus* cells to surfaces: effect of pH, ionic strength, and cation valency, *Biomaterials,* 10, 318, 1989.

25. **Lindon, M. J., Minett, T. W., and Tighe, B. J.,** Cellular interactions with synthetic polymer surfaces in culture, *Biomaterials,* 6, 396, 1985.

26. **Horbett, T. A. and Schway, B.,** Correlations between mouse 3T3 cell spreading and serum fibronectin adsorption on glass and hydroxyethylmethacrylate-ethylmethacrylate copolymers, *J. Biomed. Mater. Res.,* 22, 763, 1988.

27. **McAuslan, B. R. and Johnson, G.,** Cell responses to biomaterials. I. Adhesion and growth of vascular endothelial cells on poly(hydroxyethyl methacrylate) following surface modification by hydrolytic etching, *J. Biomed. Mater. Res.,* 21, 921, 1987.

28. **Maroudas, G.,** Adhesion and spreading of the cells on charged surfaces, *J. Theor. Biol.,* 49, 417, 1975.

29. **Jakobsen, B. S. and Ryan, U. S.,** Growth of endothelial and HeLa cells on a new multipurpose microcarrier that is positive, negative, or collagen coated, *Tissue Cult.*, 14, 69, 1982.

30. **Klein-Soyer, C., Hammendinger, S., and Cazenave, J. -P.,** Culture of human vascular endothelial cells on a positvely charged polystyrene surface, *Promaria:* comparison with fibronectin-coated tissue culture grade polystyrene, *Biomaterials,* 10, 85, 1989.

31. **Chaturvedi, N. and Picciano, P. T.,** Tissue immobilization and cell culturing system and method for affixing biologically active moieties to a substrate, European Patent Appl. 89, 111, 726.9.

32. **Nagai, T. and Machida, Y.,** Mucosal adhesive dosage forms, *Pharma. Int.,* 6, 196, 1985.

33. **Smart, J. D., Kellaway, I. W., and Worthington, H. E. C.,** An *in vitro* investigation of mucosa-adhesive materials for use in controlled drug delivery, *J. Pharm. Pharmacol.,* 36, 295, 1984.

34. **Park, K. and Robinson, J. R.,** Bioadhesive polymers as platforms for oral-controlled drug delivery: method to study bioadhesion, *Int. J. Pharm.,* 19, 107, 1984.

35. **Ponchel, G., Touchard, F., Duchene, D., and Peppas, N. A.,** Bioadhesive analysis of controlled-release systems. I. Fracture and interpenetration analysis in poly(acrylic acid)-containing systems, *J. Controlled Release,* 5, 129, 1987.

36. **Peppas, N. A., Ponchel, G., and Duchene, D.,** Bioadhesive analysis of controlled-release systems. II. Time-dependent bioadhesive stress in poly(acrylic acid)-containing systems, *J. Controlled Release,* 5, 143, 1987.

37. **Vogler, E. A. and Bussian, R. W.,** Short-term cell-attachment rates: a surface-sensitive test of cell-substrate compatibility, *J. Biomed. Mater. Res.,* 21, 1197, 1987.

38. **McAuslan, B. R., Johnson, G., Hannan, G. N., Norris, W. D., and Extner, T.,** Cell response to biomaterials. II. Endothelial cell adhesion and growth on perfluorosulfonic acid, *J. Biomed. Mater. Res.,* 22, 963, 1988.

39. **Aubert, N., Reach, G., Serne, H., and Jozefowicz, M.,** Reversible morphological and functional abnormalities of RINm5F cells cultured on polystyrene sulfonate beads, *J. Biomed. Mater. Res.,* 21, 585, 1987.

40. **Imai, Y.,** Biomedical polymers, *Kobunshi,* 21, 569, 1972.

41. **Okano, T., Nishiyama, S., Shinohara, I., Akaike, T., Sakurai, Y., Kataoka, K., and Tsuruta, T.,** Effect of hydrophilic and hydrophobic microdomains on mode of interaction between block polymer and blood platelets, *J. Biomed. Mater. Res.,* 15, 393, 1981.

42. **Kataoka, K.,** Polymers for cell separation, *CRC Crit. Rev. Biocompatibility,* 4, 341, 1988.

43. **Hasegawa, H. and Hashimoto, T.,** Morphology of block polymers near a free surface, *Macromolecules,* 18, 589, 1985.

44. **Hearn, M. J., Ratner, B. D., and Briggs, D.,** SIMS and XPS studies of polyurethane surface. I. Preliminary studies, *Macromolecules,* 21, 2950, 1988.

45. **Picha, G., Drake, R., and Mayhan, R.,** Controlling fibrous capsule formation, the soft tissue response to ion milled surface structures, *Artif. Organs,* 13, 355, 1989.

46. **Clarke, D. R. and Park, J. B.,** Prevention of erythrocyte adhesion onto porous surfaces by fluid perfusion, *Biomaterials,* 2, 9, 1981.

47. **Dunn, G. A.,** *Cell Behavior,* Bellairs, R., Curtis, A. S. G., and Dunn, G., Ed., Cambridge University Press, London, 1982, 247.

48. **Brunette, D. M., Kenner, G. S., and Gould, T. R. L.,** Grooved titanium surfaces orient growth and migration from human gingival explants, *J. Dent. Res.,* 62, 1045, 1983.

49. **Brunette, D. M.,** Fibroblasts on micromachined substrata orient hierarchically to grooves of different dimensions, *Exp. Cell Res.,* 164, 11, 1986.

50. **Brunette, D. M.,** Spreading and orientation of epithelial cells on grooved substrata, *Exp. Cell Res.,* 167, 203, 1986.

51. **Dunn, G. A. and Brown, A. F.,** Alignment of fibroblasts on grooved surfaces described by a simple geometric transformation, *J. Cell Sci.,* 83, 313, 1986.
52. **Clark, P., Connolly, P., Curtis, A. S. G., Dow, J. A. T., and Wilkinson, C. D. W.,** Topographical control of cell behavior. I. Simple step cues, *Development,* 99, 439, 1987.
53. **Wood, A.,** Contact guidance on microfabricated substrata: the response of teleost fin mesenchyme cells to repeating topographical patterns, *J. Cell Sci.,* 90, 667, 1988.
54. **Chehroudi, B., Gould, T. R. L., and Brunette, D. M.,** Effects of a grooved epoxy substratum on epithelial cell behavior *in vitro* and *in vivo, J. Biomed. Mater. Res.,* 22, 459, 1988.
55. **Chehroudi, B., Gould, T. R. L., and Brunette, D. M.,** Effects of a grooved titanium-coated implant surface on epithelial cell behavior *in vitro* and *in vivo, J. Biomed. Mater. Res.,* 24, 1067, 1988.
56. **Squier, C. A. and Collins, P.,** The relationship between soft tissue attachment, epithelial downgrowth and surface porosity, *J. Periodontal. Res.,* 16, 434, 1981.
57. **Inoue, T., Cox, J. E., Pillar, R. M., and Melcher, A. H.,** Effects of the surface geometry of smooth and porous-coated titanium alloy on the orientation of fibroblasts *in vitro, J. Biomed. Mater. Res.,* 21, 107, 1987.
58. **Bruin, P., Jonkman, M. F., Meijer, H. J., and Pennings, A. J.,** A new porous polyetherurethane wound covering, *J. Biomed. Mater. Res.,* 24, 217, 1990.
59. **Chvapil, M., Kronenthal, R. L., and van Winkle, W., Jr.,** Medical and surgical applications of collagen, *Int. Rev. Connect. Tissue Res.,* 6, 1, 1973.
60. **Huc, A.,** Collagen biomaterials characteristics and applications, *J. Am. Leather Chem. Assoc.,* 80, 195, 1985.
61. **Klopper, P. J.,** Collagen in surgical research, *Eur. Surg. Res.,* 18, 218, 1986.
62. **Pharriss, B. B.,** Collagen as a biomaterial, *J. Am. Leather Chem. Assoc.,* 75, 474, 1980.
63. **Sabelaman, E. E.,** Biology, biotechnology, and biocompatibility of collagen, in *Biocompatibility of Tissue Analogs,* Vol. 1, Williams, D. F., Ed., CRC Press, Boca Raton, FL, 1985, 27.
64. **Chvapil, M.,** Consideration on manufacturing principles of a synthetic burn dressing: a review, *J. Biomed. Mater. Res.,* 16, 245, 1982.
65. **Cervinka, F. and Krajecik, M.,** Notes on the problem of antigenicity of collagen, *Folia Biol. (Prague),* 10, 94, 1964.
66. **Bajpai, P. K. and Stull, P. A.,** Glutaraldehyde-crosslinked porcine heart valve exnografts and cell-mediated immune responses, *IRCS Med. Sci. Anat. Human Biol.,* 8, 642, 1980.
67. **Srivastava, S., Gorham, S. D., and Courtney, J. M.,** The attachment and growth of an established cell line on collagen, chemically modified collagen, and collagen composite surfaces, *Biomaterials,* 11, 162, 1990.
68. **Meade, K. R. and Silver, F. H.,** Immunogenicity of collagenous implants, *Biomaterials,* 11, 176, 1990.
69. **Woodroof, A.,** Use of glutaraldehyde and formaldehyde to process tissue heart valves, *J. Bioeng.,* 2, 1, 1978.
70. **Decker, Q. W. and Sherman, B.,** Glutaraldehyde structure and its relation to protein-crosslinking, Union Carbide Technology Series, Union Carbide, New York, April 10, 1970.
71. **Speer, D. P., Chvapil, M., Volz, and M. Holmes,** Enhancement of healing of osteochondral defects by collagen sponge implants, *Clin. Orthop.,* 144, 326, 1979.
72. **Chvapil, M., Speer, D., Mora, W., and Eskelson, C.,** Effect of tanning agent on tissue reaction to tissue implanted collagen sponge, *J. Surg. Res.,* 35, 402, 1983.
73. **Speer, D. P., Chvapil, M., Eskelson, C. D., and Ulreich, J.,** Biological effects of residual glutaraldehyde in glutaraldehyde-tanned collagen biomaterials, *J. Biomed. Mater. Res.,* 14, 753, 1980.

74. **Cooke, A., Oliver, R. F., and Edward, M.,** An *in vitro* cytotoxicity study of aldehyde-treated pig dermal collagen, *Br. J. Exp. Pathol.,* 64, 172, 1983.

75. **McPherson, J. M., Ledger, P. W., Sawamura, S., Conti, A., Wade, S., Reihanian, H., and Wallace, D. G.,** The preparation and physicochemical characterization of an injectable form of reconstituted, glutaraldehyde cross-linked, bovine corium collagen, *J. Biomed. Mater. Res.,* 20, 79, 1986.

76. **McPherson, J. M., Sawamura, S., and Armstrong, R.,** An examination of the biologic response to injectable, glutaraldehyde cross-linked collagen implants, *J. Biomed. Mater. Res.,* 20, 93, 1986.

77. **Huang-Lee, L. L. H., Cheung, D. T., and Nimni, M. E.,** Biochemical changes and cytotoxicity associated with the degradation of polymeric glutaraldehyde derived cross-links, *J. Biomed. Mater. Res.,* 24, 1185, 1990.

78. **Nimni, M. E., Cheung, D., Strates, B., Kodama, M., and Sheikh, K.,** Chemically modified collagen: a natural biomaterial for tissue replacement, *J. Biomed. Mater. Res.,* 21, 741, 1987.

79. **Rajaram, A. and Chu, C. C.,** Glutaraldehyde in collagen gel formation, *J. Biomat. Sci.,* 1, 167, 1990.

80. **Quteish, D., Singh, G., and Dolby, A. E.,** Development of a human collagen graft material, *J. Biomed. Mater. Res.,* 24, 749, 1990.

81. **Yannas, I. V. and Tobolsky, A. V.,** Cross-linking of gelatin by dehydration, *Nature (London),* 215, 509, 1967.

82. **Yannas, I. V.,** Biologically active analogues of the extracellular matrix: artificial skin and nerves, *Angew. Chem. Int. Ed. Engl.,* 29, 20, 1990.

83. **Weadock, K., Olson, R. M., and Silver, F. H.,** Evaluation of crosslinking techniques, *Biomat. Med. Dev. Artif. Organs,* 11, 293, 1983.

84. **Morykwas, M. J.,** *In vitro* properties of crosslinked, reconstituted collagen sheets, *J. Biomed. Mater. Res.,* 24, 1105, 1990.

85. **Warcquier-Clerout, R., Sigot-Luizard, M. F., and Legendre, J. M.,** F.p.l.c. and lysis of the leakage products of proteins crosslinked on Dacron R vascular prostheses, *Biomaterials,* 8, 118, 1987.

86. **Coulet, P. R. and Gautheron, D.,** Enzymes immobilized on collagen membrane: a tool for fundamental research and enzyme engineering, *J. Chromatogr.,* 215, 65, 1981.

87. **Petite, H., Rault, I., Huc, A., Menasche, Ph., and Herbage, D.,** Use of the acryl azide method for cross-linking collagen-rich tissues such as pericardium, *J. Biomed. Mater. Res.,* 24, 179, 1990.

88. **Kasai, S., Miyata, T., Kumimoto, T., and Nitta, K.,** Enhancement effect of chemically modified substratum on attachment and growth of fibroblasts in serum free media, *Biomed. Res.,* 4, 147, 1983.

89. **Gilbert, D. L. and Kim, S. W.,** Macromolecular release from collagen monolithic devices, *J. Biomed. Mater. Res.,* 24, 1221, 1990.

90. **Rosenblatt, J., Rhee, W., and Wallance, D.,** The effect of collagen fiber size distribution on the release rate of proteins from collagen matrices by diffusion, *J. Controlled Release,* 9, 195, 1989.

91. **Umemura, Y., Huskey, R. A., and Anderson, J. M.,** Human platelet interactions with surfaces of type I collagen, chondroitin-4-sulphate, and chondroitin-6-sulphate *in vitro,* *Biomaterials,* 9, 133, 1988.

92. **Zucker-Franklin, D. and Rosenberg, L.,** Platelet interaction with modified articular cartilage, *J. Clin. Invest.,* 59, 641, 1977.

93. **Teien, A. N., Abildgaard, U., and Hook, M.,** The anticoagulant effect of heparan sulphate and dermatan sulphate, *Thromb. Res.,* 8, 859, 1976.

94. **Murata, K., Nakazawa, K., and Hamai, A.,** Distribution of acidic glycosaminoglycans in the intima, media and adventitia of bovine aorta and their anticoagulant properties, *Atherosclerosis,* 21, 93, 1975.

95. **Silver, F. H., Yannas, I., and Salzman, E. W.,** Glycosaminoglycan inhibition of collagen induced platelet aggregation, *Thromb. Res.,* 13, 267, 1978.
96. **Klein, K., Sinzinger, H., Silberbauer, H., Stachelberger, H., and Leithner, Ch.,** Influence of proteoglycans (PG) and glycosaminoglycans (GAG) on ADP-. collagen- and thrombin-induced platelet aggregation, *Artery,* 8, 410, 1978.
97. **Yannas, I. V., Burke, J. F., Gordon, P. L., Huang, C., and Rubenstein, R. H.,** Design of an artificial skin. II. Control of chemical composition, *J. Biomed. Mater. Res.,* 14, 107, 1980.
98. **Dagalakis, N., Flink, J., Stasikelis, P., Burk, J. F., and Yannas, I. V.,** Design of an artificial skin. II. Control of pore structure, *J. Biomed. Mater. Res.,* 14, 511, 1980.
99. **Silverstein, M. E., Keown, K., Owen, J. A., and Chvapil, M.,** Collagen fibers as a fleece hemostatic agent, *J. Trauma,* 20, 688, 1980.
100. **Boyce, S. T., Christianson, D. J., and Hansbrough, J. F.,** Structure of a collagen-GAG dermal skin substitute optimized for cultured human epidermal keratinocytes, *J. Biomed. Mater. Res.,* 22, 939, 1988.
101. **Doillon, C. J., Wasserman, A. J., Berg, R. A., and Silver, F. H.,** Behavior of fibroblasts and epidermal cells cultivated on analogues of extracellular matrix, *Biomaterials,* 9, 91, 1988.
102. **Srivastava, S., Gorham, S. D., French, D. A., Shivas, A. A., and Courtney, J. M.,** *In vivo* evaluation and comparison of collagen, acetylated collagen and collagen/glycosaminoglycan composite films and sponges as candidate biomaterials, *Biomaterials,* 11, 155, 1990.
103. **Srivastava, S., Gorham, S. D., and Courtney, J. M.,** A comparison of the attachment and growth of an established cell line on collagen, chemically modified collagen, and collagen composite surfaces, *Biomaterials,* 11, 162, 1990.
104. **Srivastana, S., Gorham, S. D., Wu, J. C. Y., and Courtney, J. M.,** Collagen modification and cellular response, *Artif. Organs,* 14, 113, 1990.
105. **Doillon, C. J., Silver, F. H., and Berg, R. A.,** Fibroblast growth on a porous collagen sponge containing hyaluronic acid and fibronectin, *Biomaterials,* 8, 195, 1987.
106. **Doillon, C. J. and Silver, F. H.,** Collagen-based wound dressing: effects of hyaluronic acid and fibronectin on wound healing, *Biomaterials,* 7, 3, 1986.
107. **Blanchy, B. G., Coulet, P. R., and Gautheron, D. C.,** Immobilization of factor XIII on collagen membranes, *J. Biomed. Mater. Res.,* 20, 469, 1986.
108. **Amudeswari, S., Nagarajan, B., Reddy, C. R., and Joseph, K. T.,** Short-term biocompatibility studies of hydrogen-grafted collagen copolymers, *J. Biomed. Mater. Res.,* 20, 1103, 1986.
109. **Jeyanthi, R. and Rao, P.,** Controlled release of anticancer drugs from collagen-poly(HEMA) hydrogel matrices, *J. Controlled Release,* 13, 91, 1990.
110. **Cifkova, I., Stol, M., Holusa, R., and Adam, M.,** Calcification of poly(hydroxyethylmethacrylate)-collagen composites implanted in rats, *Biomaterials,* 8, 30, 1987.
111. **Gilbert, D. L. and Lyman, D. J.,** *In vitro* and *in vivo* characterization of synthetic polymer/biopolymer composites, *J. Biomed. Mater. Res.,* 21, 643, 1987.
112. **Tran, C. N. B. and Walt, D. R.,** Plasma modification and collagen binding to PTFE grafts, *J. Colloid Interface Sci.,* 132, 373, 1989.
113. **Levy, P., Nevis, A., and LaPorte, R.,** Healing potential of surgically-induced periodontal osseous defects in animals using mineralized collagen gel xenografts, *J. Periodontol.,* 52, 303, 1981.
114. **Lizarbe, M. A., Olmo, N., and Gavilanes, J. G.,** Outgrowth of fibroblasts on sepiolite-collagen complex, *Biomaterials,* 8, 35, 1987.
115. **Olmo, N., Lizarbe, M. A., and Gavilanes, J. G.,** Biocompatibility and degradability of sepiolite-collagen complex, *Biomaterials,* 8, 67, 1987.

116. **Olmo, N., Lizarbe, M. A., Turnay, J., Muller, K. P., and Gavilanes, J.,** Cell morphology, proliferation and collagen synthesis of human fibroblasts cultured on sepiolite-collagen complexes, *J. Biomed. Mater. Res.,* 22, 257, 2988.

117. **Okazaki, M., Ohmae, H., and Hino, T.,** Insolubilization of apatite-collagen composites by UV irradiation, *Biomaterials,* 10, 564, 1989.

118. **Yoshikawa, E., Murakami, U., and Yoshizato, K.,** Bilayered gelatin hybrid artificial skin consisting of thin film and porous sponge layer, *Polym. Prepr., Jpn.,* 39, 2568, 1990.

119. **Gourevitch, D., Jones, E., Crocker, J., and Goldman, M.,** Endothelial cell adhesion to vascular prosthetic surfaces, *Biomaterials,* 9, 97, 1988.

120. **Grinnell, F.,** Cellular adhesiveness and extracellular substrata, *Int. Rev. Cytol.,* 53, 65, 1978.

121. **Yamada, K. M.,** Cell surface interactions with extracellular materials, *Annu. Rev. Biochem.,* 52, 761, 1983.

122. **Hynes, R. O.,** Molecular biology of fibronectin, *Annu. Rev. Cell Biol.,* 1, 67, 1985.

123. **Ruoslahti, E. and Pierschbacher, M. D.,** New perspectives in cell adhesion: RGD and integrins, *Science,* 238, 491, 1987.

124. **Buck, C. A. and Horwitz, A. F.,** Cell surface receptors for extracellular matrix molecules, *Ann. Rev. Cell. Biol.,* 3, 179, 1987.

125. **Burridge, K., Fath, K., Kelly, T., Nuckolls, and Turner, A.,** Focal adhesions: transmembrane junction between the extracellular matrix and cytoskeleton, *Annu. Rev. Cell Biol.,* 4, 487, 1988.

126. **Culp, L. A., Lattera, J., Lark, M. W., Beyth, R. J., and Tobey, S. L.,** Heparan sulphate proteoglycan as mediator of some adhesive responses and cytoskeletal reorganization of cells on fibronectin matrices: independent versus cooperative functions, in *Functions of the Proteoglycans,* CIBA Symposium 124, Evered, E. and Whelan, J., Eds., John Wiley & Sons, London, 1986, 158.

127. **Culp, L. A., Mugnai, G., Lewandowska, K., Vallen, E. A., Kosir, M. A., and Houmiel, K. L.,** Heparan sulfate proteoglycans of Ras-transformed 3T3 or neuroblastoma cells. Differing functions in adhesion on fibronectin, *Ann. N.Y. Acad. Sci.,* 556, 194, 1989.

128. **Vroman, L.,** Methods of investigating protein interactions on artificial and natural surfaces, *Ann. N.Y. Acad. Sci.,* 516, 300, 1987.

129. **Grinnell, F. and Feld, M. K.,** Adsorption characteristics of plasma fibronectin in relationship to biological activity, *J. Biomed. Mater. Res.,* 15, 363, 1981.

130. **Grinnell, F. and Feld, M. K.,** Fibronectin adsorption on hydrophilic and hydrophobic surfaces detected by antibody binding and analyzed during cell adhesion in serum-containing medium, *J. Biol. Chem.,* 257, 4888, 1982.

131. **Grinnell, F.,** Fibronectin adsorption on material surfaces, *Ann. N.Y. Acad. Sci.,* 516, 280, 1987.

132. **Hughes, R. C., Pena, S. D. J., Clark, J., and Dourmashkin, R. R.,** Molecular requirements for the adhesion and spreading of hamster fibroblasts, *Exp. Cell Res.,* 121, 307, 1979.

133. **Haas, R. and Culp, L. A.,** Binding of fibronectin to gelatin and heparin: effect of surface denaturation and detergents, *FEBS Lett.,* 174, 279, 1984.

134. **Haas, R. and Culp, L. A.,** Properties and fate of plasma fibronectin bound to the tissue culture substratum, *J. Cell Physiol.,* 113, 289, 1982.

135. **Lewandowska, K., Balachander, N., Sukenik, C. N., and Culp, L. A.,** Modulation of fibronectin adhesive functions for fibroblasts and neural cells by chemically derivatized substrata, *J. Cell Physiol.,* 141, 334, 1989.

136. **Culp, L. A., Ansbacher, R., and Domen, C.,** Adhesion sites of neural tumor cells: biochemical composition, *Biochemistry,* 19, 5899, 1980.

137. **Akers, R. M., Mosher, D. F., and Lilien, J. E.,** Promotion of retinal neurite outgrowth by substratum-bound fibronectin, *Dev. Biol.,* 86, 179, 1981.

138. **Carbonetto, S., Gruver, M. M., and Turner, D. C.,** Nerve fiber growth on defined hydrogel substrates, *Science,* 216, 897, 1982.

139. **Rogers, S. L., Letourneau, P. C., Palm, S. L., McCarthy, J., and Furcht, L. T.,** Neurite extension by peripheral and central nervous system neurons in response to substratum-bound fibronectin and laminin, *Dev. Biol.,* 98, 212, 1983.

140. **Roufa, D. G., Johnson, M. I., and Bunge, M. B.,** Influence of ganglion age, non-neuronal cells, and substratum on neurite outgrowth in culture, *Dev. Biol.,* 99, 225, 1983.

141. **Rovasio, R. A., Delouvee, A., Yamada, K. M., Timpl, R., and Thiery, J. -P.,** Neural crest cell migration: requirements for exogenous fibronectin and high cell density, *J. Cell Biol.,* 96, 462, 1983.

142. **Waite, K. A., Mugnai, G., and Culp, L. A.,** A second cell-binding domain on fibronectin (RGDS-independent) for neurite extension of human neuroblastoma cells, *Exp. Cell Res.,* 169, 311, 1987.

143. **Mugnai, G., Lewandowska, Carnemolla, B., Zardi, L., and Culp, L. A.,** Modulation of matrix adhesive responses of human neuroblastoma cells by neighboring sequences in the fibronectins, *J. Cell Biol.,* 106, 931, 1988.

144. **Mugnai, G., Lewandowska, K., and Culp, L. A.,** Multiple and alternative adhesive responses on defined substrata of an immortalized dorsal root neuron hybrid cell line, *Eur. J. Cell Biol.,* 46, 352, 1988.

145. **van Wachem, P. B., Beugeling, T., Mallens, B. W. L., Dekker, A., Feijen, J., Bantjes, A., and van Aken, W. G.,** Deposition of endothelial fibronectin on polymeric surfaces, *Biomaterials,* 9, 121, 1988.

146. **Chinn, J. A., Horbett, T. A., Ratner, B. D., Schway, M. B., Haque, Y., and Hauschka, S. D.,** Enhancement of serum fibronectin adsorption and the clonal plating efficiencies of Swiss mouse 3T3 fibroblast and MM14 mouse myoblast cells on polymer substrates modified by radiofrequency plasma deposition, *J. Colloid Interface Sci.,* 127, 67, 1989.

147. **Truskey, G. A. and Pirone, J. S.,** The effect of fluid shear stress upon cell adhesion to fibronectin-treated surfaces, *J. Biomed. Mater. Res.,* 24, 1333, 1990.

148. **Hayman, E. G., Pierschbacher, M. D., and Rouslahti, E.,** Detachment of cells from culture substrate by soluble fibronectin peptides, *J. Cell Biol.,* 100, 1948, 1985.

149. **Suzaki, S., Pierschbacher, M. D., Hayman, E. G., Nguyen, K., Ohgren, Y., and Ruoslahti, E.,** Domain structure of vitronectin. Aligment of active sites, *J. Biol. Chem.,* 259, 15307, 1984.

150. **Barnes, D. W., Reing, J. E., and Amos, B.,** Heparin-binding properties of human serum spreading factor, *J. Biol. Chem.,* 260, 9117, 1985.

151. **Ill, C. R. and Ruoslahti, E.,** Association of thrombin-antithrombin III complex with vitronectin in serum, *J. Biol. Chem.,* 260, 15610, 1985.

152. **Collins, W. E., Mosher, D. F., Tomasini, B. R., and Cooper, S. L.,** A preliminary comparison of the thrombogenic activity of vitronectin and other RGD-containing proteins when bound to surfaces, *Ann. N.Y. Acad. Sci.,* 516, 291, 1987.

153. **Dahlback, B. and Podack, E. R.,** Characterization of human S protein, an inhibition of the membrane attach complex of complement demonstration of a free reactive thiol group, *Biochemistry,* 24, 2368, 1985.

154. **Pitt, W. G., Fabrizius-Homan, D. J., Mosher, D. F., and Cooper, S. L.,** Vitronectin adsorption on polystyrene and oxidized polystyrene, *J. Colloid Interface Sci.,* 129, 231, 1989.

155. **Hannan, G. N. and McAuslan, B. R.,** Immobilized serotonin: a novel substrate for cell culture, *Exp. Cell Res.,* 171, 153, 1987.

156. **Duval, J. L., Letort, M., and Sigot-Luizard, M. F.,** Comparative assessment of cell/substratum static adhesion using an *in vitro* organ culture method and computerized analysis system, *Biomaterials,* 9, 155, 1988.

157. **Rabaud, M., Lefebvre, F., Martin, M. T., Aprahamian, M., Schmitthaeusler, R., and Cazenave, J. P.,** Adduct formation between soluble fibrin monomers and elastin, *Biomaterials,* 8, 217, 1987.

158. **Aprahamian, M., Lambert, A., Balboni, M., Lefebvre, F., Schmitthaeusler, R., Damge, C., and Rabaud, M.,** A new reconstituted connective tissue matrix: preparation, biochemical, structural and mechanical studies, *J. Biomed. Mater. Res.,* 21, 965, 1987.

159. **Martin, M. T., Lefebvre, F., Graves, P. V., and Rabaud, M.,** Biochemical study of adduct synthesis between fibrin monomers and elastin, *Biomaterials,* 9, 519, 1988.

160. **Barbie, C., Angibaud, C., Darnis, T., Lefebvre, F., Rabaud, M., and Aprahamian, M.,** Some factors affecting properties of elastin-fibrin biomaterials, *Biomaterials,* 10, 445, 1989.

161. **Lefbvre, F., Drouillet, F., Savin de Larclause, A. M., Aprahamian, M., Midy, D., Bordenave, L., and Rabaud, M.,** Repair of experimental arteriotomy in rabbit aorta using a new resorbale elastin-fibrin biomaterial, *J. Biomed. Mater. Res.,* 23, 1423, 1989.

162. **Riquet, M., Carnot, F., Gallix, P., Callise, D., Lefebvre, F., and Rabaud, M.,** Biocompatibility of elastin-fibrin material in the rat, *Biomaterials,* 11, 518, 1990.

163. **Phillips, S. G., Lui, S., and Phillips, D. M.,** Binding of epithelial cells to lectin coated surfaces, *In Vitro,* 18, 727, 1982.

164. **Rauvala, H., Carter, W. G., and Hakomori, S.,** Studies on cell adhesion and recognition. I. Extent and specificity of cell adhesion triggered by carbohydrate reactive proteins (glycosidases and lectins) by fibronectin, *J. Cell Biol.,* 88, 127, 1981.

165. **Parhizgar, A., Galletti, P. M., and Jauregui, H. O.,** Human endothelial cell attachment receptors, *Trans. Am. Soc. Artif. Inter. Organs,* 33, 494, 1987.

166. **Muzzarelli, R. A. A., Jeuniaux, C. and Gooday, G. W., Eds.,** *Chitin in Nature and Technology,* Plenum Press, New York, 1986.

167. **Zikakis, J. P., Ed.,** *Chitin, Chitosan, and Related Enzymes,* Academic Press, New York, 1984.

168. **Nakajima, M., Atsumi, K., Kifune, K., Miura, K., and Kanamaru, H.,** Chitin is an effective material for sutures, *Jpn. J. Surg.,* 16, 418, 1986.

169. **Malette, W. G.,** Method of Altering Growth and Development and Suppressing Contamination Micro-Organisms in Cell or Tissue Culture, U.S. Patent 605, 623, 1986.

170. **Muzzarelli, R., Baldassarre, V., Conti, F., Ferrara, P., Biagini, G., Gazzanelli, G., and Vasi, V.,** Biological activity of chitosan: ultrastructureal study, *Biomaterials,* 9, 247, 1988.

171. **Pierschbacher, M. D. and Ruoslahti, E.,** Cell attachment activity of fibronectin can be duplicated by small synthetic fragments of the molecule, *Nature (London),* 309, 30, 1984.

172. **Suzuki, S., Oldberg, A., Hayman, E. G., Pierschbacher, M. D., and Ruoslahti, E.,** Complete amino acid sequence of human vitronectin deduced from cDNA. Similarity of cell attachment sites in vitronectin and fibronectin, *EMBO J.,* 4, 2519, 1985.

173. **Sasaki, M., Kleinman, H. K., Huber, H., Deutzmann, R., and Yamada, Y.,** Laminin, a multidomain protein. The A chain has a unique globular domain and homology with the basement membrane proteoglycan and the laminin B chains, *J. Biol. Chem.,* 263, 16536, 1985.

174. **Titani, K., Kumar, S., Takio, K., Ericsson, L. H., Wade, R. D., Ashida, K., Walsh, K. A., Chopek, M. W., Sadler, J. E., and Fujikawa, K.,** Amino acid sequence of human von Willebrand factor, *Biochemistry,* 25, 3171, 1986.

175. **Bernard, M. P., Myers, J. C., Chu, M. -L., Ramirez, F., Eikenberry, E. F., and Prockop, D. J.,** Structure of cDNA for the Proα2 chain of human type I procollagen. Comparison with chick cDNA for Proα2 (I) identifies structurally conserved features of the protein and gene, *Biochemistry,* 22, 1139, 1983.

176. **Watt, K. W., Cottrell, B. A., Strong, D. D., and Doolittle, R. F.,** Amino acid sequence studies on the a chain of human fibrinogen. Overlapping sequences providing the complete sequence, *Biochemistry,* 18, 5410, 1979.

177. **Lawler, J. and Hynes, R. O.,** The structure of human thrombospondin, an adhesive glycoprotein with multiple calcium-binding sites and homologies with several different proteins, *J. Cell Biol.,* 103, 1635, 1986.

178. **Oldberg, A., Franzen, A., and Heinegard, D.,** Cloning and sequence analysis of rat bone sialoprotein (osteopontin) cDNA reveals an Arg-Gly-Asp cell binding sequence, *Proc. Natl. Acad. Sci., U.S.A.,* 83, 8819, 1986.

179. **Jones, F. S., Burgoon, M. P., Hoffman, S., Crossin, K. L., Cunningham, B. A., and Edelman, G. M.,** A cDNA clone for cytotactin contains sequences similar to epidermal growth factor-like repeats and segments of fibronectin and fibrinogen, *Proc. Natl. Acad. Sci., U.S.A.,* 85, 2186, 1988.

180. **Imanishi, Y., Ito, Y., Liu, L. S., and Kajihara, M.,** Design and synthesis of biocompatible polymeric materials, *J. Macromol. Sci. Chem.,* A25, 555, 1988.

181. **Matsuda, T., Kondo, A., Makino, K., and Akutsu, T.,** Development of a novel artificial matrix with cell adhesion peptides for cell culture and artificial and hybrid organs, *Trans. Am. Soc. Artif. Organs,* 35, 677, 1989.

182. **Massia, S. P. and Hubbell, J. A.,** Covalent surface immobilization of Arg-Gly-Asp- and Tyr-Ile-Gly-Ser-Arg- containing peptides to obtain well defined cell-adhesive substrates, *Anal. Biochem.,* 187, 292, 1990.

183. **Danilov, Y. N. and Juliano, R. L.,** (Arg-Gly-Asp)$_n$-albumin conjugates as a model substratum for integrin-mediated cell adhesion, *Exp. Cell Res.,* 182, 186, 1989.

184. **Maeda, T., Oyama, R., Ichihara-Tanaka, K., Kimizuka, F., Kato, I., Titani, K., and Sekiguchi, K.,** A novel cell adhesive protein engineered by insertion of the Arg-Gly-Asp-Ser tetrapeptide, *J. Biol. Chem.,* 264, 15165, 1989.

185. **Sekiguchi, K., Maeda, T., Oyama, R., Hashino, K., Kimizuka, F., Kato, I., and Titani, K.,** Engineering of cell adhesive proteins by insertion of the Arg-Gly-Asp-Ser tetrapeptides, in *Protein Engineering,* Ikehara, M., Ed., Japan Sci. Soc. Press, Tokyo, 1990, 187.

186. **Pierschbacher, M. D. and Ruoslahti, E.,** Influence of stereochemistry of the sequence Arg-Gly-Asp-Xaa on binding specificity in cell adhesion, *J. Biol. Chem.,* 262, 17294, 1987.

187. **Ozeki, E., Matsuda, T., and Akutsu, T.,** Development of high-active peptidyl anti-platelet agent — synthesis and interaction with platelets, *Jpn. J. Artif. Organs,* 19, 1078, 1990.

188. **Saiki, I., Iida, J., Azuma, I., Nishi, N., and Matsuno, K.,** Biological activities of synthetic polypeptides containing a repetitive core sequence (Arg-Gly-Asp) of cell adhesion molecules, *Int. J. Biol. Macromol.,* 11, 23, 1989.

189. **Murata, J., Saiki, I., Iida, J., Azuma, I., Kawahara, H., Nishi, N., and Tokura, S.,** Inhibition of tumor cell adhesion by anti-metastatic polypeptide containing a repetitive Arg-Gly-Asp sequence, *Int. J. Biol. Macromol.,* 11, 226, 1989.

190. **Suzuki, K., Ito, Y., Kimura, S., and Imanishi, Y.,** Synthesis of sequential polypeptide containing cell-adhesion peptide as the repeating unit, Prep. 35th Polym. Symp. Kobe, 1989, 87.

191. **Brandley, B. K. and Schnaar, R. L.,** Covalent attachment of Arg-Gly-Asp sequence peptide to derivatizable polyacrylamide surfaces: support of fibroblast adhesion and long-term growth, *Anal. Biochem.,* 172, 270, 1988.

192. **Matsuda, T., Ozeki, E., and Akutsu, T.,** Novel artificial matrix: solution-castable cell adhesion-promoting polymers, *Artif. Organs,* 14, 110, 1990.

193. **Ito, Y. and Imanishi, Y.,** Design of bioactive materials for enhancing cell functions, in Imanishi, Y., Ed., Kagaku Dojin, Kyoto, 1991, in press.

194. **Ito, Y., Kajihara, M., and Imansihi, Y.,** Materials for enhancing cell adhesion by immobilization of cell-adhesive peptide, *J. Biomed. Mater. Res.,* in press.

195. **Streeter, H. B. and Rees, D. A.,** Fibroblast adhesion to RGDS shows novel features compared with fibronectin, *J. Cell Biol.,* 105, 507, 1987.

196. **Akiyama, S. K. and Yamada, K. M.**, Fibronectin, *Adv. Enzymol.*, 57, 1, 1987.
197. **Waite, H. J. and Tanzer,** Polyphenolic substrate of *Mytilus edulis:* novel adhesive containing L-dopa and hydroxyproline, *Science,* 212, 1038, 1981.
198. **Waite, H. J.,** Nature's underwater adhesive specialist. The lowly mussel has a lot to teach chemists about adhesives, *Chemtech,* 17, 692, 1987.
199. **Benedict, C.,** The development of unique protein polymers for surgical adhesives, U.S.A.-Italy Joint Workshop on Polymers for Biomedical Applications, Capri, Italy, June, 11–16, 1989.
200. **Yamamoto, H., Nishida, A., Takimoto, T., Ikeda, K., and Yamauchi, S.,** Synthesis and adhesive studies on the adhesive model proteins of marine invertebrates, *Polym. Prepr., Jpn.,* 39, 2643, 1990.
201. **Wood, S. A., Lemons, J. E., Prasad, K. U., and Urry, D. W.,** *In vitro* calcification and *in vivo* biocompatibility of the cross-linked polypentapeptide of elastin, *J. Biomed. Mater. Res.,* 20, 315, 1986.
202. **Schnaar, R. L., Weigel, P. H., Kuhlenschmidt, M. S., Lee, Y. C., and Roseman, S.,** Adhesion of chicken hepatocytes to polyacrylamide gels derivatized with *N*-acetyl-glucosamine, *J. Biol. Chem.,* 253, 7940, 1978.
203. **Pless, D. D., Lee, Y. C., Roseman, S., and Schnaar, R. L.,** Specific cell adhesion to immobilized glycoproteins demonstrated using new reagents for protein and glyco-protein immobilization, *J. Biol. Chem.,* 258, 2340, 1983.
204. **Schnaar, R. L.,** Immobilized glycoconjugates for cell recognition studies, *Anal. Biochem.,* 143, 1, 1984.
205. **Kobayashi, A., Akaike, T., Kobayashi, K., and Sumitomo, H.,** *Makromol. Chem., Rapid Commun.,* 7, 645, 1986.
206. **Kobayashi, K., Sumitomo, H., Kobayashi, A., and Akaike, T.,** *J. Macromol. Sci., Chem.,* A25, 655, 1988.
207. **Akaike, T., Kobayashi, A., Kobayashi, K., and Sumitomo, H.,** *J. Bioactive Compatible Polym.,* 4, 51, 1989.
208. **Ponpipom, M. M., Bugianesi, R. L., Robbins, J. C., Doebber, T. W., and Shen, T. Y.,** Saccharide receptor-mediate drug delivery, in *Receptor-Mediated Targeting of Drugs,* Gregoriadis, G., Poste, G., Senior, J., and Trouet, A., Eds., Plenum Press, New York, 1984, 53.
209. **Artursson, P., Johansson, and Sjoholm, I.,** Receptor-mediated uptake of starch and mannan microcapsules by macropharges: relative contribution of receptors for complement, immunoglobulins and carbohydrates, *Biomaterials,* 9, 241, 1988.
210. **Matsuda, T., Itoh, T., Yamaguchi, T., Iwata, H., Hayashi, K., Umezu, S., Ando, T., Adachi, S., and Nakajima, N.,** A novel elastomeric surgical adhesive: design, properties, and *in vitro* performances, *Trans. Am. Soc. Artif. Intern. Organs,* 32, 151, 1986.
211. **Ishihara, K. and Nakabayashi, N.,** Adhesive bone cement both to bone and metals: 4-META in MMA initiated with tri-*n*-butyl borane, *J. Biomed. Mater. Res.,* 23, 1475, 1989.
212. **Brauer, G. M. and Lee, C. H.,** Oligomers with pendant isocyanate groups as tissue adhesives. I. Synthesis and characterization, *J. Biomed. Mater. Res.,* 23, 295, 1989.
213. **Brauer, G. M. and Lee, C. H.,** Oligomers with pendant isocyanate groups as tissue adhesives. II. Adhesion to bone and other tissues, *J. Biomed. Mater. Res.,* 23, 753, 1989.
214. **Lee, H., Ed.,** *Cyanoacrylate Resins — The Instant Adhesives.* A Monograph of Their Applications and Technology, Pasadena Technology Press, Pasadena, 1981.
215. **Harper, M. C. and Ralston, M.,** Isobutyl 2-cyanoacrylate as an osseous adhesive in the repair of osteochondral fracture, *J. Biomed. Mater. Res.,* 17, 167, 1983.
216. **Kilpikari, J., Lapinsuo, M., Tormala, P., Patiala, H., and Rokkanen, P.,** Bonding strength of alkyl 2-cyanoacrylate to bone *in vitro, J. Biomed. Mater. Res.,* 20, 1095, 1986.

217. **Manly, R. S., Ed.,** *Adhesion in Biological Systems,* Part III, Adhesive for Soft Tissue, Academic Press, New York, 1970, 153.

218. **Peppas, N. A. and Buri, P. A.,** Surface, interfacial and molecular aspects of polymer bioadhesion on soft tissues, *J. Controlled Release,* 2, 257, 1985.

219. **Wilson, J. and Merwin, G. E.,** Biomaterials for facial bone augmentation: comparative studies, *J. Biomed. Mater. Res., Appl. Biomater.,* 22, A2, 159, 1988.

220. **Hench, L. L. and Paschall, H. A.,** Direct chemical bond of bioactive glass-ceramic materials, *J. Biomed. Mater. Res. Symp.,* 2, 5, 1973.

221. **Beckham, C. A., Greenlee, T. K., and Crebo, A. R.,** Bone formation at a ceramic implant interface, *Calcif. Tissue Res.,* 8, 165, 1971.

222. **Jarcho, M., Bolen, C. H., Thomas, M. B., Bobick, J., Kay, J. F., and Doremus, R. H.,** Hydroxyapatite synthesis and characterization in dense polycrystalline form, *J. Mater. Sci.,* 11, 2027, 1976.

223. **Tracy, B. M. and Doremus, R. H.,** Direct electron microscopy studies of the bone-hydroxyapatite interface, *J. Biomed. Mater. Res.,* 18, 719, 1984.

224. **Uchida, A., Nade, S. M. L., McCartney, E. R., and Ching, W.,** The use of ceramics for bone replacement, *J. Bone Jt. Surg.,* 66-B, 269, 1984.

225. **Blencke, B. A., Bromer, H., and Deutcher, K. K.,** Compatibility and long-term stability of glass-ceramic implants, *J. Biomed. Mater. Res.,* 12, 307, 1978.

226. **Krajewski, A. and Ravaglioli, A.,** Interpretation of difficulties in the initial adhesion of bio-active glasses to bone, *Biomaterials,* 9, 449, 1988.

227. **Greenspan, D. C. and Hench, L.,** Chemical and mechanical behavior of bioglass-coated alumina, *J. Biomed. Mater. Res.,* 7, 503, 1976.

228. **Griss, P., Greenspan, D. C., Heimke, G., Krempien, B., Buchinger, R., Hench, L. L., and Jentchura, G.,** Evaluation of bioglass-coated Al2O3 total hip prosthesis in sheep, *J. Biomed. Mater. Res. Symp.,* 7, 511, 1976.

229. **Higashi, S., Yamamuro, T., Nakamura, T., and Kitamura, Y.,** Biocompatibility of alumina ceramic implants coated with calcium phosphate compounds, *J. Jpn. Orthop. Assoc.,* 56, 1149, 1982.

230. **Kakutani, Y., Yamamuro, T., Nakamura, T., and Kotoura, Y.,** Strengthening of bone-implant interface by the use of granule coatings on alumina ceramics, *J. Biomed. Mater. Res.,* 23, 781, 1989.

231. **Krajewski, A., Ravaglioli, A., Bertoluzza, A., Monti, P., Battaglia, M. A., Pizzoferrato, A., Olmi, R., and Moroni, A.,** Structural modifications and biological compatibility of doped bioactive glasses, *Biomaterials,* 9, 528, 1988.

232. **Ravaglioli, A. and Krejewski, A.,** Bioglass ceramics covering metal prostheses: theories and prospectives, *Ceramica Informazione,* 10, 583, 1983.

233. **Ono, K., Yamamuro, T., Nakamura, T., Kakutani, Y., Kitsugi, T., Hyakuna, K., Kokubo, T., Oka, M., and Kotoura, Y.,** Apatite-wollastonite containing glass ceramic-fibrin mixture as a bone defect filler, *J. Biomed. Mater. Res.,* 22, 869, 1989.

234. **Hakkinen, L., Yli-Urpo, A., Heino, J., and Larjava, H.,** Attachment and spreading of human gingival fibroblasts on potentially bioactive glasses *in vitro, J. Biomed. Mater. Res.,* 22, 1043, 1988.

5.2 CELL GROWTH FACTOR IMMOBILIZED MATERIALS

Yoshihiro Ito

TABLE OF CONTENTS

I. INTRODUCTION

Cell functions are regulated by communications between the cell and biosignals such as those from other cells, the extracellular matrix (ECM), and biosignal molecules as shown in Figure 1. Biomaterial designs using cell-cell and cell-ECM interactions were described in Chapter 5.1.

The actions of biosignal molecules, such as hormones, cell growth factors, cell differentiation factors, and neurotransmitters, are initiated through a complex formation with their receptors. Despite extensive investigations on the intracellular processes of signal transduction, it remains unclear which process is indispensable for the signal transduction. Growth factors, such as insulin, interact with cells as illustrated in Figure 2A. After binding to its receptor, the growth factor is internalized into the target cell by receptor-mediated endocytosis. Newly formed endocytic vesicles provide the structural basis for the subsequent vesicular transport of the receptor-ligand complexes to the lysosomal compartment of the target cell. Lysosomal degradation of the receptor-ligand complexes accounts for the down-regulation of receptors and the self-regulating response of the target cell to the stimulation induced by the growth factor.

Considering the above mechanism of signal transduction, it is of interest to speculate on the events that would occur if cells interacted with growth factors immobilized on a synthetic polymer (Figure 2B) and insolubilized on a polymer matrix (Figure 2C, D, and E). Soluble growth factors (Figure 2B) or those immobilized on microspheres (diameter <1 μm, Figure 2C) would be internalized by pinocytosis or endocytosis by cells, respectively. On the other hand, insolubilized growth factors (Figure 2D, E) would not be internalized. The introduction of spacer chains (Figure 2E) will aggregate ligand-receptor complexes on the cell surface. The historical aspects of immobilization or insolubilization techniques and examples have been reviewed by Venter.[1]

One of the first uses of small molecules bound to a macromolecular support can be traced to the 1920s. "Insolubilization" techniques were first applied by Schimmer, Sato, and co-workers[2] in 1968. They reported that adrenocorticotropic hormone (ACTH) retained biological activity while covalently bound to the Sepharose beads via an azo linkage. It was argued that ACTH by nature of its being bound covalently to particles larger than cells must be exerting its action on cells by a direct interaction with membrane receptors for ACTH. Cuatrecases[3] in 1969 drew similar conclusions for the mechanism of action of insulin on isolated fat cells by using insulin covalently attached to Sepharose beads. However, the extent of immobilization was suspect.[4] In addition, it was found that cross-linked insulin extracted from the matrix was highly active.[5]

Based on recent research on biomaterials, Ito and Imanishi[6-8] investigated the effect of biosignal molecules immobilized on insoluble matrices.

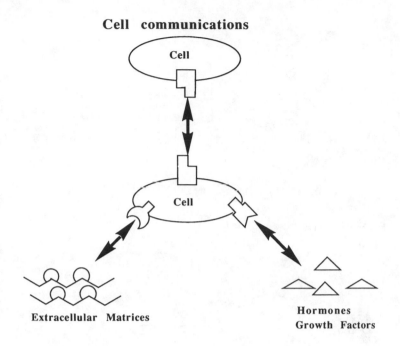

FIGURE 1.   Cell communications with other cells, extracellular matrices, and hormones.

II. IMMOBILIZATION OF GROWTH FACTORS

A. EFFECT OF IMMOBILIZATION (INSOLUBILIZATION)

Figure 3 compares the effects of soluble insulin, insulin adsorbed to a poly(methyl methacrylate) (PMMA) film, and insulin immobilized on the film on the growth of mouse fibroblast cells. The highest increase in cell growth attained by soluble and adsorbed insulin was similar at 1.4, while that by immobilized insulin was more than 2.1. Furthermore, the enhancement was increased by the introduction of polyoxyethylene spacer chains. Similar results were obtained when other matrices were used as immobilization supports.

Enhanced cell growth (1.42) was also observed on PMMA film immobilized with collagen, which is a cell-adhesion factor. Figure 4 shows the shape of cells adhered onto insulin-immobilized or collagen-immobilized PMMA. The cells are extended to a lesser extent on the insulin-immobilized than on the collagen-immobilized PMMA film. However, the cell growth was stimulated to a greater extent with exposure to insulin-immobilized (2.09) than to collagen-immobilized PMMA film (1.42). It has been reported that cell growth is closely related to the shape of the cells adhering to the matrix.[9] In the case of cell adhesion through specific ligand/receptor interactions however, cell growth may not be related simply to the shape of adherent cells.

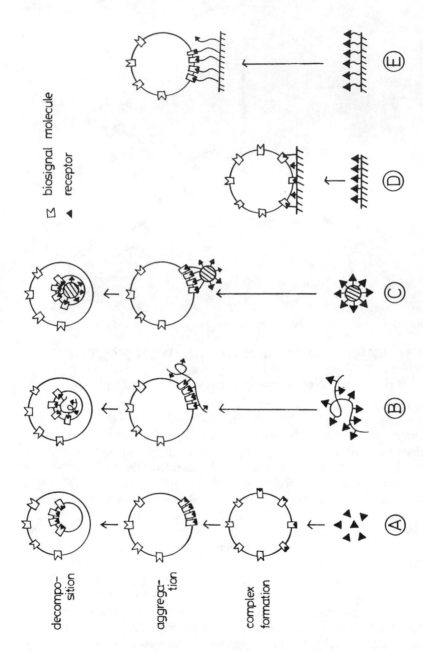

FIGURE 2.   Interactions of cells with biosignal molecules (A) Soluble; (B) immobilized on soluble polymer; (C) immobilized on microspheres (diameter <1 μm); (D) insolubilized on a solid surface; (E) insolubilized with spacer chains.

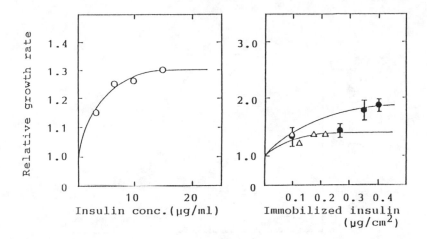

FIGURE 3. Effect of free, adsorbed, and immobilized insulin on cell growth. The insulin was adsorbed or immobilized on surface-hydrolyzed poly(methyl methacrylate) film (○), Soluble insulin; (△), adsorbed insulin; (●), immobilized insulin.

The biospecific interaction between the immobilized insulin and its receptor was demonstrated by the fact that preadsorption of the anti-insulin antibody specifically suppressed the enhancement of cell growth.[10]

Figure 5 summarizes the effects of various proteins immobilized on hydrolyzed PMMA.[11] Immobilized serum albumin did not affect either cell adhesion or growth. On the other hand, fibrinogen and fibronectin enhanced cell adhesion but cell growth only a little. Insulin exhibited the reverse effect.

B. APPLICATIONS OF IMMOBILIZED GROWTH FACTORS

The above results verified that immobilized insulin is appropriate for cell growth enhancement which prompted us to apply the technique to the development of artificial organs and cell engineering materials.

To synthesize biocompatible artificial blood vessels, insulin was immobilized on a pre-aminated polyurethane tube as described in Chapter 2.1.[12] The time needed for the tube to become covered with endothelial cells was shorter in the presence of immobilized than in free insulin.

Insulin was immobilized on poly(methyl methacrylate) microcarriers and used for large-scale cell culture.[13,14] In this application, two more advantages were revealed: cell culture without serum proteins, and the repeated use of the immobilized insulin. The re-utilization process is illustrated in Figure 6. Although the biosignal activity decreased by repeated use, immobilized insulin or transferrin was effective after use in several experiments.

III. MECHANISM OF BIOSIGNAL TRANSDUCTION

Hormones or growth factors interact with cells as shown in Figure 7. The experimental results using the immobilized insulin indicated that the mech-

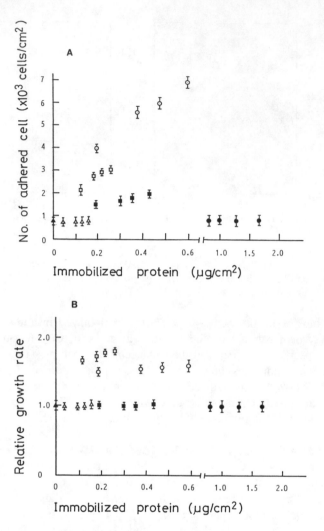

FIGURE 4.   Effect of immobilized proteins on (A) cell adhesion and (B) cell growth. (△),
Albumin; (●), immoglobulin; (■), fibrinogen, (○), collagen, (□), insulin.

anism of cell growth induced by insulin is initiated when insulin binds to its
receptor. The action mechanism of transferrin was also revealed by this immobilization method.[15] The immobilized transferrin not only transported Fe
ion into cells, but also enhanced cell growth. Pretreatment with the antitrans-
ferrin antibody suppressed these effects.

The immobilization or insolubilization method is very useful for investigating signal transduction mechanisms. Other methods have been designed
to reveal the biosignaling mechanisms. Biological responses to EGF have
been mimicked by certain classes of lectins[16] and monoclonal antireceptor

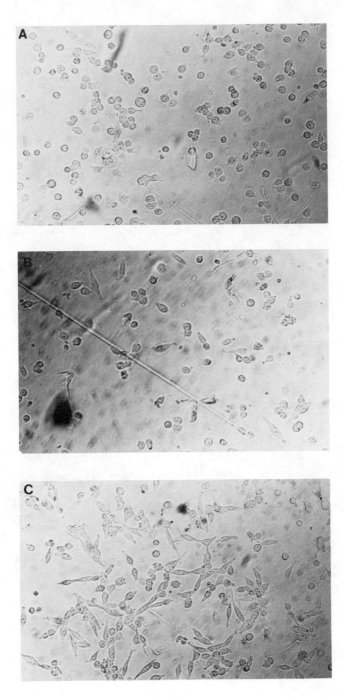

FIGURE 5.   Shape of mouse fibroblast STO cells (8h; 37°C, SFM101 medium adhered onto (A) surface-hydrolyzed poly(methyl methacrylate); (B) insulin-immobilized poly(methyl methacrylate); and (C) collagen-immobilized poly(methyl methacrylate).

Advantages of Cell Engineering Materials immobilized with Insulin

1. Non-serum, and Non-protein Cell Culture

2. Repeated Utilization

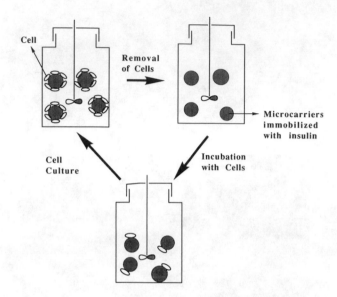

FIGURE 6.   Repeated use of insulin immobilized on microcarriers.

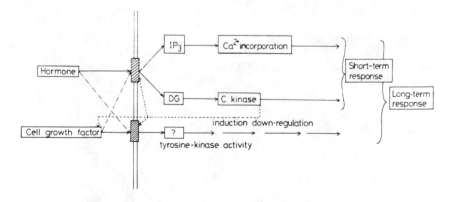

FIGURE 7.   Signal transduction pathway to cell stimulation.

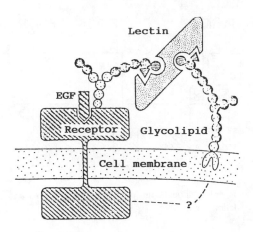

FIGURE 8. Lectin effect on EGF-receptor interaction.

antibodies,[17] suggesting that the receptor itself contains all the necessary information for signal transduction and requires only the appropriate perturbation of its conformation by ligand binding to produce biological effects. It has been proposed that microaggregation of the receptor on the cell surface is sufficient to produce a mitogenic signal,[18,19] a situation analogous to several other hormone-receptor systems.[20-23] However, other evidence that intracellular processing of the EGF-receptor complex is required for stimulation of DNA synthesis,[24-27] suggested a role for receptor internalization in inducing the mitogenic response.

Wakshull and Wharton[28] showed that incubation of cells prelabeled with ^{125}I-labeled EGF at 4°C with Con A resulted in Con A-EGF-receptor complexes that were stable to dissociation and were not internalized at 37°C, as shown in Figure 8. They used this effect of Con A on prebound EGF to test whether prolonged occupation of EGF receptors in the absence of internalization could stimulate DNA synthesis. The results showed that, although the Con A-EGF-receptor complex was functional in terms of stimulating acute effects (RNA synthesis), there was no stimulation of DNA synthesis. They concluded that internalization and processing of the hormone-receptor complex is important for inducing a mitogenic response.

Various kinds of receptors have been identified. Their structures are shown in Figure 9. Thereafter, mutant receptors were developed. Reports have suggested that specific aspects of EGF receptor function are independent of the intrinsic tyrosine kinase activity;[29,30] however, these studies used an antibody against the EGF receptor which failed to activate phosphorylation of exogenous substrates and an insertional mutation in the EGF receptor tyrosine kinase domain which had not been shown to abolish protein kinase activity in cells.

Therefore, to test the functional consequences of abolishing tyrosine kinase activity with minimal structural alterations in the EGF receptors, the

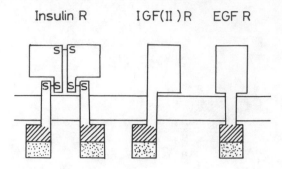

FIGURE 9.   Schematic presentation of receptors for growth factors. (▨), TPK segment;
(▣), self-phosphorylation segment.

lysine at position 721 was mutated to a methionine.[31] This study concluded
that the tyrosine kinase activity of the EGF receptor is essential for diverse
biochemical effects such as calcium ion flux, activation of gene transcription,
receptor down-regulation and ultimately stimulation of cell proliferation. Other
researchers have also reported that active kinase is essential for normal ligand-
induced receptor routing.[32]

The same conclusion has been reached for insulin using a receptor that
had alanine substituted for Lys-1018 in the ATP-binding domain of the β-
subunit.[33] However, some researchers suspect the tyrosine kinase activity of
the insulin receptor for mediation of insulin action.[34]

Wells et al.[35] used a noninternalizing receptor to investigate the EGF
mechanism. Contact-inhibited cells expressing this internalization-defective
receptor exhibited a normal mitogenic response at significantly lower ligand
concentrations than did cells expressing wild-type receptors. A transformed
phenotype and anchorage-independent growth were observed at ligand concentrations that failed to elicit these responses in cells expressing wild-type
receptors. Wells et al. concluded that activation of the protein tyrosine kinase
activity at the cell membrane is sufficient for the growth-enhancing effects
of EGF. Thus, down-regulation can serve as an attenuation mechanism, without which transformation ensures.

Membrane-anchored growth factors have been reviewed by Steele.[36] The
very existence of these growth factors indicates that internalization is not
essential for signal transduction.

IV. EFFECTS OF GROWTH FACTOR IMMOBILIZATION

The effects of immobilized biosignal molecules are considered to be
divided as follows, multivalent simultaneous activation, promotion of cross-
linking, suppression of down-regulation, and perturbation of lipid cell membranes as shown in Figure 10.[37]

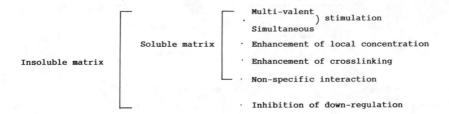

FIGURE 10. Effects of immobilized or insolubilized biosignal molecules.

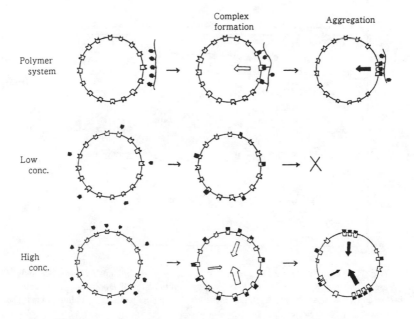

FIGURE 11. Interactions of polymeric biosignal molecules or free biosignal molecules with cells. Polymeric biosignal enhances local concentration.

The first effect is illustrated in Figure 11. Although the mobility of the signal molecule is suppressed by immobilization and the probability of binding with its receptor is decreased, a cooperative effect can be predicted.

Peptide agonists covalently attached to tobacco mosaic virus (TMV) have been synthesized by Schwyzer et al.[38,39] TMV is a crystalline, readily available "chemical," molecular weight of 40 million, with 2130 identical attachment sites. The rod-shaped, cylindrical virion (dimensions, 18 × 300 nm), lying on the surface of a target cell, would cover an area of about 0.01 μm², which would contain on the average up to five receptors. The conjugate exhibited unusual properties such as superpotency, superaffinity, enhanced resistance toward enzymic degradation, and prolonged target cell action. These observations were explained by cooperative-affinity phenomena caused by the

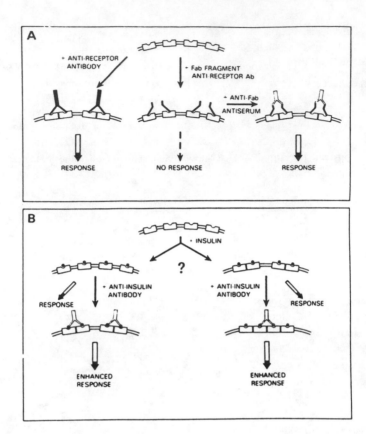

FIGURE 12. Diagrammatic representation of receptor cross-linking by various ligands. (A) Represents bivalent antireceptor antibodies or the combination of monovalent antireceptor antibodies and a second antibody. (B) Represents cross-linking of insulin upon addition of antiinsulin antibodies.

deployment in space of the agonist molecules. Schwyzer et al. considered that the degree of cooperative affinity is similar to the strong affinity of antibodies for membrane-bound antigens explained by the simultaneous binding of the two recognition sites on one antibody molecule to two antigen molecules connected through the membrane. The simultaneous interaction was considered to be stereochemically feasible. Figure 11 illustrates the reason for enhancement of stimulation. The immobilization is effective in enhancing the local concentration of biosignal molecules.

The importance of receptor cross-linking or aggregation was demonstrated by using the antibody, as shown in Figure 12.[40] This aggregation, however, was independent of microfilaments or microtubules.

The second effect is characteristic of insolubilized biosignal molecules. Since the insolubilized biosignal molecules cannot be internalized, the interaction between biosignal molecules and the receptor will continue for a long time. This phenomenon was clearly indicated by the following experiments. Liu et al.[41] added ^{125}I-labeled insulin onto cells after cultivation in the presence of free or insolubilized insulin. They reported that almost the same number of receptors remained on cell surfaces treated with insolubilized insulin, whereas the number decreased on those cells treated with free insulin as shown in Figure 13.

The third effect is based on the nonbiospecific interactions between the conjugates and cells. Bamford et al.[42,43] synthesized water soluble polymers bound to a prostaglandin analogue. They observed a remarkable synergism demonstated by the greatly enhanced activity of the mixture of a polymer coupled to the analogue with the uncoupled polymer. In some cases (e.g., with high molecular weight dextran) the effect reached two orders of magnitude. They proposed a mechanism to account for these phenomena, involving adsorption of the added (inactive) polymer onto platelet membranes, facilitating interaction of the polymer-bound drug with receptors, rendered more accessible by an alteration of the surface geometry. This situation is illustrated in Figure 14A.

It has been suggested that receptor-mediated signal transduction is accompanied by the altered calcium ion permeation through the cell membrane, as shown in Figure 14B.[44] This phenomenon is considered to be located between biospecific and nonbiospecific interactions. These interactions should be considered, when a matrix is utilized.

V. HYBRIDIZATION OF BIOSIGNAL MOLECULES

In order to enhance biosignal activity, the cooperative effect with other biocomponents was investigated by several researchers. The discovery of strong interactions between heparin and a new family of growth factors has opened a new area of investigation.[45]

Heparin binds forms of fibroblast growth factors (FGFs) and also modulates the mitogenic activity of acidic FGF (aFGF) on most cells in tissue culture.[46]

Tardieu et al.[47] found that the interaction of heparin with aFGF stabilized and protected aFGF from chemical and proteolytic degradation, and concluded that the carboxymethylbenzylamine sulfonated dextrans would be useful aFGF stabilizers with much weaker hemostatic properties than heparin for *in vivo* application of FGF.

Lowenberg et al.[48] coated titanium alloy disks with collagen and a platelet-derived growth factor to control cell migration, attachment, and orientation. They concluded that these cellular functions were modified by the coating.

The concept of hybridization described by Ito et al.[49] was sophisticated. They carried out coimmobilization of fibronectin, a cell-adhesion protein, and insulin, a cell-growth protein.[49] The coimmobilization enhanced cell attachment, but also cell growth as shown in Figure 15. The acceleration in

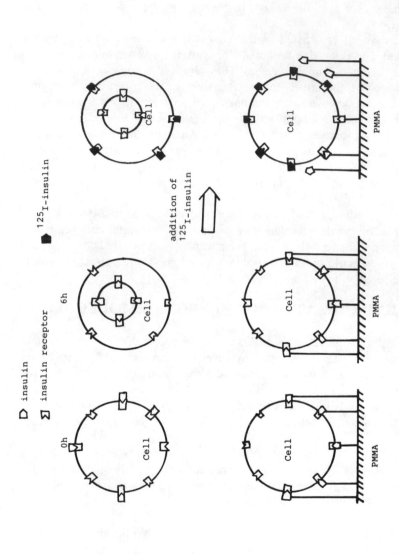

FIGURE 13.   Inhibition of down-regulation. A higher concentration of ¹²⁵I-insulin binds to cells on the insulin-immobilized PMMA than in cells incubated with free insulin. This indicates that a large amount of insulin receptor remains on the cell attached on the insulin-immobilized matrix.

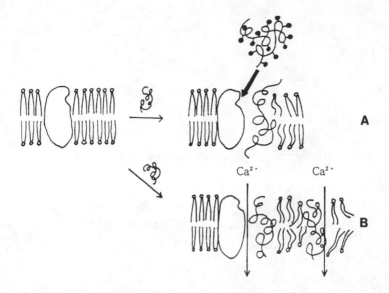

FIGURE 14.   Interaction of polymer with lipid cell membrane. (A) Disorganization of the lipid bilayer in the neighborhood of glycoprotein molecules, brought about by adsorption of a polymer; (B) Ca ion flux accompanying disorganization of lipid balayer.

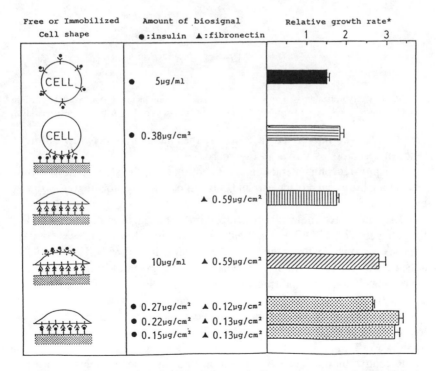

FIGURE 15.   Schematic representation of shape and effect on growth of adhered cell in the presence of free and/or immobilized insulin and/or fibronectin. * Growth rate of mouse fibroblast STO on poly(methyl methacrylate) in the absence of serum was taken as the standard.

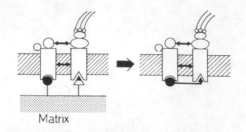

FIGURE 16.   Heterogeneous receptor-receptor interaction.

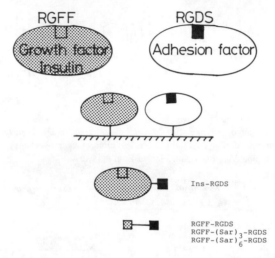

FIGURE 17.   Hybridization of a growth factor and an adhesion factor. The active site of insulin and adhesion factor is RGFF (Arg-Gly-Phe-Phe) and RGDS (Arg-Gly-Asp-Ser), respectively.

growth rate was interpreted in terms of the interactions between different kinds of receptors as shown in Figure 16.

Receptor-receptor interaction is interesting from the standpoint of cross-talk among biosignal molecules[50] and of the allosteric receptor-oligomerization hypothesis.[51]

They attempted molecular hybridization of biosignal molecules. The active-site peptide (RGDS) of a cell-adhesion protein, fibronectin, was coupled with a cell-growth factor or the active site, such as RGFF, which is considered to be an active site for insulin as shown in Figure 17. Insulin-RGDS conjugates enhanced cell adhesion and spreading, as shown in Figure 18. The hybridized peptide had no effect on cell adhesion. Although the hybridized peptide was not particularly effective in enhancing the cell growth, the protein-peptide hybrid (such as RGDS-insulin) was very effective in accelerating cell growth, as shown in Figure 19.

Recently, biomolecular hybridization and fusion of proteins have been performed using gene technology. Combination of TNF-β and interferon is an example. The hybrid molecule was reported to be very highly active.[52]

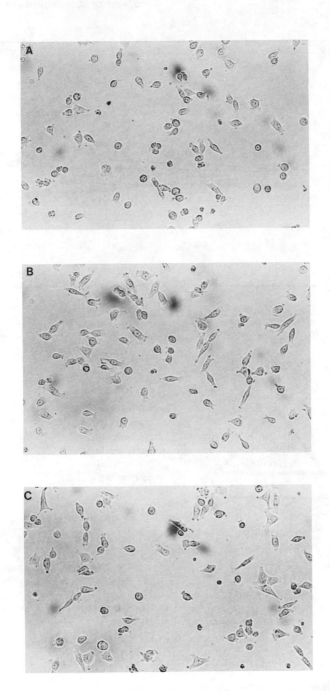

FIGURE 18.   Shape of adhered cells on polystyrene culture dish after 1 h in the presence of (A) insulin, (B) insulin-RGDS, and (C) insulin-GRGDSPC.

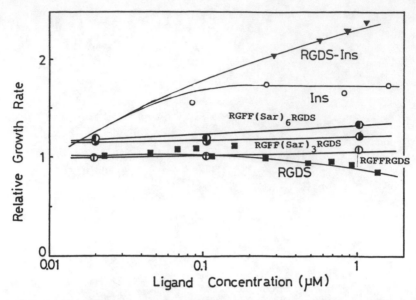

FIGURE 19.   Effect of various growth factor-adhesion factor hybrids onto cell growth.

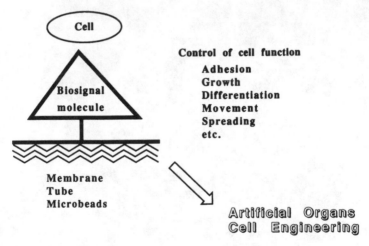

FIGURE 20.   Immobilization of biosignal molecules is a promising methodology for the development of new biomaterials controlling cell functions.

IV. FUTURE OUTLOOK

Hybridization and immobilization of biosignal molecules onto a non-biodegradable matrix are very useful tools in the development of biofunctional materials. It is expected that these biofunctional materials will be synthesized from peptide hormones, cell-differentiation factors, cytokines, and many other biosignal proteins, and may be used for regulation of cell functions such as secretion, differentiation, etc. (Figure 20).

REFERENCES

1. **Venter, J. C.,** Immobilized and insolubilized drugs, hormones, and neurotransmitters: properties, mechanisms of action and applications, *Pharm. Rev.,* 34, 1982.
2. **Schimmer, B. P., Ueda, K., and Sato, G. H.,** Site of action of adrenocorticotropic hormone (ACTH) in adrenal cell cultures, *Biochem. Biophys. Res. Commun.,* 32, 806, 1968.
3. **Cuatrecasas, P.,** Interaction of insulin with the cell membrane: the primary action of insulin, *Proc. Natl. Acad. Sci., U.S.A.,* 63, 450, 1969.
4. **Butcher, R. W., Crofford, O. B., Gammeltoff, S., Gliemann, J., Gavin, J. R., III, Goldfine, I. D., Kahn, C. R., Rodbell, M., Roth, J., Jarett, L., Larner, J., Lefkowitz, R. J., Levine, R., and Marinetti, G. V.,** Insulin activity: the solid matrix, *Science,* 182, 396, 1973.
5. **Oka, T. and Topper, Y.,** A soluble super-active form of insulin, *Proc. Natl. Acad. Sci., U.S.A.,* 71, 1630, 1974.
6. **Ito, Y., Liu, S. Q., and Imanishi, Y.,** Enhancement of cell growth-factor-immobilized polymer film, *Biomaterials,* in press.
7. **Liu, S. Q., Ito, Y., and Imanishi, Y.,** Cell growth on immobilized cell-growth factors. I. Acceleration of the grwoth of fibroblast cells on insulin-immobilized polymer matrix in culture medium without serum, *Biomaterials,* submitted.
8. **Ito, Y. and Imanishi, Y.,** Design of bioactive materials for enhancing cell functions, in Imanishi, Y., Ed., Kagaku Dojin, Kyoto, 1991.
9. **Folkman, J. and Moscona, A.,** Role of cell shape in growth control, *Nature (London),* 273, 345, 1978.
10. **Liu, S. Q., Ito, Y., and Imanishi, Y.,** The interaction between fibroblast cells and insulin immobilized on poly(methyl methacrylate), *Biochem. Biophys. Methods,* in preparation.
11. **Ito, Y., Liu, S. Q., Nakabayashi, M., and Imanishi, Y.,** Cell adhesion and growth on the poly(methyl methacrylate) immobilized with various proteins, *Biomaterials,* in preparation.
12. **Liu, S. Q., Ito, Y., and Imanishi, Y.,** Promotion of endothelial cell growth by immobilization of growth factors and adhesion factors, *J. Biomed. Mater. Res.,* in preparation.
13. **Ito, Y., Uno, T., Liu, S. Q., and Imanishi, Y.,** Cell culture on microcarrier immobilized with cell growth factor, *Biotech. Bioeng.,* in preparation.
14. **Liu, S. Q., Ito, Y., and Imanishi, Y.,** Non-protein cell culture by immobilized growth factor, *Enzyme Microb. Technol.,* in preparation.
15. **Liu, S. Q., Ito, Y., and Imanishi, Y.,** Interaction of fibroblast cells with immobilized transferrin, *Biochim. Biophys. Acta,* in preparation.
16. **Hollenberg, M. D. and Cuatrecasas, P.,** Insulin and epidermal growth factor. Human fibroblast receptors released to deoxyribonuleic acid synthesis and amino acid uptake, *J. Biol. Chem.,* 250, 3845, 1975.
17. **Schreiber, A. B., Lax, I., Yarden, Y., Eshhar, Z., and Schlessinger, J.,** Monoclonal antibodies against receptor for epidermal growth factor induce early and delayed effects of epidermal growth factor, *Proc. Natl. Acad. Sci., U.S.A.,* 78, 7535, 1981.
18. **Schechter, Y., Hernaez, L., Schlessinger, J., and Cuatrecasas, P.,** Local aggregation of hormone-receptor complexes is required for activation by epidermal growth factor, *Nature (London),* 278, 835, 1979.
19. **Maxfield, F. R., Davies, P. J. A., Klempner, L., Willingham, M. C., and Pastan, I.,** Epidermal growth factor stimulation of DNA synthesis is potentiated by compounds that inhibit its clustering in coated pits, *Proc. Natl. Acad. Sci., U.S.A.,* 76, 5731, 1979.
20. **Kahn, C. R., Baird, K. L., Jarrett, D. B., and Flier, J. S.,** Direct demonstration that receptor crosslinking or aggregation is important in insulin action, *Proc. Natl. Acad. Sci., U.S.A.,* 75, 4209, 1978.

21. **Blum, J. J. and Conn, P. M.**, Gonadotropin-releasing hormone stimulation of luteinizing hormone release: a ligand-receptor-effector model, *Proc. Natl. Acad. Sci. U.S.A.*, 79, 7307, 1982.

22. **Gregory, J., Taylor, C. L., and Hopkins, C. R.**, Luteinizing hormone release from dissociated pituitary cells by dimerization of occupied LHRH receptors, *Nature (London)*, 300, 269, 1982.

23. **Podesta, E. J., Solano, A. R., Attar, R., Sanchez, M. L., and Molina y Vedia, L.**, Receptor aggregation induced by antilutropin receptor antibody and biological response in rat testis Leydig cells, *Proc. Natl. Acad. Sci. U.S.A.*, 80, 3986, 1983.

24. **Michael, H. J., Bishayee, S., and Das, M.**, Effect of methylamine on internalization processing and biological activation of epidermal growth factor receptor, *FEBS Lett.*, 117, 125, 1980.

25. **King, A. C., Hernaez-Davis, L., and Cuatrecases, P.**, Lysosomotropic amines inhibit mitogenesis induced by growth factors, *Proc. Natl. Acad. Sci. U.S.A.*, 78, 717, 1981.

26. **King, A. C.**, *Biochem. Biophys. Res. Commun.*, 124, 585, 1984. and the increased sensitivity of EGF after heterologous down-regulation of its receptor by platelet-derived growth factor.

27. **Wharton, W., Leof, E., Pledger, W. J., and O'Keefe, E. J.**, Modulation of the epidermal growth factor receptor by platelet-derived growth factor and choleragen: effects on mitogenesis, *Proc. Natl. Acad., Sci. U.S.A.*, 79, 5567, 1982.

28. **Wakshull, E. M. and Wharton, W.**, Stabilized complexes of epidermal growth factor and its receptor on the cell surface stimulate RNA synthesis but not mitogenesis, *Proc. Natl. Acad. Sci., U.S.A.*, 82, 8513, 1985.

29. **Defize, L. H. K., Moolenaar, W. H., van der Saag, P. T., and de Laat, S. W.**, Dissociation of cellular responses to epidermal growth factor using anti-receptor mono-clonal antibodies, *EMBO J.*, 5, 1187, 1986.

30. **Prywes, R., Livneh, E., Ullrich, A., and Schlessinger, J.**, Mutations in the cytoplasmic domain of EGF receptor affect EGF binding and receptor internalization, *EMBO J.*, 5, 2179, 1986.

31. **Chen, W. S., Lazar, C. S., Poenie, M., Tsien, R. Y., Gill, G. N., and Rosenfeld, M. G.**, Requirement for intrinsic protein tyrosine kinase in the immediate and late actions of the EGF receptor, *Nature (London)*, 328, 820, 1987.

32. **Hoenegger, A. M., Dull, T. J., Felder, S., van Obberghen, E., Bellot, F., Szapary, D., Schmidt, A., Ullrich, A., and Schlessinger, J.**, Point mutation at the ATP binding site of EGF receptor abolishes protein-tyrosine kinase activity and alters cellular routing, *Cell*, 51, 199, 1987.

33. **Chou, C. K., Dull, T. J., Russell, D. S., Gherzi, R., Lebwohl, D., Ullrich, A., and Rosen, O. M.**, Human insulin receptors mutated at the ATP-binding site lack protein tyrosine kinase activity and fail to mediated postreceptor effects of insulin, *J. Biol. Chem.*, 262, 1842, 1987.

34. **Espinal, J.**, What is the role of the insulin receptor tyrosine kinase?, *TIBS*, 13, 367, 1988.

35. **Wells, A., Welsh, J. B., Lazar, C. S., Wiley, H. S., Gill, G. N., and Rosenfeld, M. G.**, Ligand-induced transformation by a noninternalizing epidermal growth factor receptor, *Science*, 247, 962, 1990.

36. **Steele, R. E.**, Membrane-anchored growth factors work!, *TIBS*, 14, 201, 1989.

37. **Ito, Y.**, Polymer systems for controlled release of drugs in response to substances and those for immobilization of drugs, in *Progress in Drug Delivery System*, Kaetsu, I., Ed., Yakugyo-Jihousha, Tokyo, 1990, 119.

38. **Schwyzer, R. and Kriwaczek, V. M.**, Tobacco mosaic virus as a carrier for small molecules: artificial receptor antibodies and superhormones, *Biopolymers*, 20, 2011, 1981.

39. **Kriwaczek, V. M., Bristow, A. F., Eberle, A. N., Gleed, C., Schulster, D., and Schwyzer, R.**, Superpotency and superaffinity phenomena in the stimulation of steroid-genesis in adrenocortical cells by adrenocorticotropin-tobacco mosaic virus conjugates, *Mol. Cell. Biochem.*, 40, 49, 1981.

40. **Kahn, C. R., Baird, K. L., Jarrett, D. B., and Flier, J. S.**, Direct demonstration that receptor crosslinking or aggregation is important in insulin action, *Proc. Natl. Acad. Sci. U.S.A.*, 75, 4209, 1978.

41. **Liu, S. Q., Ito, Y., and Imanishi, Y.**, unpublished data.

42. **Bamford, C. H., Middleton, I. P., and Al-Lamee, K. G.**, Polymeric inhibitions of platelet aggregation. Synergistic effects and proposals for a new mechanism, *Biochem. Biophys. Acta*, 886, 109, 1986.

43. **Bamford, C. H., Middleton, I. P., and Al-Lamee, K. G.**, Polymeric inhibitors of platelet aggregation. II. Copolymers of dipyridamole and related drugs with *N*-vinylpyr-rolidone, *Biochem. Biophys. Acta*, 924, 38, 1987.

44. **Ohtoyo, T., Shimagaki, M., Otoda, K., Kimura, S., and Imanishi, Y.**, Change in membrane fluidity induced by lectin-mediated phase separation of the membrane and agglutination of phospholipid vesicles containing glycopeptides, *Biochemistry*, 27, 6458, 1988.

45. **Shing, Y., Folkman, J., Sullivan, R., Butterfield, C., Murray, J., and Klagsbrun, M.**, Heparin affinity: purification of a tumor-derived capillary endothelial cell growth factor, *Science*, 223, 1296, 1984.

46. **Schreiber, A. B., Kenny, J., Kowalski, W. J., Friesel, R., Mehlman, T., and Maciag, T.**, Interaction of endothelial cell growth factor with heparin: characterization by receptor and antibody recognition, *Proc. Natl. Acad. Sci., U.S.A.*, 82, 6138, 1985.

47. **Tardieu, M., Slaoui, F., Josefonvicz, J., Courty, J., Gamby, C., and Barritault, D.**, Biological and binding studies of acidic fibroblast growth factor in the presence of substituted dextran, *J. Biomater. Sci., Polym. Ed.*, 1, 63, 1989.

48. **Lowenberg, B. F., Pilliar, R. M., Aubin, J. E., Sodek, J., and Melcher, A. H.**, Cell attachment of human gingival fibroblasts *in vitro* to porous-surfaces titanium alloy discs coated with collagen and platelet-derived growth factor, *Biomaterials*, 9, 302, 1988.

49. **Ito, Y., Inoue, M., Liu, S. Q., and Imanishi, Y.**, unpublished data.

50. **Fuxe, K. and Agnati, L. F., Ed.**, *Receptor-Receptor Interactions*, Plenum Press, New York, 1987.

51. **Schlessinger, J.**, Signal transduction by allosteric receptor-oligomerization?, *TIBS*, 13, 443, 1988.

52. **Feng, G. -S., Gray, P. W., Shepard, H. M., and Taylor, M. W.**, Antiproliferative activity of a hybrid protein between interferon-γ and tumor necrosis factor-β, *Science*, 241, 1501, 1989.

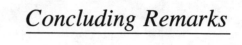

Concluding Remarks

6. CONCLUDING REMARKS

Yukio Imanishi

This book stresses the importance of chemical and genetic modification of natural polymers, in particular, proteins in the synthesis of biospecific and biosimulation materials.

In the synthesis of truly biocompatible materials, *in vivo* tissue reconstruction on a synthetic polymer membrane was found to be most promising. In order to realize tissue generation on a polymer membrane, the membrane surface should be suitably modified from the viewpoint of organ engineering. In this connection, an informative discovery is that coimmobilization of cell-adhesion factor and cell-growth factor onto a polymer membrane is very effective for cell growth on the membrane. The synergetic effect of adhesion and growth factors may be explained in terms of receptor interactions in the cell membrane and opens many research fields such as mechanistic investigation of signal transmission through the cell membrane and production of adhesion/growth hybrid proteins for biological production of useful substances. In any research field, genetic and chemical modification of cell-adhesion factor with cell-growth factor or vice versa is important. Somewhat different types of modifications are also important where cell-growth factors are immobilized onto the membrane surface which is coated with cell-adhesion proteins or is suitable for cell adhesion.

In the synthesis of artificial enzymes, which are useful for chemical production and new therapeutic systems, protein engineering has been found to be an effective tool. Protein engineering is an extremely attractive technology and offers a variety of possibilities for the production of protein materials. However, it is accompanied by considerable difficulties in experimental procedures. A good way to avoid these difficulties is through the utilization of monoclonal antibodies. The efficient production of a monoclonal antibody from a hybridoma enabled the generation of a specific substrate-binding site such as the one found in enzymes. On the basis of this fact, antibodies have recently been given the position of the most important receptor molecule either in biology or in medicine. Many kinds of applications of an antibody have been explored recently, and the most interesting application is the production of a catalytic antibody.[2,3] Catalytic antibodies are produced on the basis of the principle: an antibody that is produced by immunization against a hapten, which is the transition state analogue of a substrate, catalyzes the reaction of the substrate. Therefore, by designing the structure of the antigen, an antibody corresponding to the antibody, a new species of protein is biologically produced. In other words, we can produce a tailor-made enzyme protein using an antibody-production mechanism of hybridoma only by the design of antigen structure.

In reference to chemical and biological modifications of natural polymers (the subtitle of this book) the present author emphasizes the importance of combining the two types of modifications. Genetic engineering offers an effective way of biological production of new proteins. However, it cannot be effective to incorporate an unnatural α-amino acid into a protein molecule.[4] This conflict becomes real in the synthesis of a photoactive cytochrome. Cytochrome is an electron-mediating protein and can be modified to a photoelectron-mediating protein by substituting a phenylalanyl unit at the active site with an alanine derivative carrying a fused-ring aromatic substituent such as the naphthyl, anthryl, or pyrenyl group. These fused-ring aromatic groups can generate photoelectrons, and photoirradiation of mutant cytochrome may generate and transmit photoelectrons. In the synthesis of mutant cytochrome, the original protein should be broken into several portions. The peptide fragment including the site for substitution with a nonnatural α-amino acid should be chemically synthesized. Other peptide fragments can be synthesized either chemically or biologically. The mutant cytochrome can be synthesized by reorganization of the fragments. The importance of the semisynthetic method, a combination of chemical and biological syntheses, in the synthesis of biocomposite materials should be kept in mind.

REFERENCES

1. **Inoue, Y., Liu, S. Q., Ito, Y., and Imanishi, Y.,** Cell growth on growth factor/adhesion factor-coimmobilized polymer films, *Polym. Prepr., Jpn.,* 39, 608, 1990.
2. **Riechmann, L., Clark, M., Waldmann, H., and Winter, G.,** Reshaping human antibodies for therapy, *Nature (London),* 332, 323, 1988.
3. **Tramontano, A., Janda, K. D., and Lerner, R. A.,** Catalytic antibodies, *Science,* 234, 1566, 1986.
4. **Ueda, T., Ueda, M., Tanaka, A., Sisido, M., and Imanishi, Y.,** Biosynthesis of mutant β-D-galactosidases containing nonnatural aromatic amino acids by *Escherichia coli, Bull. Chem. Soc. Jpn.,* 64, 1576, 1991.

INDEX